A GARDEN OF FLOWERS IN FULL BLOOM.

THE GARDEN
Month by Month

Describing the appearance, color, dates of bloom, height and cultivation of all desirable

HARDY HERBACEOUS PERENNIALS

for the formal or wild garden, with additional lists of aquatics, vines, ferns, etc.

By

MABEL CABOT SEDGWICK

Assisted by

ROBERT CAMERON

Gardener of the Harvard Botanical Gardens

WITH OVER TWO HUNDRED HALF-TONE
ENGRAVINGS FROM PHOTOGRAPHS
OF GROWING PLANTS, AND
A CHART IN COLORS.

GARDEN CITY PUBLISHING COMPANY, INC.
GARDEN CITY, NEW YORK

Printed in the United States of America

TO

W. C. C. AND E. R. C.

ERRATA

pp. 200 and 304, *Leontopòdium alpìnum* or *Edelweiss* now under *Yellows,* should be under *Whites.*

pp. 458 and 466, under *Ìris lævigàta* omit *water during flowering season.*

PREFACE

O Perpetui Fiori del Eterna Letizia

IT is prudent perhaps for many of us to have our pleasure gardens shaped for us by an expert wiser than we may hope to be. A trained eye, long study of old forms, and that knowledge which is born only of experience, make possible a beauty of outline and insure a perfection of detail in a project which in the minds of most of us is a desire rather than a definite conception. Yet he who truly loves his garden will not relinquish altogether the happy task of creating it. For him it is the centre of bright imaginings. He dreams of it asleep and awake, until from among the multitude of his thoughts there flashes some happy vision finished in all things, like the completed picture which the painter sees on the white canvas before him. Quickly before it fades he rushes to his task. But to the amateur, garden catalogues are often a snare and most books a delusion. Search as he may, these helpers serve him little, and as he struggles to find the appropriate flowers with which to paint his picture, the gay vision fades and confusion and discouragement ensue.

It is for this gardener that I have made this book and offer it as a full palette, to enable him the more readily to paint the picture as he sees it, and save him the discouragement of looking in a thousand places for a thousand bits of information. However small a part of the garden it may be that he himself plans, he will look upon that portion with a kindlier eye, and find more in it to love and enjoy than all the rest of the garden has to offer.

With this book I wish my gardener joy of his experiments, and if he fails to make his garden altogether as he has imagined it, may he have a fancy quick to suggest new visions; for in the possibilities of change lies the imperishable charm of gardens. Forever through past experience shine the bright alluring pictures of the future.

PREFACE

The plan of this book is simple. The plants are arranged in the order of the months in which they bloom, while for the reader's convenience a plant which flowers in more months than one is listed afresh in each appropriate month, though the full description of its habit and the directions for its culture are given only where it makes its first appearance.

In the great majority of cases, the dates of bloom are taken from personal observations in the vicinity of Boston. The season about New York is, generally speaking, about ten days earlier. A rough and ready calculation allows six days' difference to every degree of latitude.

Yet in this matter of the date of bloom the reader must understand that nothing like exactness is possible. All that can be claimed is the representation of a fair average. The season of bloom is very irregular, often varying as much as a fortnight in the spring. But though early dates may vary, by June first all irregularities seem to disappear, and the reader can be confident that whatever are the dates of bloom, the succession of bloom remains invariable.

As the plants are divided according to the months in which they first bloom, so they are subdivided according to color. In each month's list of blooming plants there are nine color groups, including "parti-colored," *i.e.*, those plants in which each blossom is variegated, and "various," *i.e.*, those in which the color of the blossoms vary.

Since color is the chief glory of a garden, much stress has been laid upon it throughout the preparation of the book. Almost every flower mentioned has been accurately compared with the appended color chart, and in the column devoted to that purpose it bears its appropriate color number, while above this in quotation marks is the color ascribed to it by some reliable authority.

The reader must remember, however, that with matters of color it is much as with matters of taste. One may call the wood violet purple and another insist that it is blue, while red fades so insensibly into pink, and yellow blends so imperceptibly into orange, that he is an artist indeed who can define the precise point where one becomes the other. It must also be borne in mind that the same flower may vary in color in different localities and the same plant may put forth blossoms of varying shades. And yet, though you may quarrel with the division lines, they are just in the main and are not further wrong than others might well be.

A word or two is needed concerning the comprehensiveness of this book. Annuals have *not* been included, as their dates depend altogether upon the time when their seed is sown. Of biennials but few are mentioned; but of hardy perennials it can fairly be said that all are included which deserve a place in the garden proper, in the rock or wild garden, or which are worthy of naturalization; and of tender perennials a few that should find a place in every garden have been added. Large estates as well as small gardens are increasing so rapidly throughout the country and so many new and unfamiliar plants have come to enlarge the gardener's choice, that it has seemed best to make this book offer all which the most varied taste could wish for beautifying a great estate; but it is selection rather than variety which the small gardener needs, and for his sake such plants as are especially serviceable or exceptionally to be desired are marked with a single or double asterisk.

The details of this book have been almost infinite in number, and my best thanks are due to all who have helped me: To Mr. Robert Cameron, of the Harvard Botanical Gardens, for revision of all facts mentioned in the book and for much valuable assistance; to Prof. Benjamin Watson, of Harvard, for advice and encouragement; to Mr. L. T. Ernst for observing many of the dates of flowering; to Miss Louisa B. Stevens and Miss Turner for the preparation of the color chart and, with the assistance of Miss Edith May, for the skilful comparison of colors; to Miss Rose Standish Nichols, who has conducted the book through the press; to Messrs. John L. Gardner, J. S. Lee and J. Woodward Manning for many admirable photographs which I should have gone far to secure elsewhere; and to Miss Elizabeth Dean for much devoted work. To all of these I feel greatly indebted for their real interest and assistance, and in a special sense I wish to record the gratitude with which I shall always remember the unselfish and unremitting assistance and encouragement given me by Miss Pauline Brigham.

For the rest I only hope that this book may help to make more gardens lovely and more gardeners content.

M. C. S.,

Brookline, Mass., Nov., 1906.

CONTENTS

ILLUSTRATIONS

(For English Names of Flowers, see Index)

ILLUSTRATIONS

ILLUSTRATIONS

ILLUSTRATIONS

ILLUSTRATIONS

THE GARDEN MONTH BY MONTH

COMMON SNOW DROP AND WINTER ACONITE. *Galanthus nivalis and Eranthis hyemalis.*

MARCH

WHITE

Color	English Name	Botanical Name and *Synonyms*	Description	Height and *Situation*	Time of Bloom
"White"	WHITE GLORY OF THE SNOW	*Chionodóxa Lucíliæ var. álba	Pretty bulbous plant somewhat resembling the Squill. Ten or twelve flowers on each stem. Lance-shaped leaves. Plant in border or rock-garden. Prop. by seed and offsets. Well-drained loam. Asia Minor.	3-6 in. *Sun or half shade*	Mid. Mar. to early May
"White"	CROCUS	**Cròcus vars.	See color "various," page 8.		Mid. Mar. to late Apr.
"White"	GIANT SNOWDROP	**Galánthus Élwesii	Large handsome species; flowers more globular than G. Nivalis. Bulbous. Prop. by offsets in autumn. Any ordinary garden soil. Mts. of Asia Minor. See Plate, page 5.	6-12 in. *Sun or half shade*	Mar., Apr.
"White"	COMMON SNOWDROP	**Galánthus nivàlis	A charming plant with drooping blossoms. Blooms whenever the snow leaves the ground. Plant in quantity on the lawn, in borders or under the shade of trees. Bulbous. Prop. by offsets in autumn. Any garden soil. Pyrenees to Caucasus. See Plate, page 2.	4-6 in. *Sun or half shade*	"
"White"	PLAITED SNOWDROP	**Galánthus plicàtus	Excellent Snowdrop, larger in all its parts than G. Nivalis and blossoming later. Solitary bell-shaped flowers on long graceful stems. Good for rock-garden or border. Bulbous. Prop. by offsets in autumn. Any ordinary garden soil. Crimea.	4-8 in. *Sun or half shade*	Mar. to early May
"White"	CHRISTMAS ROSE	*Helléborus nìger	Cup-shaped flowers about 3 in. across, sometimes purple-tinted. Good for cutting. Attractive evergreen foliage. Protect in winter and do not disturb. Plant in border or rock-garden. Prop. by division. Any well-drained rich soil. Europe. See Plate, page 6.	9-15 in. *Half shade best*	Mar., early Apr.
"White"	LONG-LEAVED CHRISTMAS ROSE	*Helléborus nìger var. altifòlius *H. n. var. màjor, var. máximus*	Largest flowered variety, with several blossoms, 3-5 in. across, on a stem. See H. niger. Hort.	6-18 in. *Half shade best*	Mar. to early Apr.
"White"	OLYMPIC HELLEBORE	*Helléborus orientàlis var. Olýmpicus	Profusion of small spreading flowers; good for cutting. Protect in winter. Plant in rock-garden and border. Prop. by division. Well-drained rich soil. Bithynia.	12-15 in. *Half shade best*	"

Color	English Name	Botanical Name and *Synonyms*	Description	Height and *Situation*	Time of Bloom
"White"	WHITE SIBERIAN SQUILL	**Scílla Sibírica var. álba.	Very pretty early-flowering species. One to three flowers droop from slender stems. Long narrow leaves. Plant in clumps in rock-garden or border. Bulbous. Prop. by offsets in autumn. Sandy soil; give an occasional top-dressing. Hort.	3-6 in. *Sun or half shade*	Mid. Mar. to early May
"Yellow" 6	CROCUS	**Cròcus vars.	See color "various," page 8.		Mid. Mar. to late Apr.
"Deep yellow" 6	CLOTH OF GOLD CROCUS	**Cròcus Susiànus	One of the earliest Crocuses. Flowers tinged or striped with brown. Variegated leaves in tufts. Naturalize in the grass or plant under trees and in rows or clumps in the border. Bulbous. Prop. by offsets in late Sept. or Oct. Deep well-drained soil. Caucasus; Crimea.	3 in. *Sun or half shade*	"
"Yellow" 4	COMMON WINTER ACONITE	**Eránthis hyemàlis	Pretty cup-shaped flower, surrounded by a whorl of leaves, appears at the same time as the Snowdrop and does well under trees, as it prefers partial shade. Bulbous. Any garden soil. S. Europe. See Plate, page 2.	3-8 in. *Sun or half shade*	Mar., Apr.
"Purple"	CROCUS	**Cròcus vars.	See color "various," page 8.		Mid. Mar. to late Apr.
"Reddish purple" 42	DARK-RED EASTERN HELLEBORE	*Hellébor̀us orientàlis var. atrórubens *H. o. var. Cólchicus*	Large cup-shaped flowers in clusters. Leaves evergreen, one on each flower-stem. Good for rock-garden and border. Prop. by division. Any well-drained rich soil. Hungary. Vars. *F. C. Heineman* and *Gretchen Heineman* are good and somewhat similar in color.	9-15 in. *Half shade best*	Mar. to early Apr.
"Purple" 55	NETTED IRIS	*Ìris reticulàta	Bulbous Iris. Fragrant yellow-crested flowers on short stalks. Leaves stiff and narrow, growing to full height after the flowers are gone, disappear when bulbs ripen. Plant in masses in the border in a sheltered spot. Well-drained sandy or fibrous soil. Several blue vars. Caucasus; Palestine.	1 ft. *Sun*	Mid. Mar.
"Plum" 42	KRELAGE'S NETTED IRIS	*Ìris reticulàta var. Krèlagei	Flowers almost scentless; color varies, yellow markings not distinct. See Iris reticulata. Caucasus. See Plate, page 9.	1 ft. *Sun*	Late Mar.
"Blue" 52	ALLEN'S GLORY OF THE SNOW	*Chionodóxa Álleni *Chionoscilla Álleni*	Similar to C. Luciliæ, the white centre being less distinct. Border and rock-garden. Bulbous. Prop. by seed and offsets. Any good soil. Requires moisture while growing. Hort.	3-6 in. *Sun or half shade*	Mid. Mar. to early May

(A) PLAITED SNOWDROP. *Leucojum vernum.*

(B) GIANT SNOWDROP. *Galanthus elwesii.*

CHRISTMAS ROSE. *Helleborus niger.*

CROCUS. *Crocus vars.*

MARCH

Color	English Name	Botanical Name and *Synonyms*	Description	Height and *Situation*	Time of Bloom
"Blue" 52 & white	**GLORY OF THE SNOW**	****Chionodóxa Lucíliæ**	Pretty, bulbous plant somewhat resembling the Squill. Ten or twelve flowers on each stem; petals tipped with blue, shading to white at the centre. Lance-shaped leaves. Border or rock-garden. Bulbous. Prop. by seed and offsets. Any good soil. Likes moisture while growing. Asia Minor; Crete. See Plate, page 9.	3-6 in. *Sun or half shade*	Mid. Mar. to early May
"Blue" 52 or 63	**GIANT GLORY OF THE SNOW**	****Chionodóxa Lucíliæ var. gigantèa**	Large form of C. Luciliæ, of different habit. Groups of this plant are very effective in early spring. For cultivation see C. Luciliæ. Hort.	3-8 in. *Sun or half shade*	"
"Blue" 52	**TMOLUS' GLORY OF THE SNOW**	***Chionodóxa Lucíliæ var. Tmolùsi** *C. Tmolùsi*	Large flowers, bright blue and white; blooms later than the species. Choice rockery plant; rather rare. For cultivation see C. Luciliæ. Asia Minor.	3-6 in. *Sun or half shade*	Late Mar. to early May
"Gentian-blue" 52 deeper	**SARDIAN GLORY OF THE SNOW**	***Chionodóxa Sardénsis**	Flowers borne on branching flower-stems smaller than C. Luciliæ, deeper blue and without the white markings. Two vars., one having white stamens, the other black. Bulbous. Prop. by seed and offsets. Any good soil. They like moisture while growing. Asia Minor.	3-6 in. *Sun or half shade*	Mid. Mar. to early May
"Blue" 61	**STAR HYACINTH**	***Scílla amœna**	Blooms freely, but not so attractive as other Squills. Several flower-stems bear 4-8 star-shaped flowers. Bulbous. Enrich the soil occasionally with a top-dressing of manure. Tyrol.	6-9 in. *Sun or shade*	"
"Dark blue" 62	**EARLY SQUILL**	***Scílla bifòlia**	Earliest species. Spikes of 4-6 starry flowers. Graceful lance-shaped leaves. Desirable in wild garden, rock-garden, or border. Vars. *alba* and *rosea;* (color no. 25). Treat like Scilla Sibirica. Europe; Asia Minor.	4-6 in. *Sun or shade*	"
"China-blue" 61 deeper	**SIBERIAN SQUILL**	****Scílla Sibírica** *S. amœna var. præcox*	Early flowering species. One to three flowers of beautiful deep blue droop from slender stalks. Long narrow leaves. Charming in clumps in rock-garden or border, or for edging beds. Bulbous. Prop. by offsets in autumn. Sandy soil is preferable; give an occasional top-dressing. Several vars. are grown. Russia; Asia Minor. See Plate, page 9.	2-6 in. *Sun or half shade*	"
Parti-colored		****Cròcus vars.**	See color "various," page 8.		Mid. Mar. to late Apr.

Color	English Name	Botanical Name and *Synonyms*	Description	Height and *Situation*	Time of Bloom
Cream & 42 outside, 44 inside	**SCOTCH CROCUS, CLOTH OF SILVER CROCUS**	****Cròcus biflòrus**	Stemless funnel-form yellowish white flowers, tinged and striped with purple. Variegated grass-like leaves rise above the flowers. Pretty in the grass, in borders, or under trees. Bulbous. Prop. by seed and offsets. Plant in late Sept. or Oct. in deep, well-drained soil. S. Europe.	6-8 in. *Sun or half shade*	Mid. Mar. to late Apr.
Parti-colored	**EASTERN HELLEBORE FRAU IRENE HEINEMANN**	***Hellèborus orientàlis var. "Frau Irene Heinemann"**	Large cup-shaped flowers, rose-purple outside, and greenish, streaked and dotted, within. Good for cutting. Evergreen foliage. Slightly protect in winter. Should not be disturbed. Plant in border or edge of shrubbery. Prop. by division. Any rich well-drained soil. Hort.	12-15 in. *Half shade best*	Mar. to early Apr.
Green & 35	**PURPLISH GREEN HELLEBORE**	***Hellèborus víridis var. purpuráscens**	Drooping green flowers purple tinted, and grayish foliage. Needs slight protection in winter. Should not be disturbed. Plant in border or edge of shrubbery. Prop. by division. Any rich well-drained soil. Hungary.	12-18 in. *Half shade best*	"
Various	**CROCUS**	****Cròcus vars.**	Large funnel-shaped stemless flowers. Stiff grass-like leaves. Naturalize in masses in the grass and under trees, or plant in clumps in the border. Bulbous. Prop. by offsets. Plant in light well-drained soil in late Sept. or Oct. The following are some of the best garden vars.:— **White vars.**—*Caroline Chisholm;* free bloomer. *Mammoth White;* very large. *Mont Blanc;* large and snow white. **Yellow vars.**—*Golden Yellow;* (color no. 6), deep yellow. **Purple vars.**—*King of the Blues;* deep purple blue, very large. *Purpurea grandiflora;* large deep purple flowers. **Parti-colored vars.**—*Albion;* deep violet striped with white. *Cloth of Silver;* silvery white lilac-striped. *La Majestueuse;* large lilac flowers striped white. *Sir Walter Scott;* white striped with lilac; large and free blooming. See Plate, page 6.	6-8 in. *Sun or half shade*	Mid. Mar. to late Apr.
White or 46	**SPRING CROCUS**	****Cròcus vérnus**	The commonest species. Large stemless funnel-shaped flowers, white or lilac, sometimes striped with purple. Stiff grass-like leaves. Mass in border or scatter in the grass or under trees. Bulbous. Prop. by offsets. Plant in late Sept. or Oct. in light, well-drained soil. Europe.	4-5 in. *Sun or half shade*	"

KRELAGE'S NETTED IRIS. *Iris reticulata var. Krelagei.*

GLORY OF THE SNOW. *Chionodoxa Luciliæ.* SIBERIAN SQUILL. *Scilla Sibirica.*

AN APRIL LANDSCAPE

APRIL

WHITE TO GREENISH

Color	English Name	Botanical Name and *Synonyms*	Description	Height and *Situation*	Time of Bloom
"Cream white"	**WHITE BANE-BERRY**	**Actæa álba** *A. rùbra*	Strong-growing plant of the Buttercup order. Flowers in showy clusters. Berries white. Foliage finely cut. Suitable for rock and wild garden. Prop. by seed and division. Loose soil.	1-1½ ft. *Shade*	Early Apr. to July
"Greenish"	**MOUNTAIN LADY'S MANTLE**	**Alchemílla alpìna**	Drooping inconspicuous flowers and digitate leaves. Easily cultivated. Suitable for the rock-garden. Prop. by seed and division. Europe.	6-8 in. *Sun or shade*	Late Apr., May
"White" turns to 29 pale	**WILD ONION**	***Állium mutábile**	Flowers change to rose-pink and are borne in many-flowered umbels above the narrow leaves. Effective rock-garden plant. Easily cultivated. Bulbous. Prop. by seed and offsets. Well-drained soil. N. America.	1-2 ft. *Sun or half shade*	Apr., May
"White" often 36	**WOOD ANEMONE**	**Anemòne nemoròsa**	Common Anemone. Slender stemmed plant, bearing a solitary flower, often pink or purplish, 1 in. across. Attractive delicate foliage. Pretty in wild garden and in masses under trees. Prop. by division; of easy culture in good soil. U. S. A.; Europe; Siberia.	4-8 in. *Shade*	Late Apr. to early June
"Cream white" tinged with 39	**SNOWDROP WIND-FLOWER**	****Anemòne sylvéstris**	Large nodding sweet-scented flowers, tinged with lavender, solitary or two together; pretty drooping buds. Fine deeply cut foliage. Good in border or under trees. Prop. by division. Europe. See Plate, page 13.	1-1½ ft. *Sun or half shade*	Late Apr. to mid. July
"Cream white"	**DOUBLE SNOWDROP WIND-FLOWER**	***Anemòne sylvéstris var. flòre-plèno**	Large double-flowered var. of A. sylvestris. Good border plant; also suitable for open position in the rock-garden. Europe.	12-15 in. *Sun or shade*	Apr., May
"White"	**MUN-STEAD'S WHITE COLUM-BINE**	***Aquilègia vulgàris var. álba.** *A. v. var. nívea*	Abundance of large flowers, many on a stem. Good for sheltered spots, in the wild garden or border. Prop. by seed, sown as early as possible, or division. Deep sandy loam is best. Hort.	2-3 ft. *Sun*	"
"White"	**WHITE ROCK CRESS**	****Árabis álbida** *A. Caucásica*	Heads of fragrant flowers form a sheet of pure white. Good for the spring garden, for rockwork and edgings, and for covering bare and rocky places. Prop. by seed, division and cuttings. Thrives even in poor soil. Europe. See Plate, page 13. Var. *variegata;* var. with gold and green foliage. Var. *flore-pleno;* double-flowered var.	6-8 in. *Sun*	Early Apr. to June

Color	English Name	Botanical Name and *Synonyms*	Description	Height and *Situation*	Time of Bloom
"White"	ALPINE ROCK CRESS	*Árabis alpìna	Resembles A. albida but has smaller flowers. Excellent rock-plant. Prop. usually by division, also by seed and cuttings. Thrives in any ordinary soil. Europe. Var. *flore-pleno;* a double var. Var. *variegata* has green and yellow foliage.	6 in. *Sun or half shade*	Early Apr. to late May
"White"	RUNNING ROCK CRESS	*Árabis procúrrens	Useful trailing rock-plant with small flowers. Prop. by stolons. Any soil. Europe.		"
"White"	WHITE GLORY OF THE SNOW	*Chionodóxa Lucíliæ var. álba	See page 3.		Mid. Mar. to early May
"Green-ish"	GOLDEN SAXIFRAGE, WATER CARPET	Chrysosplè-nium Americànum	Creeping plant with inconspicuous flowers. Stems covered with small pulpy leaves. For bog-garden. Prop. by division. Wet soil. N. Amer.	6 in. *Sun*	Late Apr., May
"White"	CROCUS	**Cròcus vars.	See page 8.		Mid. Mar. to late Apr.
"White"	PEPPER-ROOT, TWO-LEAVED TOOTH-WORT	Dentària diphýlla *Cardamìne diphýlla*	Petals pale purplish beneath. Two compound leaves. Rootstocks edible. Suitable for margin of border or shrubbery. Prop. by division and by bulblets. Easily cultivated. Light peaty soil, well enriched. Eastern N. Amer.	1 ft. *Half shade or shade*	Apr., May
"White"	SQUIRREL CORN	*Dicéntra Canadénsis *Diélytra Canadénsis*	A few pendent pink-tipped flowers terminate leafless stems. Pretty grayish fern-like foliage. Good plant for border or rock-garden. Prop. in spring by crown-division or root-cuttings. Light rich soil. Eastern N. Amer.	6-12 in. *Half shade*	"
"White" white & 3	DUTCH-MAN'S BREECHES	*Dicéntra Cucullària *Diélytra Cucullària*	Racemes of drooping yellow-tipped flowers on leafless stems. Deeply-divided leaves. Suitable for rock-garden. Prop. in the spring by crown-division or root-cuttings. Light rich soil. Eastern N. Amer. See Plate, page 14.	3-6 in. *Half shade*	"
"White"	LARGE WHITE-FLOWERED BARREN-WORT	*Epimèdium macránthum var. níveum *E. níveum*	Dainty interesting plant with curiously shaped flowers. New leaves reddish. Suitable for rock-garden or border. Prop. by division. Any garden soil. Japan.	8-10 in. *Half shade best*	Late Apr. to late May

SNOWDROP WINDFLOWER. *Anemone sylvestris.*

WHITE ROCK CRESS. *Arabis albida.*

WHITE GUINEA-HEN FLOWER. *Fritillaria Meleagris var. alba.*

DUTCHMAN'S BREECHES. *Dicentra Cucullaria.*

Color	English Name	Botanical Name and *Synonyms*	Description	Height and *Situation*	Time of Bloom
"White" white & 6	**WHITE DOGTOOTH VIOLET**	****Erythrònium álbidum**	Solitary flower with recurved petals, yellow at base, and two long narrow unmottled leaves, rising from the root. Pretty on grassy banks and extremely good for planting in Rhododendron beds. Bulbous. Any light fibrous soil with good drainage. N. J.; Western U. S. A.	6 in. *Half shade*	Late Apr., May
"Greenish"	**LILY-LIKE FRITILLARY**	***Fritillària liliàcea**	One to six lily-like flowers, veined with green, droop from each stalk. Leaves mostly in a whorl near the ground. Border or rock-garden. Bulbous. Prop. by offsets. Rich loam. Cal.	6-12 in. *Half shade*	Apr., May
"White"	**WHITE GUINEA-HEN FLOWER OR CHECKERED LILY**	****Fritillària Meleàgris var. álba**	Solitary drooping bell-shaped flowers, slightly checkered. Delicate grass-like leaves which disappear after the bulb ripens. Charming for the border and rock-garden. Bulbous. Prop. by offsets; division every 3 or 4 years necessary. Grows best in deep sandy loam. Europe; Caucasus. See Plate, page 14.	1 ft. *Sun or shade*	Late Apr., May.
"Greenish white"	**CAUCASIAN SNOWDROP**	***Galánthus Caucásicus** *G. Redoùtei*	Later blooming than G. nivalis. Plant in masses in the rock-garden and leave undisturbed for years. Bulbous. Prop. by offsets. Any ordinary garden soil. Caucasus.	8-9 in. *Half shade best*	Apr., May
"White"	**GIANT SNOWDROP**	****Galánthus Élwesii**	See page 3.		Mar., Apr.
"White"	**COMMON SNOWDROP**	****Galánthus nivàlis**	See page 3.		"
"White"	**PLAITED SNOWDROP**	****Galánthus plicàtus**	See page 3.		Mar. to early May
"White"	**CHRISTMAS ROSE**	***Helléborus nìger**	See page 3.		Mar., early Apr.
"White"	**LONG-LEAVED CHRISTMAS ROSE**	***Helléborus nìger var. altifòlius** *H. n. var. màjor, var. máximus*	See page 3.		Mar. to early Apr.
"White"	**OLYMPIC HELLEBORE**	***Helléborus orientàlis var. Olýmpicus**	See page 3.		"

Color	English Name	Botanical Name and *Synonyms*	Description	Height and *Situation*	Time of Bloom
"White"	**DUTCH HYACINTH**	****Hyacínthus orientàlis vars.**	See color "various," page 59.		Late Apr., May
"Greenish white"	**ORANGE ROOT, GOLDEN SEAL**	**Hydrástis Canadénsis**	Grown for its bright red berries and attractive foliage rather than for its small solitary flowers. Root of medicinal value. Wild garden. Prop. by seed and division. Moist loam with leaf-mold. Eastern N. Amer.	12-15 in. *Half shade*	"
"White"	**WHITE SPRING BITTER VETCH**	****Láthyrus vérnus var. álbus** *Órobus vèrnus var. álbus*	Tufted habit. Drooping pea-like flowers in clusters of 5 or 7. Pale green compound leaves. Excellent for rock-garden and border. Prop. by seed or division. Any good garden soil. Central Europe.	1-2 ft. *Sun or half shade*	"
"Transparent white"	**SAND LILY** of Colorado	**Leucocrìnum montànum**	Clusters of fragrant funnel-shaped stemless flowers with pale anthers rise among the narrow leaves. Blooms for several weeks. Good for the rock-garden. Bulbous. Pacific Coast.	4-8 in. *Sun*	Apr.
"White"	**SUMMER SNOW-FLAKE**	***Leucòjum æstìvum**	Clusters of fragrant bell-shaped flowers, with green tips, somewhat resembling Snowdrops, droop among the narcissus-like foliage. Attractive in the border or for edging shrubbery. Bulbous. Any rich soil. Central and S. Europe. See Plate, page 17.	1 ft. *Half shade*	Apr., May
"White"	**SPRING SNOW-FLAKE**	***Leucòjum vérnum**	Resembles L. æstivum, but is a smaller plant with solitary flowers. Pretty in the border. Bulbous. Any rich soil. Central and S. Europe. See Plate, page 5.	6-12 in. *Half shade*	"
"White"	**WHITE OR EVENING CAMPION**	***Lýchnis álba**	Star-shaped flowers in showy clusters. Foliage thick at the base of the plant. Plant in border or rock-garden. Prop. by seed or division. Ordinary light garden soil. N. Asia; Europe; Amer.	6-12 in. *Sun*	"
"White"	**COMMON WHITE GRAPE HYACINTH**	****Muscàri botryoìdes var. álbum**	Pretty bulbous plant. Dense clusters of small globular scentless flowers. Narrow stiff leaves. Plant in clumps in the border or rock-garden. Prop. by offsets. Ordinary garden soil. Europe.	9 in. *Sun or half shade*	"
"White"	**PRIMROSE PEERLESS**	**Narcíssus biflòrus**	Two flowers on each stem; pure white with pale yellow short cups. Plant in masses in the border, in the grass or the edge of shrubbery. Bulbous. Prop. by offsets. Plant in late Sept. or Oct., 6 or 8 inches deep, 3 in. apart. Any good soil. S. Europe.	1-2 ft. *Half shade best*	Mid. Apr. to late May

SPRING SNOWFLAKE. *Leucojum æstivum.*

DROOPING STAR-OF-BETHLEHEM. *Ornithogalum nutans.*

WHITE MOSS PINK. *Phlox subulata var. alba.*

Color	English Name	Botanical Name and *Synonyms*	Description	Height and *Situation*	Time of Bloom
"White"	**ALGERIAN WHITE HOOP-PETTICOAT DAFFODIL**	***Narcíssus Bulbocò-dium var. Monophýllus**	Small delicate plant with several solitary flowers and slender leaves. Not very hardy. Plant in Sept. or Oct., 5 in. deep and 3 in. apart. Bulbous. Prop. by offsets. A good well-drained soil. Give an occasional top-dressing. Algiers.	5-8 in. *Half shade best*	Late Apr., early May
"White"	**BUR-BIDGE'S NARCISSUS**	**Narcíssus Burbidgei**	A short-cupped Narcissus similar to the Poet's N. Yellow cups edged with scarlet, and white base-petals. For cultivation, etc., see N. biflorus.	12-15 in. *Half shade best*	Mid. Apr. to late May
"White"	**WHITE CREEPING FORGET-ME-NOT**	***Omphalòdes vérna var. álba**	Trailing plant. Flowers like Forget-me-nots in clusters on an erect stem; less pretty than the blue var. Good for rock-garden or for fringing walks. Prop. by division. Moist soil. Europe. Var. **plena-alba* has double white flowers.	6 in. *Sun or half shade*	Apr., May
"Cream white"	**DROOPING STAR-OF-BETHLE-HEM**	**Ornithógalum nùtans**	Star-shaped flowers, white backed with green, in racemes of 3 to 12. Foliage pale green, long and narrow. Pretty plant, good for covering bare places in wild gardens as it increases rapidly. Bulbous. Prop. by offsets. Europe; Asia Minor. See Plate, page 17.	8-12 in. *Sun*	"
"White"	**WHITE ICELAND POPPY**	****Papàver nudicaùle var. álbum**	Charming large cup-shaped flowers on graceful erect and leafless stems. Buds nodding. Good for cutting. Treat as an annual. Rock-gardens or borders. Prop. by seed. Light fairly rich soil. Arctic and Alpine Regions.	9-15 in. *Sun*	Late Apr. to mid. June; late Aug., Sept.
"White"	**WHITE MOSS OR GROUND PINK**	***Phlóx subulàta var. álba**	Creeping evergreen plant thickly covered with small flowers which rise from a moss-like bed of foliage. Favorite rock-plant; good for carpeting rock-gardens or borders. Prop. by seed, division or cuttings. Light dry soil. N. Y., West and South. See Plate, page 18.	4-6 in. *Sun*	Late Apr. to late May
"White"	**MOSS OR GROUND PINK, THE BRIDE**	****Phlóx subulàta var. "The Bride"**	Sheets of pink-centred flowers cover the thick moss-like bed of foliage. Excellent for carpeting. Prop. by seed, division or cuttings. Light dry soil. Hort.	4-6 in. *Sun*	"
"White"	**NELSON'S MOSS OR GROUND PINK**	***Phlóx subulàta var. Nelsoni**	Resembles the preceding in habit, use and cultivation. Hort.	4-6 in. *Sun*	"

Color	English Name	Botanical Name and *Synonyms*	Description	Height and *Situation*	Time of Bloom
"Bluish white"	**STRIPED SQUILL**	***Puschkínia scilloìdes** *Adámsia scilloìdes*	Resembles the Squill family. Flowers marked with blue, in clusters of 1-10. Foliage long and narrow. Good border or rock-garden plant, but cannot be crowded. Bulbous. Light soil. Asia Minor.	4-12 in. *Sun*	Late Apr., early May
"White" buds 23	**FLOWERING MOSS, PYXIE**	**Pyxidanthèra barbulàta** *Diapênsia barbulàta*	Minute evergreen creeping plant growing in dense cushions. Buds rose-color, flowers white. Rock-garden. Prop. by division. Sand and leaf-mold. N. J. to N. C.	2 in. *Sun or half shade*	Apr., May
"White"	**BLOODROOT, RED PUCCOON**	***Sanguinària Canadénsis**	Large star-shaped flowers with yellow centres, surrounded by handsome dark foliage which springs from the root. Gives a charming effect in masses under deciduous trees. Prop. by division. Light, preferably moist soil. Eastern N. Amer. See Plate, page 21.	8 in. *Sun or shade*	Early Apr. to mid. May
"Dull white"	**EARLY SAXIFRAGE**	**Saxífraga Virginiénsis**	Downy tufted plant. Numerous insignificant flowers in erect spikes. Leaves in rosette at base of plant. Protect with leaves. Suitable for rock-garden and used sometimes for border. Prop. by division. N. Eastern U. S. A.	4-9 in. *Sun*	Apr. to late June
"White"	**WHITE SIBERIAN SQUILL**	****Scílla Sibírica var. álba**	See page 4.		Mid. Mar. to early May
"White"	**EASTER DAISY** of Colorado	**Townséndia serícea**	Low-growing, almost stemless plant with white or purple-tinted flowers, resembling Asters. Leaves very narrow. Western N. Amer.	6-12 in. *Sun*	Apr., May
"White"	**NODDING WAKE-ROBIN**	**Tríllium cérnuum**	Solitary flowers with wavy reflexed petals, somewhat hidden by the broad leaves. Pretty for moist borders among ferns or for naturalization under trees. Prop. by seed or division. Rich moist soil is best. Eastern N. Amer.	1 ft. or more *Half shade*	Late Apr. to early June
"White"	**WHITE ILL-SCENTED WAKE-ROBIN**	***Tríllium eréctum var. álbum**	Showy. Charming flowers on graceful stems from which spring broad dark leaves. For use and cultivation see T. cernuum. N. Amer.	1 ft. or more *Half shade*	"
"Cream white"	**EARLY WAKE-ROBIN**	**Tríllium nivàle**	Dwarf species. Flowers solitary on erect or drooping stalks. Foliage narrow. Pretty for the wild garden or border. Prop. by seed or division. Moist rich soil is best. U. S. A.	5 in. *Half shade*	Late Apr., May
"White"	**PAINTED WAKE-ROBIN**	**Tríllium úndulàtum** *T. erythrocárpum, T. píctum*	Large handsome flowers marked with crimson. Large leaves on the flower-stalks. Pretty in moist borders, among ferns, or naturalized under trees. For cultivation see T. nivale.	1 ft. or more *Half shade*	"

BLOOD-ROOT. *Sanguinaria Canadensis.*

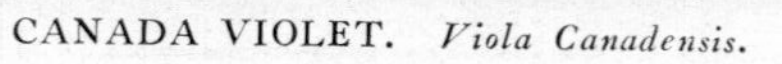

CANADA VIOLET. *Viola Canadensis.*

Color	English Name	Botanical Name and *Synonyms*	Description	Height and *Situation*	Time of Bloom
"White"	**TULIP**	****Tùlipa vars.**	**Single Early Bedding Tulips.** See color "various," page 64.		Late Apr. to late May
"White"	**TULIP**	****Tùlipa vars.**	**Double Early Bedding Tulips.** See color "various," page 64.		"
"White"	**WHITE GENTIAN-LEAVED SPEED-WELL**	***Verónica gentianoìdes var. álba**	Pretty tufted plant. Small flowers borne in erect spikes on leafy stems above the carpet of foliage. Good for border or rock-garden and for covering bare places. Prop. by division. Any soil. Hort.	6-24 in. *Sun or half shade*	"
"White"	**COMMON WHITE PERI-WINKLE OR RUNNING MYRTLE**	***Vínca mìnor var. álba**	Trailing evergreen. Large solitary flowers and dark glossy foliage. Spreads rapidly forming a thick carpet for bare places, under trees or in rock-garden. Prop. by division or cuttings. Easily cultivated in any ordinary soil. Europe.	2-6 in. *Sun or shade*	Late Apr. to June
"White"	**SWEET WHITE VIOLET**	**Vìola blánda**	Common species. Small flowers somewhat veined with lilac and faintly fragrant. Round leaves minutely downy. Prop. by division. Found in low wet places. N. Amer.	2-6 in. *Sun*	Apr., May
"White" tinged with 43	**CANADA VIOLET**	**Vìola Canadénsis**	Common in rich woods. Flowers tinged with purple. Pointed heart-shaped leaves. Vigorous grower; good under shade of trees. Prop. by division. N. Amer. See Plate, page 22.	3-14 in. *Half shade*	Late Apr. to Mid. June
"White"	**WHITE DOG VIOLET**	**Vìola canìna var. álba**	Petals pinkish outside. Plant in masses in a shady spot of rock or wild garden. Prop. by division. Europe.	3-5 in. *Half shade*	Late Apr. to late May
"White"	**WHITE HORNED VIOLET OR BED-DING PANSY**	****Vìola cornùta var. álba**	Vigorous tufted plant bearing a continuous profusion of sweet-scented flowers the size of small Pansies. Clean bright foliage. Good for cutting and excellent for border or rock-garden and for spring bedding. Prop. by seed, division or cuttings. Any good soil. Pyrenees.	5-8 in. *Sun or half shade*	Late Apr. until frost
"White"	**SWEET VIOLET**	**Vìola odoràta var. álba**	Tufted Violet with creeping runners. Flowers very fragrant. Heart-shaped leaves. Should be naturalized in large quantities. Prop. by seed and division. Loose, rich sandy soil preferable. Europe.	6 in. *Half shade*	Late Apr. to late May
"Yellow" 5	**SPRING ADONIS, OX EYE**	****Adònis vernàlis** *A. Apennìna, A. Davùrica*	Alpine plant. Dense tufts bear large buttercup-like flowers. Very finely cut foliage in whorls. Useful for rock-garden or border. Prop. by seed sown when gathered, or by division. Light sandy soil. Europe.	8-15 in. *Sun or half shade*	Mid. Apr. to June

Color	English Name	Botanical Name and *Synonyms*	Description	Height and *Situation*	Time of Bloom
"Yellow" 4	GOLDEN GARLIC	**Àllium Mòly	Flowers in round clusters surmounting a slender stem. Leaves lance-shaped. Easily cultivated. Good in masses. Bulbous. Prop. by offsets and seed. Hungary; the Pyrenees.	1 ft. *Sun or shade*	Mid. Apr. to June
"Lemon yellow" 2	AUSTRIAN MADWORT	*Alýssum Gemonénse	Shrubby habit. Less hardy than A. saxatile, but with larger flowers. Rock-plant. Prop. by seed, division and cuttings. Europe.	9-12 in. *Sun*	Apr., May
"Golden yellow" 5	ROCK MADWORT, GOLDEN-TUFT	**Alýssum saxátile	Excellent rock-plant of spreading habit covered with numerous clusters of small flowers which form an effective mass of color in border or rock-garden. Good for edgings. If cut after flowering it will bloom again, even after frost. Prop. by seed, division or cuttings. Of easy culture. Prefers a well-drained soil. Europe. See Flate, page 25. *A. saxatile var. variegatum* is a variegated form. **A. saxatile var. flore-pleno* is a double var., a splendid rock-plant and good for the front of borders.	1 ft. *Sun*	Mid. Apr. to late May
"Yellow" 5	COMPACT ROCK MADWORT OR GOLDEN-TUFT	**Alýssum saxátile var. compàctum	A compact var. of A. saxatile. Fragrant flowers in clusters. Foliage silvery. Used for rockwork or border. See A. saxatile. Europe.	1 ft. *Sun*	Mid. Apr. to June
"Golden yellow" bet. 1 & 2	YELLOW WOOD ANEMONE	*Anemòne ranuncu-loídes	Plant resembles the Apennine Anemone. Flowers single or semi-double, usually solitary. Leaves deeply cut. Good rock-garden plant. Rich light soil. Europe; Siberia.	3-8 in. *Sun or shade*	Apr., May
"Yellow" 2	YELLOW CANADIAN COLUMBINE	*Aquilègia Canadénsis var. flaviflòra *A. C. var. flavéscens. A. cærùlea var. flavéscens*	Pretty variety. Flowers droop over the grayish foliage. Plant in sheltered situation in the rock-garden or border. Prop. by seed sown as early as possible, or by division. Deep sandy loam. N. Amer.	1-2 ft. *Sun*	Late Apr. to early July
"Yellow" 5	PROPHET FLOWER	Arnèbia echioìdes *Macrotòmia echioìdes*	Slightly curving spikes of flowers with purple spots which gradually disappear. Spreading leaves. Does well in a northern exposure. Suitable for rock-gardens. Prop. by seed, division or cuttings. Moist, but well-drained soil. Caucasus.	3-12 in. *Half shade*	Apr., May
"Yellow" 5	DOUBLE MARSH MARIGOLD	Cáltha palústris var. flòre-plèno *C. p. var. monstròsa-plèna*	Double form of our native plant. Flowers 1½ in. broad. Good for cutting. Prefers wet places, but does well in a rich border. Hort.	1-2 ft. *Sun or half shade*	"

ROCK MADWORT. *Alyssum saxatile.*

26 COMMON DOGTOOTH VIOLET. *Erythronium Americanum.*

Color	English Name	Botanical Name and *Synonyms*	Description	Height and *Situation*	Time of Bloom
"Greenish yellow" green & 2	BLUE COHOSH	**Caulophýllum thalictroìdes**	Native plant of the Barberry family. Racemes of small flowers, succeeded by deep blue berries. Foliage dark and finely cut. Good for wild garden. Prop. by division. Grows well in rich soil or peat. N. Amer.	1-2½ ft. *Half shade*	Apr., May
"Yellow" 6	CROCUS	****Cròcus vars.**	See page 8.		Mid. Mar. to late Apr.
"Deep yellow" 6	CLOTH OF GOLD CROCUS	***Cròcus Susiànus**	See page 4.		"
"Yellow" 4	EVER-GREEN WHITLOW GRASS	**Dràba Aìzoon**	Tiny Alpine plant bearing a profusion of small flowers rising from dense rosettes of stiff dark green linear foliage. Plant on walls or in masses in rock-garden. Prop. by fall-sown seed, generally by division. Well-drained soil. Europe.	3 in. *Sun*	Late Apr., May
"Yellow" 4	ALPINE WHITLOW GRASS	**Dràba alpìna**	Flower-stems hairy; leaves lance-shaped. For description see D. Aizoon.	3 in. *Sun*	"
"Yellow" 4	COMMON WINTER ACONITE	****Eránthis hyemàlis**	See page 4.		Mar., Apr.
"Sulphur yellow" 4	ALPINE WALL-FLOWER	**Erýsimum ochroleùcum**	Resembles the Wallflower. Flowers cover the plant thickly. Foliage forms a dense tuft. Good for front of border, dry banks and rock-gardens. "On level ground it is likely to lose lower leaves and to perish on heavy soils in a hard winter." Prop. by division and cuttings. Spain.	6-12 in. *Sun*	Late Apr., May
"Yellow" 1	COMMON ADDER'S TONGUE OR DOGTOOTH VIOLET	****Erythrònium Americànum**	Native plant. Charming lily-like drooping flowers with recurved petals, sometimes purple-tinged. Long narrow mottled leaves. Excellent for rock-garden, to naturalize or to plant in Rhododendron beds. Mulch in winter. Bulbous. Prop. by seed and offsets. Light peaty soil. Eastern U. S. A.; Canada. See Plate, page 26.	6 in. *Half shade*	Late Apr. to late May
"Yellow" 15 yellower & lighter	LEMON-COLORED DOGTOOTH VIOLET	***Erythrònium citrìnum**	Pretty native plant. Flowers with broad petals much reflexed, tipped with pink, 1-3 on the stem. Leaves mottled. For location and cultivation see E. Americanum. N. Amer.	6 in. *Shade*	Late Apr. to mid. May
"Yellow" 5 light	LARGE-FLOWERED DOGTOOTH VIOLET	****Erythrònium grandiflòrum** *E. gigantèum*	Attractive native plant. One to six flowers on slender stems. Handsome unmottled leaves. Blooms early. Naturalize under deciduous trees. For location and cultivation see E. Americanum. N. Western Amer.	6 in. *Half shade or shade*	"

Color	English Name	Botanical Name and *Synonyms*	Description	Height and *Situation*	Time of Bloom
"Yellow" 2 pale & greenish	MYRSINITES-LIKE SPURGE	*Euphórbia Myrsinìtes	Low prostrate plant. Flowers in umbels surrounded by leaflets. Pulpy whitish green leaves. Good for rock-gardens. Prop. by division. Any soil. S. Europe.	1 ft. *Sun*	Mid. Apr. to late May
"Bright yellow" 5	GOLDEN FRITILLARY	*Fritillària aùrea	Solitary drooping bell-shaped flowers checkered with brown. Deep green lance-shaped leaves generally in whorls of three. Replanting every 3 or 4 years necessary. Bulbous. Prop. by offsets. Deep sandy loam. Cilicia.	6-12 in. *Shade*	Mid. Apr. to June
"Greenish yellow" 2 deep & greenish	PALE-FLOWERED FRITILLARY	*Fritillària pallidiflòra	Large pale bell-shaped flowers. Bluish green foliage. Pretty for the border and rock-garden. For cultivation see F. aurea. Siberia.	6-15 in. *Shade*	"
"Yellow" 5	SHY FRITILLARY	*Fritillària pùdica	A charming plant and graceful species with fragrant bell-shaped uncheckered flowers, solitary and drooping. Long and narrow grayish leaves. Replanting every 3 or 4 years necessary. Give slight shelter. Bulbous. Prop. by offsets. Deep sandy loam mixed with leaf-mold. Also a purple var. N. Western Amer.	6-12 in. *Half shade*	Mid. Apr. to mid. May
"Yellow"	DUTCH HYACINTH	Hyacínthus orientàlis vars.	See color "various," page 59.		Late Apr., May
"Yellow" 5	HOOP-PETTICOAT DAFFODIL	*Narcíssus Bulbocòdium	Small delicate species with several solitary flowers and slender leaves. Pretty but not very hardy. Plant in Sept. or Oct., 5 in. deep and 3 in. apart. Prop. by offsets. Any good well-drained soil. Give an occasional top-dressing. S. France; Spain.	5-8 in. *Half shade best*	Late Apr. to late May
"Yellow mixed"	STAR DAFFODIL	**Narcíssus incomparábilis & vars.	A vigorous species which increases rapidly. Solitary scentless flowers. Plant in rows or masses in the border or edge of shrubbery in Sept. or Oct., 6 in. deep and 6 in. apart. Cover in winter. In 5 or 6 years the bulbs become crowded and should be separated. Bulbous. Prop. by offsets. Any good well-drained soil. Give an occasional top-dressing. S. Spain; France to the Tyrol. See Plate, page 29. The following are some of the best garden vars. They are hardy and increase rapidly. **Medium Trumpet Narcissi, Single vars.**—***Barrii Conspicuus;* (petals color no. 2, cup 5), yellow crown, edged with orange-scarlet, and broad spreading primrose base-petals; unexcelled for cutting, lasting long in water. **Incomparabilis cynosure;* (color no. 2, crown 5), large base-petals with	12-15 in. *Half shade best*	Mid. Apr. to mid. May

STAR DAFFODIL "SIR WATKIN." *Narcissus incomparabilis "Sir Watkin."*

STAR DAFFODIL VAR. LEEDSII. *Narcissus Leedsii.*

JONQUIL. *Narcissus Jonquilla.*

Color	English Name	Botanical Name and *Synonyms*	Description	Height and *Situation*	Time of Bloom
			bold orange crown. ***Incomparabilis Sir Watkin,* Giant Welsh Daffodil; (petals color no. 3, cup 6), rich golden yellow cup, beautifully fringed, very large primrose base-petals. **Incomparabilis Stella;* (white and color no. 5), pure white base-petals, with bright yellow crown; blooms early. **Leedsii;* (color nos. 2 pale & 3 light), white base-petals, pale primrose cup which becomes white; fragrant. See Plate, page 30. ***Leedsii Amabilis;* (color nos. 2 pale & 3 light), pure white base-petals, and long lemon-yellow crown; a very beautiful var. ***Leedsii Duchess of Brabant;* (color nos. 2 pale & 3), white base-petals small canary-yellow cup; a pretty var. **Nelsoni Major;* (color nos. 2 pale & 4), large white base-petals, and long fluted bright yellow cup. **Double vars.**—These bloom somewhat later and are not so charming as the single vars. *Incomparabilis Orange Phœnix,* Eggs and Bacon; (color no. 7 & white), very large double cream white and orange flower of great beauty. *Incomparabilis plenus,* Butter and Eggs; (color no. 2, centre 5), large double yellow flowers with orange centres. *Incomparabilis Silver Phœnix,* Codlins and Cream; (color no. 3 & white), one of the finest double sorts, with large pale creamy white fragrant flowers.		
"Yellow" 3	**JONQUIL**	***Narcissus Jonquílla & vars.**	Delicate and graceful species; 2-6 flowers on a stalk. Plant in Sept. or Oct., 5 in. deep and 3 in. apart. Prop. by offsets. Any good well-drained soil. Give an occasional top-dressing. S. Europe; Algeria. See Plate, page 30. **Vars.—Double** and **Single,** Sweet scented; deep yellow. *Campernelle;* (color no. 5), large deep yellow flowers, excellent for cutting. *Rugulosus;* (color no. 4), very fragrant deep yellow flowers; the largest Jonquil.	8-15 in. *Half shade best*	Apr., early May
"Yellow mixed"	**COMMON OR TRUMPET DAFFODIL, LENT LILY**	****Narcíssus Pseùdo-Narcíssus vars.**	Hardy and common species. Usually a single Trumpet Narcissus, but having also double forms in which the trumpet disappears. Plant in Sept. or Oct., 6-8 in. deep and 4 in. apart. Cover in winter and separate after 5 or 6 years. Plant in rows or masses in the border or on the edge of shrubbery. Prop. by offsets. Any good	12-18 in. *Half shade best*	Late Apr., May

Color	English Name	Botanical Name and *Synonyms*	Description	Height and *Situation*	Time of Bloom
			well-drained soil. Give an occasional top-dressing. The following is a list of hybrids. This is an especially beautiful group, but not so hardy as the Medium Trumpet Narcissus.		
"Yellow mixed"			**Large Trumpet Narcissi. Single vars.**—***Ard Righ;* (color nos. 3 & 4), large flowers blooming early. ***Emperor;* (color no. 3, cup 4), immense flowers with golden yellow trumpets, and deep primrose base-petals; hardy. **Golden Spur;* (color no. 3, cup 4), large flowers with immense trumpets. **Henry Irving;* (color no. 4), very handsome golden yellow flowers with large trumpet recurved at the tip, and broad overlapping base-petals. **Johnstoni,* Queen of Spain; (color no. 2), sulphur-yellow flowers; blossoms early. **Nanus;* (color no. 3), about 6 in. high, bearing clear yellow flowers which bloom early. **Obvallaris,* Tenby Daffodil; (color nos. 4 & 3 bright), bright yellow flowers with wide trumpets and broad base-petals; quite hardy. **Princeps;* (color no. 2, cup 3), large yellow trumpets and pale primrose base-petals. ***Rugilobus;* (color no. 4 to 3 light), large golden yellow trumpets and broad primrose base-petals; a free bloomer. ***Spurius;* (between color nos. 3 & 5), one of the finest Trumpet Daffodils, clear yellow flowers with large wide-mouthed trumpets. ***Trumpet Major;* (color no. 5), very effective golden yellow flowers. ***Trumpet Maximus;* (color no. 2, cup 5), golden yellow flowers with fringed trumpets and twisted base-petals. **Trumpet Minor;* is a dwarf form of *T. Maximus,* about 5 in. high. **Empress;* (color no. 2, cup 4), one of the best Daffodils, rich yellow trumpet and broad cream-white base-petals. ***Grandee Maximus;* (color no. 2, cup 5), large golden yellow trumpet and broad cream-white finely imbricated base-petals. ***Horsfieldii,* King of the Daffodils; (color no. 2, cup 5), large golden yellow trumpet and cream-white base-petals. **Scoticus;* (color no. 4), deep yellow trumpet beautifully serrated, white base-petals; not quite hardy. **Double vars.**—*Telemonius plenus, Van Sion;* "Old Double Yellow Daffodil"; (color nos. 5 & 6), very large deep golden yellow flowers. See Plate, page 33.	12-15 in. *Half shade best*	Apr. to mid. May

DOUBLE TRUMPET DAFFODIL. *Narcissus Pseudo-Narcissus double var.*

ENGLISH COWSLIP. *Primula officinalis.*

ORANGE GLOBE FLOWER. *Trollius Asiaticus.*

APRIL

Color	English Name	Botanical Name and *Synonyms*	Description	Height and *Situation*	Time of Bloom
"Yellow" 5 or 6 brilliant	ICELAND POPPY	**Papàver nudicaùle	Lovely plant. Large solitary cup-shaped flowers on graceful leafless stems; nodding buds. Not enduring; should be treated as an annual. If cut back flowers almost continuously from spring until frost. Mass in border or rock-garden. Prop. by seed. Light rather rich soil. Arctic Regions.	9-15 in. *Sun*	Late Apr. to July, late Aug. to Oct.
"Pale yellow" 2	COMMON EUROPEAN OR TRUE PRIMROSE	*Prímula acaùlis *P. grandiflòra* *P. vulgàris*	Several solitary flowers, 1 in. across on naked stems, rise above the pretty tuft of foliage. Slight protection needed in winter. Very pretty in sheltered positions in the rock-garden or border. Prop. by seed and offsets. Rich light soil, not dry. England; Europe.	6-9 in. *Half shade*	Mid. Apr. to June
"Yellow" 3	ENGLISH COWSLIP	**Prímula officinàlis	Six to twelve small cup-shaped flowers drooping in one-sided clusters which rise high above the rosette of foliage. Charming in sheltered positions in the rock-garden or border. Protect in winter. Prop. by seed. Rich light soil, not dry. N. and Central Europe. See Plate, page 34.	6-12 in. *Half shade*	Late Apr. to late May
"Orange yellow" 6	ORANGE GLOBE FLOWER	**Tróllius Asiáticus	Large solitary flowers, 1-2 in. across, less globular than T. Europæus, on leafy stems. Foliage bronze-green and finely divided. Good for cutting. Popular border plant. Prop. by division or seed which will flower the second year. Prefers a damp peaty soil. Siberia. See Plate, page 34.	1½-2 ft. *Sun or half shade*	Late Apr. to late May, early Aug. to Oct.
"Yellow"	TULIP	**Tùlipa vars.	**Single Early Bedding Tulips.** See color "various," page 64.		Late Apr. to late May
"Yellow"	TULIP	**Tùlipa vars.	**Double Early Bedding Tulips.** See color "various," page 64.		"
"Yellow" 3	VARIEGATED COLTSFOOT	Tussilàgo Fárfara var. variegàta	Somewhat coarse spreading plant. Dandelion-like flowers. Grown chiefly for its large downy leaves margined and spotted with white. Apt to spread too readily. Wild garden. Associate with ferns, etc. Prop. by runners. Grows in stiff clay. Hort.	6-8 in. *Shade*	Apr.
"Yellow" 6	YELLOW HORNED VIOLET OR BEDDING PANSY	**Vìola cornùta var. lùtea màjor	One of the many charming yellow varieties of this species. Excellent for border or rock-garden. Prop. by seed, division or cuttings. Any good soil.	5-8 in. *Sun or half shade*	Late Apr. until frost
"Yellow" effect 2 light	HAIRY YELLOW VIOLET	Vìola pubéscens	Common in our woods. Softly downy. Lower petals purple veined. Broad heart-shaped leaves. Plant in a shady part of rock or wild garden. Prop. sometimes by seed, generally by division. N. Amer.	6-12 in. *Half shade*	Late Apr. to late May

Color	English Name	Botanical Name and *Synonyms*	Description	Height and *Situation*	Time of Bloom
"Yellow" 4	EARLY OR GOLDEN MEADOW PARSNIP	Zízia aùrea	Flowers in flat-topped clusters. Plant in masses in the wild garden. N. Amer.	1-2½ ft. *Sun*	Late Apr. to late May
"Orange yellow" 5	GOLDEN BLOOMERIA	Bloomèria aùrea *Nothoscordúm aúreum*	Round clump of flowers surmounting a slender stem. Foliage grass-like springing from the root. Winter protection of leaves or litter. Bulbous. Flowers from seed in 4 years. Warm sandy soil. Cal.	6-18 in. *Sun*	Apr., May
"Orange"	DUTCH HYACINTH	**Hyacínthus orientàlis vars.	See color "various," page 59.		Late Apr., May
"Deep orange" bet. 12 & 17	ORANGE ICELAND POPPY	*Papàver nudicaùle var. aurantìacum	Charming large cup-shaped flowers on graceful erect and leafless stems; buds nodding. Treat as an annual. If cut back flowers almost continuously from spring until frost. Rock-garden or border. Prop. by seed. Light soil, fairly rich. Arctic Regions.	9-15 in. *Sun*	Late Apr. to July, late Aug. to Oct.
"Deep orange" bet. 12 & 17	SMALL ICELAND POPPY	*Papàver nudicaùle var. miniàtum	Very showy var. See P. nudicaule var. aurantiacum.	9-15 in. *Sun*	Late Apr. to July, mid. Aug. to Oct.
"Scarlet" 19 & 4	WILD COLUMBINE	*Aquilègia Canadénsis	Compact native plant of slender growth, with many flowers. Effect is scarlet, but petals are yellow, with spurs of brilliant scarlet and protruding stamens. Pretty foliage. Useful for wild or rock-garden. Prop. preferably by seed. Any ordinary soil. N. Amer. See Plate, page 37.	1-2 ft. *Sun or shade*	Late Apr. to mid. June
"Orange scarlet" 17	CALIFORNIAN COLUMBINE	*Aquilègia truncàta *A. Califórnica, A. eximea*	Drooping yellow-tinged flowers with short blunt petals and stout spurs. Border plant. Prop. by seed or careful division. Sandy loam, deep and well-drained. Cal.	2-3 ft. *Sun or half shade*	Apr. to early June
"Red"	DUTCH HYACINTH	**Hyacínthus orientàlis vars.	See color "various," page 59.		Late Apr., May
"Brownish red" 28 neutral & lighter	ILL-SCENTED WAKE-ROBIN	*Tríllium eréctum *T. pèndulum T. purpùreum T. fœtidum*	Charming flowers on graceful stems from which spring broad dark leaves. Pretty for damp borders, to plant among ferns, or to naturalize under trees. Prop. by seed or division. A rich moist soil is best. N. Amer.	1 ft. or more *Half shade*	Late Apr. to early June
"Red"	TULIP	**Tùlipa vars.	**Single Early Bedding Tulips.** See color "various," page 64.		Late Apr. to late May
"Red"	TULIP	**Tùlipa vars.	**Double Early Bedding Tulips.** See color "various," page 64.		"

WILD COLUMBINE. *Aquilegia Canadensis.*

GREIG'S TULIP. *Tulipa Greigi.*

APRIL

Color	English Name	Botanical Name and *Synonyms*	Description	Height and *Situation*	Time of Bloom
"Red" 19	GREIG'S TULIP	**Tùlipa Greìgi	Very showy. Large flowers 5 in. across with dark spots at base of petals. Leaves marked with purple. Effective in masses. Plant in late Sept. or Oct. at a depth of 4 in. to the bottom of the bulb, allowing 4 in. between each. Best to protect in winter and to give an occasional top-dressing. Prop. by offsets. Light soil. If drainage is poor, put sand around the bulb. Turkestan. See Plate, page 38.	3-8 in. *Sun*	Late Apr. to mid. May
"Intense pink" 27 lighter	LAUCHE'S THRIFT OR SEA PINK	**Armèria marítima var. Laucheàna *A. Laucheàna*	Flowers in dense heads spring from cushions of foliage. Easily grown and blooms profusely. Excellent for rock-garden or for edging. Prop. by seed and division. Any garden soil. Hort.	3-6 in. *Sun*	Late Apr. to mid. June
"Pink" 31 deeper & redder	LEICHTLIN'S ROCK CRESS	*Aubriètia deltoìdea var. Leìchtlini	Creeping plant, usually evergreen, covered in spring with a profusion of flowers. Leaves grayish. Attractive when trailing over rocks and when used for edgings. Prop. by seed, division or cuttings. Easily grown. S. Europe.	2-10 in. *Sun or half shade*	Late Apr., May
"Pink" 29 & white	ENGLISH DAISY	*Béllis perénnis	A plant of neat effect. Rays white, tipped with pink or red, so numerous as frequently to conceal the yellow centre. Should be protected in winter. Desirable for edging and border. Prop. by seed in spring or division in Sept. Rich soil. W. Europe. See Plate, page 41.	3-6 in. *Sun*	Mid. Apr. to mid. June
"Rose" 36 & 23	CAROLINA SPRING BEAUTY	Claytònia Caroliniàna	Spreading dwarf plant. Flowers in loose racemes on slender stems, pretty, but of few days' duration. Good for the wild or rock-garden. Tuberous plant. Prefers moist soil. Minn. to the Atlantic and S. to the Mts. of N. C.	6 in. *Half shade*	Late Apr., May
"Light pink" 36	SPRING BEAUTY	Claytònia Virgínica	Flowers are larger and more numerous than in C. Caroliniana. Slender stem bears loose clusters of star-shaped flowers, veined with deeper pink, which pass quickly. Plant six to twelve together in wild or rock-garden. Tuberous plant. Moist rich soil. Middle and Eastern U. S. A.	6-12 in. *Half shade*	Apr., May
"Deep pink" 29	GARLAND FLOWER	**Dáphne Cneòrum	Charming dense trailing evergreen shrub with close clusters of deliciously fragrant flowers and crimson buds. Attractive dark glossy foliage. Often blooms again in summer. Plant in sheltered position in front of shrubs or in the rock-garden. Prop. by layers or cuttings of half ripe wood. Light loam well-drained and enriched. Europe.	6-12 in. *Sun or half shade*	Late Apr., May

Color	English Name	Botanical Name and *Synonyms*	Description	Height and *Situation*	Time of Bloom
			Var. *majus;* (color no. 30), "deep pink." Of stronger growth than the type. Hort.		
"Rose" 30	**BLEEDING HEART**	**Dicéntra spectábilis *Diélytra spectábilis*	The best species of Dicentra. An excellent perennial. Arching stems bear drooping heart-shaped flowers with white protuberances. Handsome deeply-cut foliage. Good old-fashioned border plant. Prop. by division of crown or roots. Light loam not too rich. Japan.	1-2 ft. *Half shade best*	Late Apr. to mid. July
"Pinkish" 36 or 29	**MAY-FLOWER, TRAILING ARBUTUS, GROUND LAUREL**	*Epigǽa rèpens	The well-known and lovely plant, native of the woods, with fragrant flowers in clusters amidst dark glossy foliage. Difficult of culture. Plant in the wild garden. Prop. by layers, cuttings or division. Moist soil, sand and leaf-mold. U. S. A. east of the Mississippi.	2-4 in. *Shade*	Apr., early May
"Pink"	**DUTCH HYACINTH**	**Hyacínthus orientàlis vars.	See color "various," page 59.		Late Apr., May
"Pink" 29 pale	**RED ALPINE CAMPION**	*Lýchnis alpìna *Viscària alpìna*	Star-like flowers in showy clusters. Foliage thick at base of plant. Plant in the border or rock-garden. Prop. by seed or division. Ordinary light garden soil. N. Asia; Europe; Amer.	6-12 in. *Sun*	"
"Magenta" bet. 44 & 38 dull	**DARK PURPLE MOSS PINK**	*Phlóx subulàta var. atropurpùrea	Creeping evergreen plant covered with masses of small flowers which rise from a moss-like bed of foliage. Good carpeting plant for rock-garden or border. Prop. by seed, division or cuttings. Dry light soil. Hort.	4-6 in. *Sun*	"
"Rose"	**LEAFY MOSS PINK**	*Phlóx subulàta var. frondòsa	For description and cultivation see P. subulata var. atropurpurea. Hort.	4-6 in. *Sun*	"
"Pink" bet. 29 & 32	**THICK-LEAVED SAXIFRAGE**	*Saxífraga crassifòlia	Showy and spreading. Drooping masses of flowers high above the large clustered leaves. Protect in winter. One of the most popular rock-plants. Prop. by division or offshoots. Any soil. Siberia.	12-15 in. *Half shade*	Late Apr. to late May
"Bright pink" 29	**WILD PINK**	*Silène Pennsylvánica	Small attractive plant. Flowers in nodding clusters of 6 or 8 about 1 in. across. Foliage mostly about the root. Good rock-garden plant. Prop. by seed or cuttings. Sandy soil. Eastern U. S. A.	6-9 in. *Sun*	Late Apr., May
"Pale flesh-color" white turns 36	**TINY TRILLIUM**	Tríllium pusíllum	Flowers small, about an inch long, on erect stems. Leaves lance-shaped or oblong. Pretty in moist borders among ferns and naturalized under trees. Prop. by seed and division. Rich moist soil. N. Amer.	1 ft. *Half shade*	Late Apr. to late May
"Pink"	**TULIP**	**Tùlipa vars.	**Single Early Bedding Tulips.** See color "various," page 64.		"

ENGLISH DAISY AND POET'S NARCISSUS. *Bellis perennis and Narcissus poeticus.*

ALPINE THRIFT. *Armeria alpina.*

Color	English Name	Botanical Name and *Synonyms*	Description	Height and *Situation*	Time of Bloom
"Pink"	TULIP	**Tùlipa vars.	**Double Early Bedding Tulips.** See color "various," page 64.		Late Apr. to late May
"Rosy" bet. 36 & 37	DOUBLE PINK PERIWINKLE OR MYRTLE	Vìnca mìnor var. ròsea plèna	A rare form with double flowers of unusual and conspicuous color. Its habit of growth is the same as that of the Common Myrtle. Prop. by division or cuttings. Any ordinary garden soil. Hort.	2-6 in. *Sun*	Late Apr. to June
"Pale purple" white to 44 bluer	CAROLINA WIND-FLOWER	Anemòne Caroliniàna *A. decapètala*	Single flowers 1-1½ in. across on slender stems. Color varies to white. Foliage much divided. Wild garden. Tuberous. Prop. by seed and division. Rather rich light sandy loam. U. S. A.	6-12 in. *Sun*	Apr., May
"Lilac" 50	AMERICAN PASQUE FLOWER, WILD PATENS	*Anemòne pàtens var. Nuttalliàna *Pulsatìlla hirsutìssima* (Brit.)	Star-shaped flowers on erect stems. Fine foliage. Good for the wild garden. Prop. by seed or division. U. S. A.; Siberia.	4-9 in. *Shade*	"
"Purple" 44 or 37	PASQUE FLOWER	*Anemòne Pulsatìlla *Pulsatìlla vulgàris* *A. acutipètala*	Flowers varying to lilac, 1½-2½ in. across, erect on hairy stems. Leaves much divided. Adapted for rockwork or border. Prop. by seed or division. Well-drained or rocky soil. Europe; Siberia.	9-12 in. *Sun or shade*	Early Apr. to late May
"Pinkish lavender" 37	ALPINE THRIFT	*Armèria alpìna	Flowers in compact heads spring from cushions of needle-like evergreen foliage. Blooms freely and is of easiest culture. Excellent for rock-garden and for edging. Prop. by seed and division. Any garden soil. Europe. See Plate, page 42.	6-9 in. *Sun*	Late Apr. to mid. June
"Dark violet" 48 or 47	PURPLE ROCK CRESS	**Aubrìetia deltoìdea	Creeping Alpine plant usually evergreen and covered in blooming season with a profusion of small flowers. Leaves grayish. Pretty for rock-garden, border and edging. Prop. by seed, layers and cuttings. Any garden soil. S. Europe.	2-10 in. *Sun or half shade*	Early Apr. to late May
"Dark violet" 48 lighter	EYRE'S PURPLE ROCK CRESS	Aubrìetia deltoìdea var. Eỳrei	Large flowered and branching var. Rock plant of trailing habit thickly covered with flowers when in bloom. Good for edging and rock-garden. Prop. by layers and cuttings. Thrives in stony places or any garden soil. S. Europe.	2-10 in. *Sun*	Late Apr., May
"Dark violet" 46 & white	GRECIAN PURPLE ROCK CRESS	*Aubrìetia deltoìdea var. Græca	A compact large-flowered var. Hort.	2-10 in. *Sun*	Mid. Apr. to late May

APRIL

Color	English Name	Botanical Name and *Synonyms*	Description	Height and *Situation*	Time of Bloom
"Dark violet" 56	OLYMPIAN PURPLE ROCK CRESS	*Aubriètia deltoìdea var. Olýmpica	Large flowers. Similar to var. Eyrei. Hort.	2-10 in. *Sun*	Late Apr., May
"Purple" 47 light	DEEP PURPLE ROCK CRESS	*Aubriètia deltoìdea var. purpùrea	More erect than the type, with larger flowers borne on leafy stalks, covering the plant with bloom. Pretty for edging or for carpeting. Prop. by seed, cuttings or layers. Rich deep loam is best. Europe.	2-10 in. *Sun or half shade*	Mid. Apr. to late May
"Rosy purple" 39	SPRING MEADOW-SAFFRON	**Bulbocòdium vérnum	One to three flowers resembling Crocuses spring from each bulb. Large broad leaves appear after the flowers. Rock-garden or border. Bulbs should be divided every 3 years. Prop. by offsets. Plant in light sandy soil. Alps of Europe.	4-6 in. *Sun*	Apr.
"Rosy purple" 39	BEAR'S EAR SANICLE	Cortùsa Matthioli	Choice but delicate downy Alpine plant resembling Primula cortusoides. Drooping flowers in clusters. Winter protection necessary. Rock-garden plant. Prop. by seed sown when gathered. Moist loam, sand and peat. Swiss Alps.	6 in. *Half shade or shade*	Apr., May
"Purple"	CROCUS	**Cròcus vars.	See page 8.		Mid. Mar. to late Apr.
"Lilac" often bet. 43 & 45	COMMON DOGTOOTH VIOLET of Europe	**Erythrònium Dens-Cànis	Pretty solitary drooping flowers; color varies from rosy purple to white. Leaves mottled with reddish brown. Perfectly hardy. Excellent in rock-crannies or for naturalization. Bulbous. Prop. by offsets. Light fibrous soil. Central Europe. See Plate, page 45.	4-6 in. *Half shade*	Late Apr., May
"Deep purple" 42 lighter	PURPLE FRITILLARY	*Fritillària atropurpùrea	One to six bell-shaped flowers faintly checkered with green droop from leafy stems. Division every 3 or 4 years necessary. Bulbous. Prop. by offsets. Sandy and open loam mixed with leaf-mold.	12-15 in. *Shade*	Mid. Apr. to June
"Purple black" 56	TWO-FLOWERED FRITILLARY	*Fritillària biflòra	Two to ten open bell-shaped flowers with green tints, droop from leafy stems. Leaves in whorls. Replanting every 3 or 4 years necessary. Bulbous. Prop. by offsets. Rich sandy loam.	1-2 ft. *Half shade*	"
"Deep purple" 21 purpler	BLACK LILY	*Fritillària Camtschatcénsis *Lílium Camtschatcênse*	One to three drooping bell-shaped flowers on a stem. Deep purple lance-shaped leaves. Plant in the border. Division every 3 or 4 years necessary. Bulbous. Prop. by offsets. Rich deep and sandy soil. Siberia; Western N. Amer.	6-18 in. *Shade*	"

COMMON DOGTOOTH VIOLET (OF EUROPE). *Erythronium Dens-Canis.*

TRAILING PHLOX. *Phlox procumbens.*

APRIL

Color	English Name	Botanical Name and *Synonyms*	Description	Height and *Situation*	Time of Bloom
"Livid purple" 33	RUSSIAN FRITILLARY	*Fritillària Ruthénica	Drooping bell-shaped flowers, 1-3 on a stem. Lance-shaped leaves. For location and cultivation see F. Camtschatcensis. Caucasus Regions.	1-2 ft. *Shade*	Mid. Apr. to mid. May
"Reddish purple" 42	DARK-RED EASTERN HELLEBORE	*Hellèborus orientàlis var. atrórubens *H. o. var. Cólchicus*	See page 4.		Mar., early Apr.
"Purplish" 44 deeper	SWAMP OR STUD PINK	Helònias bullàta	Very showy. Twenty to thirty flowers crowded in terminal racemes. Long dark green leaves. Excellent bog plant; also good for edge of stream or moist sandy border. Bulbous. Prop. by division or offsets. Multiplies rapidly. Any soil, preferably moist. Eastern U. S. A.	1-2 ft. *Sun or shade*	Apr., May
"Purple & lilac"	DUTCH HYACINTH	**Hyacínthus orientàlis vars.	See color "various," page 59.		Late Apr., May
"Blue violet" 55	SLENDER DWARF IRIS	*Iris vérna	Sweet-scented species. Stems very short, giving a dwarfed effect; free blooming. Grassy leaves. Spreads rapidly. Plant in masses in sheltered spots. Ohio and southward.	6 in. *Half shade*	"
"Bluish violet" 39	SPRING BITTER VETCH	**Láthyrus vérnus *Órobus vérnus*	Tufted plant with small pea-like flowers; good for cutting. Pale green foliage. Grows well in almost any situation. Excellent for rock-garden and border. Prop. by seed and division. Any good garden soil. S. and Central Europe.	12-15 in. *Sun*	Mid. Apr. to late May
"Pinkish purple" 27 lighter & more purple	TRAILING PHLOX	*Phlóx procúmbens	A choice dwarf plant closely related to P. amœna but lower growing. Stems prostrate and terminated with panicles of flowers. Excellent for front row of border or rock-garden. Dr. Asa Gray regarded this plant as a hybrid between P. amœna and P. subulata. Prop. by division. See Plate, page 46.	3-5 in. *Sun*	Late Apr. to late May
"Pale lilac" 50	LILAC MOSS PINK	*Phlóx subulàta var. lilacìna	Masses of small flowers cover the mossy bed of foliage. Good for carpeting rock-garden or border. Prop. by seed, division or cuttings. Soil light and dry. Hort.	4-6 in. *Sun*	Late Apr., May
"Lilac" 39	HEART-LEAVED SAXIFRAGE	*Saxífraga cordifòlia	Flowers in thick panicles half hidden by the large dark and heart-shaped foliage. Protect in winter. Beautiful foliage plants for rocky places. Prop. by division or offshoots. Any ordinary garden soil. Siberia. See Plate, page 49.	12-15 in. *Half shade*	Late Apr. to late May

APRIL

Color	English Name	Botanical Name and *Synonyms*	Description	Height and *Situation*	Time of Bloom
"Purple" 53 brighter	PURPLE OR MOUNTAIN SAXIFRAGE	Saxífraga oppositifòlia	Flowers solitary on leafy stalks, so numerous as almost to cover the purplish mossy evergreen foliage. For carpeting rock-gardens or for edging. Prop. by division. Sandy moist soil. Europe; N. Amer. There are several vars.	2-6 in. *Sun or shade*	Late Apr., May
"Violet"	TULIP	**Tùlipa vars.	**Single Early Bedding Tulips.** See color "various," page 64.		Late Apr. to late May
"Violet" 49	COMMON PERIWINKLE, BLUE, RUNNING OR TRAILING MYRTLE	*Vínca mìnor	Evergreen trailer. Large solitary flowers. Dark glossy foliage. Spreads rapidly forming a thick carpet for bare places, under trees and for rock-gardens. Prop. by division or cuttings. Easily cultivated in any ordinary soil. Europe. Var. *cærulea* (*V. cærulea*), has bluer flowers than the type. Hort. Var. *plena,* has double flowers. Var. *aurea variegata,* foliage variegated with yellow.	8-10 in. *Sun or shade*	Late Apr. to June
"Violet" 47 or 49	HORNED VIOLET, BEDDING PANSY	**Vìola cornùta	Charming tufted plant of strong growth; bears a profusion of faintly scented flowers the size of a small Pansy. Good for cutting. Pretty bright foliage. Excellent for border or rock-garden or for spring bedding. Prop. by seed, division or cuttings. Any good soil. Pyrenees.	5-8 in. *Sun or half shade*	Late Apr. until frost
"Violet" 48 or 47	SWEET VIOLET	Vìola odoràta	Parent of the cultivated Violet; tufted, with creeping runners. Flowers very fragrant, varying to reddish purple. Heart-shaped leaves. Should be naturalized in large quantities. Prop. by seed or division. Loose rich sandy soil preferable. Europe; Africa; N. Asia.	6 in. *Half shade*	Late Apr. to late May
"Violet" 53	EARLY BLUE VIOLET	Vìola palmàta *V. cucullàta var. palmàta*	Flowers sometimes striped with white, growing in very dense clumps intermingled with foliage. Leaves heart-shaped. Good to naturalize in shady spots. Prop. by division. A rather heavy sandy loam is best. Eastern U. S. A.	2-6 in. *Half shade*	Late Apr., May
"Violet" 46	COMMON BLUE OR LARGE AMERICAN VIOLET	Vìola palmàta var. cucullàta	Commonest species in the east. Thick clumps of leaves and flowers. Good for naturalizing. For cultivation see V. palmata. Atlantic States.	2-6 in. *Half shade*	Late Apr. to late May
"Sky-blue" 52	APENNINE WIND-FLOWER	*Anemòne Apennìna	Flowers 1½ in. across. This and white var. well suited for the wild garden. Prop. by division. Any good garden soil, though a well-drained rich sandy loam is best. Somewhat tender. Italy.	4-9 in. *Shade*	Apr., May

HEART-LEAVED SAXIFRAGE. *Saxifraga cordifolia.*

 VIRGINIAN COWSLIP. *Mertensia pulmonarioides.*

Color	English Name	Botanical Name and *Synonyms*	Description	Height and *Situation*	Time of Bloom
"Sky-blue" 52 intense	BLUE WINTER WIND-FLOWER OR ANEMONE	*Anemòne blánda	Flowers star-like. Foliage divided and toothed. Excellent for rock-garden and to naturalize. Prop. by seed and division. Rich sandy and well-drained loam. Asia Minor; Greece.	4-6 in *Sun*	Apr., May
"Blue" 44 tinged blue	ROBINSON'S WOOD ANEMONE	*Anemòne nemoròsa var. Robinsoniàna *A. n. var. cærùlea*	Superior to A. nemorosa. Solitary flowers 1 in. or more across. Attractive foliage. Good for rockwork, for naturalizing and for border. Prop. by seed and division. Rich well-drained loam. Hort.	6-12 in. *Half shade*	Mid. Apr. to June
"Blue" 52	ALLEN'S GLORY OF THE SNOW	*Chionodóxa Álleni *Chionoscílla Álleni*	See page 4.		Mid. Mar. to early May
"Blue" 52 & white	GLORY OF THE SNOW	**Chionodóxa Lucíliæ	See page 7.		"
"Blue" 52 or 63	GIANT GLORY OF THE SNOW	**Chionodóxa Lucíliæ var. gigantèa	See page 7.		"
"Blue" 52	TMOLUS' GLORY OF THE SNOW	*Chionodóxa Lucíliæ var. Tmolùsi *C. Tmolùsi*	See page 7.		"
"Gentian blue" 52 deeper	SARDIAN GLORY OF THE SNOW	*Chionodóxa Sardénsis	See page 7		"
"Light blue" 60 deeper & brilliant	VERNAL GENTIAN	*Gentiàna vérna	Beautiful Alpine plant of tufted habit. Solitary salver-form flowers. Rock-garden. Prop. by seed when just ripe. Deep sandy loam, moist but well-drained. Alps; Caucasus; Great Britain.	3 in. *Sun*	Apr. to early June
"Bluish" near 44	SHARP-LOBED OR HEART LIVER LEAF	*Hepática acutíloba *H. tríloba var. acùta, Anemòne acutíloba*	Resembles H. triloba, but the lobes of the leaves are more pointed. Eastern U. S. A.	4-6 in. *Half shade or shade*	Apr., early May
"Bluish" 53 lighter	FIVE-LOBED HEPATICA	*Hepática angulòsa *Anemòne angulòsa*	Flowers over 2 in. across, sometimes varying to reddish or white. For cultivation see H. triloba. Hungary.	4-6 in. *Half shade or shade*	Late Apr., early May
"Bluish" near 44 or 46	ROUND-LOBED OR KIDNEY LIVER LEAF	*Hepática tríloba *H. Hepática, Anemòne Hepática, A. tríloba*	Our common Hepatica. Solitary flowers about half an in. across, varying to pinkish and white. Leaves three lobed. Good for eastern and northern exposures in the rock-garden or for naturalizing under trees. Do not disturb. Prop. by seed and division. Well-drained soil and enriched with leaf-mold. Eastern U. S. A.; Asia; Europe.	4-6 in. *Half shade or shade*	"

Color	English Name	Botanical Name and *Synonyms*	Description	Height and *Situation*	Time of Bloom
"Blue"	DUTCH HYACINTH	**Hyacìnthus orientàlis vars.	See color "various," page 59.		Late Apr., May
"Blue" 57 deeper turns to 29	VIRGINIAN COWSLIP, BLUE BELLS	*Mertènsia pulmona-rioìdes *M. Virgìnica*	The handsomest species of Mertensia. Pretty tubular flowers, which turn pink, in pendent clusters, the effect being a mixture of pink and blue. Large bluish-gray leaves. Charming for the border and for rocky places. Prop. by seed when just gathered, and by division. Rich loose soil. Eastern U. S. A. See Plate, page 50.	1-2 ft. *Sun or half shade*	Late Apr. to late May
"Blue" 63 or paler	COMMON GRAPE HYACINTH	**Muscàri botryoìdes *Hyacìnthus botryoìdes*	Pretty bulb. Small globular scentless flowers in dense clusters. Rather narrow leaves spring from the root. Plant in masses in the border or rock-garden with Narcissi, etc. It makes a pretty edging to beds of bulbs. Prop. by offsets. Ordinary garden soil. Europe; Orient. See Plate, page 53.	6-9 in. *Sun or half shade*	Apr., May
"Dark blue" 49 redder	DARK PURPLE GRAPE HYACINTH	*Muscàri commutàtum	Flowers small and scentless, in short racemes. Narrow leaves. Plant in masses in rock-garden or margin of border. Bulbous. Prop. by seed and offsets. Any ordinary soil. Sicily.	6-10 in. *Sun or half shade*	"
"Blue" 63 duller	TUFTED GRAPE HYACINTH, PURSE TASSELS, TUZZY-MUZZY	*Muscàri comòsum *Hyacìnthus comòsus*	Flowers 40-100 on a long stalk; the lower ones brownish green, projecting outwards, the upper ones blue or violet, on upcurved stems. Very long and narrow leaves. Bulbous. Prop. by seed or offsets. Ordinary soil. Orient; Mediterranean Region.	1 ft. *Sun or half shade*	"
"Blue" 63 duller	FEATHERED, FAIR-HAIRED OR TASSELED HYACINTH	**Muscàri comòsum var. monstròsum *M. plumòsum* *M. p. var. monstròsum*	A distinct form with round fluffy flowers which have a feathery effect. Much preferred to the type. Plant in masses. Bulbous. Prop. by seed and offsets. Ordinary soil. Hort.	1 ft. *Sun or half shade*	"
"Dark blue" 61	STARCH GRAPE HYACINTH	*Muscàri racemòsum *Hyacìnthus racemòsus*	Clusters of flowers with plum-like fragrance rise above long and almost prostrate leaves. Good for border. Bulbous. Prop. by seed and offsets. Any ordinary soil. Mediterranean Region; Caucasus.	4-8 in. *Sun or half shade*	"
"Deep sky-blue" 58 brighter	EARLY FORGET-ME-NOT	*Myosòtis dissitiflòra	Of tufted habit. Flowers in small clusters. Foliage large, bright green. Biennial, but self perpetuating. Good for open places in the rock-garden or for naturalizing on grassy banks. Prop. by division and cuttings. Prefers moist soil but will grow almost anywhere. Switzerland.	6-12 in. *Sun or half shade*	Late Apr. to July

COMMON GRAPE HYACINTH. *Muscari botryoides.*

CHICKWEED PHLOX. *Phlox stellaria.*

BETHLEHEM SAGE. *Pulmonaria saccharata.*

Color	English Name	Botanical Name and *Synonyms*	Description	Height and *Situation*	Time of Bloom
"Blue" 58 deeper	**CREEPING FORGET-ME-NOT**	***Omphalòdes vérna**	Trailing plant. Pretty flowers like Forget-me-nots borne in clusters on an erect stem. Excellent for rock-garden or for fringing walks. Prop. by division. Cool moist loam. Europe.	6 in. *Sun or half shade*	Apr., May
"Pale blue" 50	**CHICK-WEED PHLOX**	***Phlóx stellária**	A tufted or creeping Phlox. Flowers varying to bluish white somewhat scattered. Rock-garden. Prop. by seed, division and cuttings. Light well-drained soil. U. S. A. See Plate, page 54.	5-6 in. *Sun*	Late Apr., May
"Light blue" bet. 46 & 52	**GREEK VALERIAN**	***Polemònium réptans**	Dwarf bushy plant. Flowers in drooping lax panicles. For rock-garden or border. Prop. by fall-sown seed or division. Deep rich loamy soil. Eastern U. S. A.	6-8 in. *Half shade*	Late Apr. to early June
"Blue" 51	**BETHLE-HEM SAGE**	**Pulmonària saccharàta**	Tufted plant. Tubular flowers in clusters, turn to pink. Smooth leaves. Good in semi-wild spot or in the border. Prop. by division. Light soil not too dry is advisable. Europe. See Plate, page 54.	6-18 in. *Half shade or shade*	Late Apr. to late May
"Blue" 61	**STAR HYACINTH**	***Scílla amœna**	See page 7.		Mid. Mar. to early May
"Dark blue" 62	**EARLY SQUILL**	***Scílla bifòlia**	See page 7.		"
"China-blue" 61 deeper	**SIBERIAN SQUILL**	****Scílla Sibírica** *S. amœna var. præcox*	See page 7.		"
"Blue" 46 or 53	**GENTIAN-LEAVED SPEED-WELL**	****Verónica gentian-oìdes**	Charming tufted species. Light blue flowers with dark streaks borne in erect spikes on leafy stems above the glossy foliage. Good to cover bare places in shade or for border and rock-garden. Prop. by division. Any soil. S. Eastern Europe.	½-2 ft. *Sun or half shade*	Late Apr. to late May
Parti-colored	**CROCUS**	****Cròcus vars.**	See page 8.		Mid. Mar. to late Apr.
Cream & 42 out-side, 44 inside	**SCOTCH OR CLOTH OF SILVER CROCUS**	***Cròcus biflòrus**	See page 8.		"
Effect 60 lighter	**TROLLIUS-LEAVED LARKSPUR**	***Delphínium trollifòlium**	Long lax racemes of blue and white flowers surmount leafy stalks. Useful for cutting. Second bloom possible if flowers are cut when faded. Border plant. Prop. in spring or autumn by seed division or cuttings; transplant every 3 or 4 years. Deep	2-5 ft. *Sun*	**Apr.**

Color	English Name	Botanical Name and *Synonyms*	Description	Height and *Situation*	Time of Bloom
			rich soil, sandy and loamy. Western U. S. A.		
26, 43 & white	LARGE-FLOWERED BARREN-WORT	*Epimèdium macránthum	Dainty plant as interesting as many Orchids. "Outer sepals bright red; inner sepals violet; spurs white." New leaves reddish. Suitable for the rock-garden or border. Prop. by division. Any garden soil. Japan.	10-15 in. *Half shade*	Late Apr. to late May
1, markings 17	TWISTED-LEAVED FRITIL-LARY	*Fritillària obliqua	Bell-shaped flowers checkered with brown and purple. Leaves grayish, oblique, numerous. Suitable for mixed border or rock-garden. Bulbous. Prop. by offsets. Rich soil, well-drained. Greece.	1 ft. *Sun*	Mid. Apr. to mid. May
Parti-colored	EASTERN HELLEBORE FRAU IRENE HEINEMANN	*Hellèborus orientàlis var. "Frau Irene Heinemann"	See page 8.		Mid. Mar. to mid. Apr.
Green & 35	PURPLISH GREEN HELLEBORE	*Hellèborus víridis var. purpu-ráscens	See page 8.		"
Parti-colored	TULIP	**Tùlipa vars.	**Single Early Bedding Tulips.** See color "various," page 64.		Late Apr. to late May
Various	CROCUS	**Cròcus vars.	See page 8.		Mid. Mar. to late Apr.
Various	SPRING CROCUS	*Cròcus vérnus	See page 8.		"
Often near 22	FLESH-COLORED HEATH	Erìca cárnea	Branching shrub. Numerous small bell-shaped flowers, rose, pale red or white. Needle-like foliage. Protect in winter. Plant in rock or wild garden. Prop. by division and cuttings. Soil of peat and sand is best. Europe.	6-9 in. *Sun*	Apr., May
4, 13, or 19 more orange	CROWN IMPERIAL	**Fritillària Imperiàlis	Showy plant. Bell-shaped flowers, red, yellow or orange, droop under an upright crown of leaves. Can be cut down when foliage turns brown. Disagreeable odor. Divide every 2 or 3 years. Handsome in groups among shrubs or in border. Bulbous. Prop. by offsets. Deep rich well-drained soil. See Plate, page 57. Var. **aurea marginata* (*F. I. coronata var. aurea marginata*); foliage striped with gold. Var. *variegata*, foliage marked with white and gold.	2-3 ft. *Sun, half shade best*	Mid. Apr. to mid. May

CROWN IMPERIAL. *Fritillaria imperialis.*

POLYANTHUS. *Primula Polyantha.*

 GUINEA-HEN FLOWER. *Fritillaria Meleagris.*

Color	English Name	Botanical Name and *Synonyms*	Description	Height and *Situation*	Time of Bloom
Various	**GUINEA-HEN FLOWER, SNAKE'S HEAD CHECKERED LILY**	****Fritillària Meleàgris**	Large attractive, bell-shaped flowers in light colors checkered with green and purple, also white and purple Good for cutting. Foliage inconspicuous. Plant in border and rock-garden. It is well to have them spring through the green of some low carpeting plant like Creeping Phlox. Divide every 2 or 3 years. Bulbous. Deep sandy loam. Europe; Caucasus. See Plate, page 58.	10-12 in. *Sun or shade*	Late Apr. to late May
Various	**DUTCH HYACINTH**	****Hyacínthus orientàlis vars.**	Fragrant flowers in large dense spikes surrounded by narrow stiff leaves. Protect with stable manure or dry litter. Most effective when planted in rows. Plant bulbs 6 in. deep, 5 in. apart in Sept. or early Oct. Bulbous. Prop. by offsets. Light rich well-drained soil. Give an occasional top-dressing. S. Eastern Europe; Levant. The following are some of the best horticultural vars.:— **Single white vars.**—**Mont Blanc;* snow white with large bells. **Queen of the Netherlands;* pure white, of great substance. **Snowball;* pure white with large bells. *Grand Vainqueur;* pure white. ***La Grandesse;* the finest pure white. **Lady Plimsoll;* white, large truss, very fine. *L'Innocence;* pure white extra large bells. *Madame van der Hoop;* pure white, very large bells. **Alba Maxima;* pure white, large bells and spike. **Alba superbissima;* snow white. **Baroness van Thuyll;* white with primrose eye. **Paix de l'Europe,* large bells, long truss. **Double white vars.**—*Bouquet Royal;* white, very double. *Grootvorstin;* pure white. *Jenny Lind;* white with purple eye. **Miss Nightingale;* pure white. **Prince of Waterloo;* pure white, large double bells. *Sir Bulwer-Lytton;* white, large bells. **Single yellow vars.**—**Bird of Paradise;* (color no. 2), beautiful clear yellow, magnificent spike; one of the best. **Ida;* (color no. 1), clear primrose yellow, large and superb spike. **King of the Yellows;* (color no. 3). pure yellow, large bells in long symmetrical spike. *L'Or d'Australie;* (color no. 2), pure deep yellow, long and very handsome truss. **Obelesque;* (color no. 2 deeper), pure yellow, large truss. *Yellow Hammer;* (color no. 2 warmer), golden yellow.	8-18 in. *Sun*	Late Apr., May

Color	English Name	Botanical Name and *Synonyms*	Description	Height and *Situation*	Time of Bloom
			Double yellow vars.—*Goethe;* (color no. 3 light), light yellow almost white, large spike. *Minerva;* orange yellow, extra large and fine flower. *Ophir d'Or;* (color no. 1), citron yellow with purple centre. *Sunflower;* (color no. 3), buff yellow. **Double orange vars.**—*Bouquet d'Orange;* (color no. 15 light), reddish orange. **Single red vars.**—*Mr Stanley;* (color no. 27 light and brilliant), "dark red." *Robert Steiger;* (color no. bet. 26 deep & 33), deep crimson, compact spikes. *Roi des Belges;* (color no. bet. 26 deep & 33), dark red. **Double red vars.**—**Princess Louise;* (color no. 26 darker), "dark red." **Single pink vars.**—*Baron van Thuyll;* (color no. 36), beautiful delicate rose in large spikes; early. *Charles Dickens;* (color no. 29 warmer), salmon-rose striped carmine; early. *Cosmos;* (color no. 29), deep rose with light eye, large bells. *Fabiola;* (color no. 22), pink striped with bright rose. **Gertrude;* (color no. 23), bright pink, in large spikes. **Gigantea;* (color no. 36 deep), bright rose, very large spikes. **Lord Macaulay;* (color no. 26), deep carmine pink, striped darker, very attractive. *Norma;* (color no. 22), coral pink, large waxy flowers. *Von Schiller;* deep salmon pink, striped with crimson. See Plate, page 61. **Double pink vars.**—*Czar Nicholas;* (color no. 36 warm), delicate rose. *Kohinoor;* (color no. 30), bright rose, large spikes. **Lord Wellington;* (color no. 36), the finest double pink, large bells and massive spike. **Single purple and lilac vars.**—*Grand Lilas;* (color no. 57), "porcelain lilac," large bells and fine spike. *King of the Blacks;* (color no. 56), "black purple." **Single blue vars.**—*Argus;* (color no. 49 & white), "dark blue, white eye." **Charles Dickens;* (color no. 46 deeper), dark porcelain blue, large bells and spike. **Czar Peter;* (color no. bet. 50 & 57), light porcelain blue, large bells, magnificent spike. **General Havelock;* (color no. 63 darker), "deep purple shaded with black," large bells and truss. **Queen of the Blues:* (color no. 46), sky-blue, fine bells, stately spike. **Sir J. Lawrence;* dark blue, white eye. **King of the Blues;* (color no.		

SINGLE HYACINTHS.

POLYANTHUS. *Primula Polyantha.*

Color	English Name	Botanical Name and *Synonyms*	Description	Height and *Situation*	Time of Bloom
			54 darker), clear dark blue, large bells and fine spike. **Lord Derby;* (color no. 44), light porcelain blue, large. See Plate, page 61. **Double blue vars.**—**Bloksberg;* (color no. 46 lighter), porcelain blue, striped. **Garrick;* (color no. 55), "deep lilac blue," very large spike. *Lord Raglan;* (color no. 46), "dark blue." *Lord Wellington;* (color no. 42), dark stripes. **Van Speyk;* pale blue, large bells, compact truss.		
Often 31 duller	**HAIRY PHLOX**	****Phlóx amœna** *P. procúmbens,* (Gray)	Profusion of flowers varying from purple through pink to white in flat-topped clusters. Leaves in tufts at the base of flower-stalks. Good for carpeting border or rock-garden. Prop. by division and cuttings. Light dry soil. S. Eastern U. S. A.	4-6 in. *Sun*	Late Apr., May
40, 36, 43, white & others	**MOSS OR GROUND PINK**	****Phlóx subulàta** *P. nivàlis*	Creeping evergreen plant thickly covered by pink, blue, or white flowers which rise from a mossy bed of foliage, no. 40 being the most common color. Favorite rock-plant, good for carpeting in border or rock-garden. Prop. by seed, division or cuttings. Any light soil. N. Y., West and South.	4-6 in. *Sun*	"
34 & others	**AURICULA**	***Prímula Aurícula** *Auricula*	Not very hardy. Round bright flowers in clusters terminating a leafless stem. Usually yellow-flowered in the wild state; ranging from white to purple in garden vars. Broad thick foliage in a rosette at the base of the plant. Protect in winter. Charming in a sheltered position in the rock-garden or border. Prop. best by offsets or division; also by seed. Rich light soil not dry. Europe.	6-9 in. *Sun or half shade*	"
Various	**POLYANTHUS**	***Prímula Polyántha**	Mixed yellow and red or yellow flowers in umbels rise above the long leaves which spring from the root. Because of its richness in coloring it is good for beds or borders. Prop. by seed, generally by division. Rich moist soil. Hort. See Plate, page 62.	6-10 in. *Half shade best*	"
Mixed white & 48 & others	**VON SIEBOLD'S PRIMROSE**	****Prímula Sièboldi** *P. cortusoìdes var. amœna, var. grandiflòra & var. Sièboldi*	Among the largest and most showy of Primroses. Flowers 1-2 in. across ranging from white to dark purple-rose, in clusters. Leaves in a rosette at the base of the flower-stem. Rock-garden or border, in sheltered position. Winter protection of leaves. Prop. by seed or root division. Light rather rich soil with leaf-mold. Has many pretty vars. Japan.	6-12 in. *Sun or half shade*	"

Color	English Name	Botanical Name and *Synonyms*	Description	Height and *Situation*	Time of Bloom
Various	**TONGUE-LEAVED SAXIFRAGE**	***Saxífraga ligulàta** *S. Schmídtii*	Vigorous showy plant. Flowers white to light purple, globe-shaped, in spreading panicles. Foliage broad, springing from the root. Winter protection of leaves is necessary. Plant in the rock-garden. Prop. by division and offshoots. Any soil. Var. **ciliata* has roundish hairy leaves. Himalayan Region.	1 ft. *Half shade*	Late Apr., May
34 or green	**SESSILE-FLOWERED WAKE-ROBIN**	**Tríllium séssile**	Showy plant with small purple or green flowers surrounded by whorls of broad spotted leaves. Effective in moist borders among ferns and in masses under trees. Prop. by seed or division. Moist rich soil is best. Eastern U. S. A.	8-12 in. *Half shade*	"
Various	**LARGE CALIFORNIAN WAKE-ROBIN**	**Tríllium séssile var. gigantèum** *T. s. var. Califórnicum*	Larger than the type. Petals 4 in. long and leaves often 6 in. Flowers purple pink, or white. For use and cultivation see T. sessile. Pacific States.	8-12 in. *Half shade*	"
Various	**TULIP**	****Tùlipa vars.**	**Early Bedding Tulips.** The following is a carefully selected list of some of the best single and double early Tulips of garden origin. There are many others in the trade also desirable. Plant in late Sept. or Oct. at a depth of 4 in. to the bottom of the bulb, allowing 4 in. between each. It is best to protect in winter and to give an occasional top-dressing. Prop. by offsets. Light soil. If drainage is not good put sand around the bulb. See Plate, page 65. **Single white vars.**—**Joost Van Vondel, White;* very large pure white flowers, height 10 in. *Le Matelas;* (white & color no. 30), white, shaded with bright rose, large blossoms, height 12 in. **Pottebakker, White;* large showy pure white blossoms, height 14 in. *White Swan;* pure white large egg-shaped flowers, height 14 in. **Double white vars.**—*Alba Maxima;* very large white blossoms, height 12 in. *Le Blason:* white shaded with rose, height 11 in. **Single yellow vars.**—*Canary Bird;* (color no. 4), clear rich yellow, height 11 in. **Chrysolora;* (color no. 4), large pure yellow flowers, height 11 in. **Mon Trésor;* (color no. 4), rich golden yellow, extra large, height 11 in. *Ophir d'Or;* (color no. 5), very large pure golden yellow, height 11 in. **Pottebakker, Yellow;* (color no. 5), pure yellow, height 14 in. **Yellow*	10-14 in. *Sun*	Late Apr. to late May

TULIPS.

TULIPA "YELLOW PRINCE"

Color	English Name	Botanical Name and *Synonyms*	Description	Height and *Situation*	Time of Bloom
			Prince; (color no. 4), fragrant bright yellow flowers, height 10 in. See Plate, page 66. **Double yellow vars.**—*Crown of Gold;* (color no. 7), rich golden yellow, very large and very double, height 11 in. *Tournesol Yellow;* (fire color mixed 20 lighter & 3), golden yellow, shaded with orange, height 11 in., especially fine. **Single red vars.**—*Artus;* (color no. 18), bright scarlet, height 10 in. *Belle Alliance;* (color no 20), rich crimson, height 10 in. *Bizard Verdict;* brilliant carmine, gold striped. **Couleur Cardinal;* (color no. 20 brilliant), rich crimson, height 11 in. *Duchess of Parma;* (color no. 19 deeper & brighter), orange-red, yellow margined, height 12 in. **Keizer Kroon;* (color no. 18 bordered with 5), "crimson-scarlet, bordered with yellow," extra large, height 14 in. *Pottebakker Scarlet;* (color no. 20), large bright scarlet, height 14 in *Prince of Austria;* (color no. 17), "rich orange-red," fragrant flowers, height 12 in. *Stanley;* (color no. 30 deep & brilliant), large "dark rich crimson," height 10 in. *Vermilion Brilliant;* (color no. 18 redder & more brilliant), "vermilion," height 12 in. **Double red vars.**—*Gloria Solis;* (color no. 19 bordered with 5), "deep crimson, with yellow border," height 12 in. *Imperator Rubrorum;* (color no. 18), bright scarlet, much doubled and large, height 12 in *Le Matador;* (color no. 14), brownish red, height 12 in. *Rex Rubrorum;* (color no. 20 more brilliant), brilliant scarlet. *Tournesol Scarlet;* (effect color no. 17), bright scarlet, edged with yellow, height 11 in. **Single pink vars.**—*Cottage Maid;* (color no 29 & white), "rose shaded with white," height 9 in. *Proserpine;* (color no. 27), very large "rich deep rose flowers," height 14 in.; it turns bluish. *Queen of the Netherlands;* (color no. 36), beautiful satiny pink, very large blossoms, height 13 in. *Rachel Ruisch;* (color no. 23), "rose," height 10 in. **Double pink vars.**—*Crown of Roses;* (color no. 30), "delicate rose," very double flowers, height 11 in. *Murillo;* (color no. 36 deeper), blush white, shaded with rose, height 10 in.		

Color	English Name	Botanical Name and *Synonyms*	Description	Height and *Situation*	Time of Bloom
			Single violet vars.—*Van der Neer;* (color no. 33 lighter), large purplish violet flowers. **Single parti-colored vars.**—*Grand Duke of Russia;* white striped deep violet *Joost van Vondel, Striped;* (color no. 33 & white), cherry and white, height 10 in. *Rose Grisdelin;* (color no. 23 & white), rose and white, height 10 in. *Wapen Van Leiden;* (white striped with color no. 26), "white striped with rose," height 13 in. See Plate, page 69.		
Various	**DUC VAN THOL TULIP**	****Tùlipa "Duc van Thol"**	The earliest blooming of all Tulips. Dwarf habit. Large fragrant flowers from white to vermilion Plant in masses in late Sept. or Oct. at a depth of 4 in. to the bottom of the bulb, allowing 4 in between each. It is best to protect in winter and to give an occasional top-dressing. Prop. by offsets. Light soil. If drainage is not good put sand around the bulb. Hort. Among the most desirable are:—*Maximus;* (color no. 20 brighter edged with 5), red with yellow edge, fine form. *Rose;* (white tinged with color nos. 36 & 29), milky white shading to rose. *Scarlet;* (color no. 18), "reddish scarlet." *Crimson;* (color no. 20 lighter), bright crimson. *Orange;* (effect color no. 17), very fine. **Vermilion;* (color no. 18), bright vermilion.	6 in. *Sun or half shade*	Late Apr., early May
46, etc.	**DOG VIOLET**	**Vìola canìna**	Flowers blue, lilac, grayish or white, on leafy stems, petals pinkish outside. Plant in shady part of rock or wild garden. Prop. by division. Europe.	3-5 in. *Half shade*	Late Apr. to late May
Various	**PANSY, HEARTSEASE**	****Vìola trícolor**	Our hybrid pansies come under this head as well as the little Heartsease which is often parti-colored, upper petals color no. 42, lower and centre no. 5. This plant has a wonderful length of blooming season and flowers with great richness of coloring, being purple, yellow, whitish, or parti-colored. Best to treat as an annual. Plant in border or rock-garden. Prop. by seed or division. Any garden soil. Europe.	6-8 in. *Sun or half shade*	Mid. Apr. to mid. Sept.

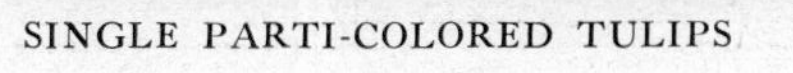

SINGLE PARTI-COLORED TULIPS

A MAY BORDER

MAY

WHITE TO GREENISH

Color	English Name	Botanical Name and *Synonyms*	Description	Height and *Situation*	Time of Bloom
"Cream white"	**WHITE BANE-BERRY**	**Actæa álba** *A. rùbra*	See page 11.		Early Apr. to July
"White"	**COHOSH, HERB CHRISTO-PHER**	**Actæa spicàta**	Vigorous plant of somewhat coarse habit. Large racemes of small flowers tinged with blue which pass quickly. Foliage good. Showy purple berries in autumn. Suitable for the wild garden. Prop. by seed and division in spring. Japan; Europe. Var. *rubra* (*A. rubra*); flower clusters larger; showy red berries. Northern U. S. A.	1-2 ft. *Shade*	May, June
"White"	**WHITE BUGLE**	***Ajùga réptans var. álba**	Dense creeping plant. Spreads rapidly. Numerous spikes of flowers. Shiny leaves. Good for carpeting a shady spot. Not so often cultivated as the blue var. Prop. by seed and division. Europe.	3-4 in. *Sun or shade*	May to mid. June
"Green-ish"	**MOUNTAIN LADY'S MANTLE**	**Alchemílla alpìna**	See page 11.		Late Apr., May
"White" turns to 29 pale	**WILD ONION**	***Állium mutábile**	See page 11.		Apr., May
"White"	**ALPINE ANEMONE**	***Anemòne alpìna** *A. acutipétala*	Beautiful alpine plant. Large flowers solitary or few on the stem, varying to purple and cream outside, rise above the large handsome finely cut foliage. Good for cutting and pretty for naturalization and for the border. Prop. by seed and division. Rich deep soil. Europe.	9-24 in. *Half shade*	May
"White" 39 under petals	**NARCISSUS-FLOWERED ANEMONE**	**Anemòne narcissiflòra** *A. umbellàta*	Flowers ¾ in. across, with yellow centres, in umbels. Foliage deeply divided. Pretty for the rock-garden or border. Prop. by seed and division. Rich soil. Northern Hemisphere.	6-18 in. *Half shade*	May to Aug.
"White" often 36	**WOOD ANEMONE**	**Anemòne nemoròsa**	See page 11 and Plate, page 73.		Late Apr. to early June
"White"	**CANADA ANEMONE**	***Anemòne Pennsyl-vánica** *A Canadén-sis,* *A. dichótoma*	Erect plant. Cup-shaped flowers 1-2 in. across on strong stems. Leaves much divided. Prop. by seed and division. Any good soil; a well-drained good loam is best. N. Amer. See Plate, page 73.	1-2 ft. *Sun or shade*	Mid. May to early July

MAY

Color	English Name	Botanical Name and *Synonyms*	Description	Height and *Situation*	Time of Bloom
"Cream white" tinged with 39	SNOWDROP WIND-FLOWER	**Anemòne sylvéstris	See page 11.		Late Apr. to mid. July
"Cream white"	DOUBLE SNOWDROP WIND-FLOWER	*Anemòne sylvéstris var. flòre-plèno	See page 11.		Apr., May
"Cream white"	RUE ANEMONE	*Anemonélla thalictroìdes *Thalíctrum anemonoìdes*	A pretty native found in woods. Flowers nearly 1 in. across. Prominent stamens. Fern-like foliage. Delights in a shady spot in the rock-garden. Prop. by division. Prefers sandy peat. U. S. A.	3-6 in. *Shade*	May
"White"	ST. BRUNO'S LILY	*Anthéricum Liliástrum *Paradísea Liliástrum*	Lily-like flowers with green tips, in few-flowered racemes. Rush-like foliage springs from the root. Protect in winter. Charming in the border. Prop. by seed and division. S. Europe. Var. *major;* 2-3 ft. An improvement on A. Liliastrum. Flowers larger and more numerous, 2 in. long, 2½ in. across. Good border or rock-garden plant. Hort.	1-2 ft. *Sun*	Late May to early July
"White"	MUNSTEAD'S WHITE COLUMBINE	*Aquilègia vulgàris var. álba *A. v. var. nívea*	See page 11.		Apr., May
"White"	WHITE COLUMBINE	**Aquilègia vulgàris var. nívea	A vigorous plant of compact habit with an abundance of large flowers. Excellent in border or rock-garden. Prop. by seed. Rich soil. Europe.	2-3 ft. *Sun*	Mid. May to July
"White"	WHITE ROCK CRESS	**Árabis álbida *A. Caucásica*	See page 11.		Early Apr. to June
"White"	ALPINE ROCK CRESS	*Árabis alpìna	See page 12.		Early Apr. to late May
"White"	RUNNING ROCK CRESS	*Árabis procúrrens	See page 12.		"
"White"	VERNAL SANDWORT	Arenària vérna *Alsìne vérna*	A dwarf alpine plant with star-shaped flowers on slender stems. Foliage erect. Excellent for rockwork and for carpeting. Prop. by seed, division and rarely by cuttings. Any soil. Europe; Asia; Rocky Mts.	1-3 in. *Sun*	May
"White"	WHITE PLANTAIN-LIKE THRIFT	*Armèria plantagínea var. leucántha *A. dianthoìdes*	Flowers in round compact heads spring from cushions of evergreen foliage. Easily grown. Excellent for rock-garden and edging. Prop. by seed and division. Any garden soil.	6-8 in. *Sun*	Late May to late June

CANADA ANEMONE. *Anemone Pennsylvanica.*

WOOD ANEMONE. *Anemone nemerosa.*

LILY-OF-THE-VALLEY. *Convallaria majalis.*

MAY

Color	English Name	Botanical Name and *Synonyms*	Description	Height and *Situation*	Time of Bloom
"White"	SWEET WOOD-RUFF, HAY PLANT	Aspérula odoràta	Delicate flowers in tight clusters rise from whorls of leaves which, when dry, have a fragrance like hay. Spreads rapidly. Good for edging and carpeting. Prop. by seed and division. Moist loam. Europe; the Orient.	6-8 in. *Half shade or shade*	Early May to mid. June
"White"	WHITE ALPINE ASTER	**Áster alpìnus var. álbus	Similar to the type, but less desirable and less vigorous. Large solitary single flowers with yellow centres. Leaves in clusters at the base of plant. Good in rock-garden or border. Prop. generally by division of clumps. Easily cultivated in ordinary soil and exposure. Europe.	3-10 in. *Sun or half shade*	Late May to late June
"White"	WHITE MOUNTAIN BLUET OR KNAP-WEED	**Centaurèa montàna var. álba	Dwarf compact plant. Large showy flowers resemble Bachelor's Buttons; produced plentifully. Good for border. Prop. by division. Ordinary garden soil. Hort.	9-15 in. *Sun*	Late May to early July
"White"	STARRY GRASS-WORT, FIELD CHICK-WEED	Cerástium arvénse	Mat-like habit; flowers abundant. Good for bedding or half shady places under trees. Europe; Asia; U. S. A.	4-6 in. *Sun or half shade*	May
"White"	LARGE-FLOWERED MOUSE-EAR CHICK-WEED	*Cerástium grandiflòrum	Prostrate creeper. Profusion of flowers on erect stalks. Silvery leaves forming a dense mat. Good for edging or rock-garden, also used in the border. Prop. by division or cuttings. Any soil. Hungary; Iberia.	6-8 in. *Sun*	May, June
"White"	WHITE GLORY OF THE SNOW	*Chionodóxa Lucíliæ var. álba	See page 3.		Mid. Mar. to early May
"Green-ish"	GOLDEN SAXIFRAGE, WATER CARPET	Chrysosplè-nium Americànum	See page 12.		Late Apr., May
"White"	LILY-OF-THE-VALLEY	**Convallària majàlis	A favorite plant. Many small, globular, fragrant flowers droop from the flower stalks. Clean and attractive foliage. Plant in masses under trees where it will spread rapidly. Prop. by division in fall or early spring. Moderately rich soil. Europe; Asia; S. Alleghanies. See Plate, page 74.	8 in. *Half shade or shade*	Mid. May to mid. June
"Green-ish white"	DWARF CORNEL, BUNCH-BERRY	Córnus Canadénsis	Native plant. Involucres white varying to pink, supported by a whorl of dark foliage and succeeded by bunches of red berries. Good for boggy places and wild gardens. Prop. by layers. Any soil. N. Amer.	4-8 in. *Half shade*	May. June

Color	English Name	Botanical Name and *Synonyms*	Description	Height and *Situation*	Time of Bloom
"White" 36 in effect	**SMALL WHITE LADY'S SLIPPER**	***Cypripèdium cándidum**	Leafy stem bears a solitary flower; the sac is striped and spotted with purple. Protect in winter. Suitable for rock or wild garden. Porous moist peaty soil. U. S. A.	6-12 in. *Shade*	May to early June
"White"	**MOUNTAIN LADY'S SLIPPER OR MOCCASIN FLOWER**	***Cypripèdium montànum**	Handsome species. Fragrant flowers with wavy brownish petals, the sac being dull white, veined with purple. Large leaves. Winter protection necessary. Forms clumps in the rock or wild garden. Porous moist peaty soil. Pacific States.	1-2 ft. *Shade*	"
"White"	**PEPPER-ROOT, TWO-LEAVED TOOTH-WORT**	**Dentària diphýlla** *Cardamìne d.*	See page 12.		Apr., May
"White"	**PINK MISS SIMKINS**	****Dianthus "Miss Simkins"**	Dwarf Pink of compact habit with numerous showy flowers. Good for cutting. Excellent border plant. Grows in any ordinary garden soil. Hort.	4-6 in. *Sun*	Late May to late June
"White"	**SQUIRREL CORN**	***Dicéntra Canadénsis** *Dièlytra C.*	See page 12.		Apr., May
"White" white & 3	**DUTCH-MAN'S BREECHES**	***Dicéntra Cucullària** *Dièlytra C.*	See page 12.		"
"White"	**WHITE BLEEDING HEART**	***Dicéntra spectábilis var. álba** *Dièlytra spectábilis var. álba*	Less vigorous in appearance than the type. Heart-shaped flowers in drooping racemes hang above the pretty foliage. Often found in old-fashioned gardens. Prop. by crown or root division. Good light soil, not too rich. Japan.	1-2 ft. *Half shade*	Late May, June
"White"	**UMBRELLA-LEAF**	**Diphyllèia cymòsa**	Flowers in round loose clusters, followed by dark blue berries. One or two very large umbrella-like leaves, heart or shield-shaped. Border. Prop. in spring by division. Dry peaty soil. Mts., Va. to Ga.	1 ft. *Shade*	May
"White"	**MOUNTAIN AVENS**	***Drỳas octopétala**	Evergreen creeper with large rose-like flowers and dense clumps of glossy foliage. Protect from winter sun with boughs. Excellent for the rock-garden. Prop. by seed, division or cuttings. Loose well-drained soil. Northern Hemisphere.	3-4 in *Sun*	May, June
"White"	**LARGE WHITE-FLOWERED BARREN-WORT**	***Epimèdium macránthum var. níveum** *E. níveum*	See page 12.		Late Apr. to late May

Color	English Name	Botanical Name and *Synonyms*	Description	Height and *Situation*	Time of Bloom
"Creamy white"	**MUSSCHE'S BARREN-WORT**	***Epimèdium Musschiànum**	Not so ornamental as the other species, but desirable. Flowers have curious spurs. Foliage leathery, changing in color in the autumn. Prop. by division. Any garden soil. Japan.	1 ft. *Half shade*	May
"White"	**LARGE WHITE MOUNTAIN DAISY**	***Erígeron Coúlteri**	Large feathery daisy-like flowers 1-2 on a stem. Good to mass in the wild garden or plant in the border. Prop. by seed or division. Any ordinary garden soil. Sierras; Rocky Mts.; Col.	6-20 in. *Sun*	Late May to late June
"Yellow-ish white"	**RUNNING FLEABANE**	***Erígeron flagellàris**	Yellow-centred solitary aster-like flowers on half trailing leafy stems. Color varies to pink. Good for front of border, and for naturalizing in wild garden. Prop. by seed and division. Any ordinary soil. Western U. S. A.	4-6 in *Half shade*	Mid. May to late June
"Bluish white" white & 6	**WHITE DOGTOOTH VIOLET**	****Erythrònium álbidum**	See page 15.		Late Apr., May
"Green-ish"	**LILY-LIKE FRITILLARY**	***Fritillària liliàcea**	See page 15.		Apr., May
"White"	**WHITE GUINEA-HEN FLOWER OR CHECK-ERED LILY**	****Fritillària Meleàgris var. álba**	See page 15.		Late Apr., May
"Green-ish white"	**CAUCASIAN SNOWDROP**	***Galánthus Caucásicus** *G. Redoùtei*	See page 15.		Apr., May
"White"	**PLAITED SNOWDROP**	****Galánthus plicàtus**	See page 3.		Mar. to early May
"White"	**WHITE BLOOD CRANES-BILL**	***Geránium sanguíneum var. álbum**	Good species. Large flowers borne on branching stems. Pretty foliage. Border or rock-garden. Prop. by seed and division. Any good soil. Europe.	1 ft. *Sun or half shade*	Late May to mid. July
"White" often near 41	**WHITE WATER AVENS OR WATER FLOWER**	***Gèum rivàle var. álbum**	Single flowers 1-2 in. across, sometimes vary to blue. Leaves mostly at the base of plant. Good for the rock-garden. Prop. by seed and division. Light moist soil. N. Amer.	1-2 ft. *Half shade*	Late May, June
"Yellow-ish white"	**ALPINE HUTCH-INSIA**	***Hutchinsia alpìna**	Minute alpine plant with clustered flowers and glistening leaves. Attractive mass of bloom in rock-garden. Prop. by seed and division. Sandy soil. Pyrenees.	2-3 in. *Sun*	May, June
"White"	**DUTCH HYACINTH**	****Hyacínthus orientàlis vars.**	See page 59.		Late Apr., May

Color	English Name	Botanical Name and *Synonyms*	Description	Height and *Situation*	Time of Bloom
"Greenish white"	**ORANGE ROOT, GOLDEN SEAL**	**Hydrástis Canadénsis**	See page 16.		Late Apr., May
"White"	**GARREX'S CANDYTUFT**	***Ibèris Garrexiàna**	Spreading evergreen plant with clusters of blossoms which are good for cutting, rising above clumps of dark foliage. Mass in border or rock-garden. Leave undisturbed. Prop. by seed, division or cuttings. Any rich soil. S. Europe.	6-9 in. *Sun or shade*	May, June
"White"	**ROCK CANDYTUFT**	***Ibèris saxátilis**	Spreading evergreen plant with clusters of flowers borne above masses of fine dark green foliage. Effective in clumps, drooping over rocks or edging the border. Prop. by seed, division or cuttings. Any ordinary garden soil, not too moist. S. Europe.	6-9 in. *Sun*	May, early June
"White"	**LEATHERY-LEAVED ROCK CANDYTUFT**	***Ibèris saxátilis var. corifòlia** *I. corifòlia*	Excellent alpine plant, dwarf var. Flowers in round clusters on slightly drooping stems. See I. saxatilis. Sicily.	3-6 in. *Sun*	May, June
"White"	**EVER-GREEN CANDYTUFT**	****Ibèris sempérvirens**	Excellent shrubby plant with numerous dense clusters of flowers and handsome dark evergreen foliage. Very hardy and vigorous, and may prove troublesome by spreading. Splendid rock-garden or border plant and good for edging shrubberies. Prop. by seed, division or cuttings. Any soil. S. Europe. Var. *pleno;* double flowered form not so desirable. Hort. Var. *foliis variegatis,* variegated leaves. Hort.	9-15 in. *Sun or half shade*	May, early June
"White"	**TENORE'S CANDYTUFT**	****Ibèris Tenoreàna**	Dwarf evergreen shrubby plant. Flowers turn purple. Foliage very hairy. Liable to perish in heavy soils in winter. Pretty in rock-garden or margin of border. Prop. by seed. Sandy soil. S. Italy.	9-12 in. *Half shade*	"
"White"	**FLORENTINE FLAG, ORRIS ROOT**	****Ìris Florentìna**	Large delicate scented flowers tinted with blue, with purple veining and orange-yellow beards, stand above the dark green leaves. Root-stock fragrant, (Orris-root). Good border plant in soil not too dry. Prop. by division. S. Europe. See Plate, page 79.	1-2 ft. *Half shade*	"
"White"	**WHITE-FLOWERED DWARF FLAG**	***Ìris pùmila var. álba**	One of the best dwarf Iris. Short-lived flowers borne close to the ground. Sword-shaped leaves. Spreads rapidly. Good for rock-garden or margin of border. Prop. by division. Any garden soil. Europe.	4-8 in. *Sun or half shade*	May

POET'S NARCISSUS. *Narcissus poeticus.*

ORRIS ROOT. *Iris Florentina.*

LEUCOJUM AESTIVUM AND TULIPA "WHITE SWAN"

Color	English Name	Botanical Name and *Synonyms*	Description	Height and *Situation*	Time of Bloom
"White"	**WHITE SIBERIAN FLAG**	****Ìris Sibírica var. álba**	Attractive species, distinguished by its tall slender stalks and early season. Forms dense tufts; many flower-stems bear clusters of pretty though rather small flowers, which rise above the grassy foliage. Good border plant. Prop. by division. Rich soil. Europe; E. Siberia.	2-3 ft. *Sun*	Late May to mid. June
"White"	**TWIN-LEAF**	***Jeffersònia binàta** *J. diphýlla*	Dwarf plant, similar to the Bloodroot. Solitary flowers 1 in. across on leafless stems. Round two-parted leaves. Bog or rock-garden plant. Prop. by seed and division. Peaty soil. S. Eastern U. S. A.	16-18 in. *Shade*	May
"White"	**WHITE VARIEGATED NETTLE**	**Làmium maculàtum var. álbum** *L. álbum*	Semi-trailing plant. Flowers in whorls on leafy stalks. Heart-shaped foliage blotched with white. Good for border or for covering barren places. Prop. by division. Preferably sandy open soil. Europe.	6-8 in. *Sun*	Mid. May to late July
"White"	**WHITE SPRING BITTER VETCH**	****Láthyrus vérnus var. álbus** *Órobus v. var. a.*	See page 16.		Late Apr., May
"White"	**SUMMER SNOW-FLAKE**	***Leucòjum æstìvum**	See page 16 and Plate, page 80.		Apr., May
"White"	**SPRING SNOW-FLAKE**	***Leucòjum vérnum**	See page 16.		"
"White"	**WHITE PERENNIAL FLAX**	***Lìnum perénne var. álbum**	Branching plant. Flowers on slender leafy stems A free and continuous bloomer. Good border and rock-garden plant. Prop. by seed and division. Any garden soil. Europe.	1-2 ft. *Sun or half shade*	Mid. May to Aug.
"White"	**WHITE OR EVENING CAMPION**	***Lýchnis álba**	See page 16.		Apr., May
"White"	**MITRE-WORT, TWO-LEAVED BISHOP'S-CAP**	**Mitélla diphýlla**	Small flowers in racemes on erect unbranching stalks. Fine large heart-shaped leaves on the flower-stalks and in a clump at base of plant. Rock or wild garden. Prop. by division. Moist soil somewhat peaty. Eastern States.	6-12 in. *Shade*	May
"White"	**COMMON WHITE GRAPE HYACINTH**	****Muscàri botryoìdes var. álbum**	See page 16.		Apr., May
"White"	**WHITE WOOD FORGET-ME-NOT**	***Myosòtis sylvática var. álba**	Loose clusters of small flowers with yellow eyes. Good for spring bedding, for fringing walks, or in garden beds. Prop. by seed. Any good soil. Europe; Asia.	1-2 ft. *Sun or shade*	May, early June

Color	English Name	Botanical Name and *Synonyms*	Description	Height and *Situation*	Time of Bloom
"White"	**PRIMROSE PEERLESS**	**Narcíssus biflòrus**	See page 16.		Mid. Apr. to late May
"White"	**ALGERIAN WHITE HOOP-PETTICOAT DAFFODIL**	***Narcíssus Bulbocòdium var. monophy̆llus**	See page 19.		Late Apr., early May
"White"	**BUR-BIDGE'S NARCISSUS**	**Narcíssus Burbidgei**	See page 19.		Mid. Apr. to late May
"White"	**PHEASANT'S EYE, POET'S NARCISSUS**	****Narcíssus poéticus**	A lovely and vigorous species. Fragrant wide-open flowers, the short cups having red edges. An old favorite. Plant in masses in the border, in the grass, or the edge of shrubbery beds, where it will endure for years. Bulbous. Prop. by offsets. Plant the bulbs in late Sept. or Oct., 6 or 8 in. deep, 3 in. apart. Any good soil. Mediterranean Region. See Plate, p. 79. Vars.—*P. Grandiflorus;* the largest of the type, pure white base-petals, cup suffused with crimson. *P. Ornatus.* Resembles N. poeticus though larger. It blooms earlier but is not so fragrant. *P. Poetarum;* large flowers, a fine var.	12-15 in. *Half shade best*	May
"White" or 2	**WHITE SPANISH TRUMPET DAFFODIL**	**Narcíssus Pseudo-Narcíssus var. moschàtus**	Very fragrant large trumpet flowers. These Narcissi unfortunately die out after a few years. Plant in rows or masses in the border or on the edge of shrubbery. Plant in late Sept. or Oct., 6-8 in. deep and 4 in. apart. Cover in winter. Bulbous. Prop. by offsets. Any good well-drained soil. Give an occasional top-dressing. See Plate, page 83.	6-9 in. *Half shade best*	Early to mid. May
"White" turns to 36	**STEMLESS EVENING PRIMROSE**	***Œnothèra acaùlis** *Œ. Taraxacifòlia*	Tender perennial or biennial; treat as an annual. Trailing. Flowers 2½-3½ in. across, pure white, becoming rose-color. Dandelion-like leaves. Rock-garden. Prop. by seed and cuttings. Any well-drained soil. Chili.	6 in. *Sun*	Late May to Aug.
"White"	**TUFTED EVENING PRIMROSE**	***Œnothèra cæspitòsa** *Œ. exímea, Œ. marginàta*	Tender perennial or biennial; treat as an annual. Large fragrant flowers 3-4 in. across. Color changes to rose; scent like the Magnolia. Long jagged leaves. Good for rock-garden and border. Prop. by seed, cuttings or suckers. Any soil. S. Western U. S.A	4-12 in. *Sun*	Late May, June

WHITE SPANISH TRUMPET DAFFODIL. *Narcissus Pseudo-Narcissus var. moschatus.*

COMMON STAR-OF-BETHLEHEM. *Ornithogalum umbellatum.*

WHITE SPIDERWORT. *Tradescantia Virginiana var. alba.*

MAY

Color	English Name	Botanical Name and *Synonyms*	Description	Height and *Situation*	Time of Bloom
"White"	WHITE CREEPING FORGET-ME-NOT	*Omphalòdes vérna var. álba	See page 19.		Apr., May
"Cream white"	DROOPING STAR-OF-BETHLE-HEM	Ornithógalum nùtans	See page 19.		"
"White"	COMMON STAR-OF-BETHLE-HEM, TEN-O'CLOCK	Ornithógalum umbellàtum	Star-shaped flowers in clusters of 12-20. Leaves long and narrow, spotted with white. Pretty in grass and for wild gardens. Bulbous, increases rapidly. Any soil. Mediterranean Region. See Plate, page 84.	½ ft. *Sun*	May, June
"White"	ALLEGHANY MOUNTAIN SPURGE	Pachysándra procúmbens	Dwarf shrubby plant with spikes of flowers sometimes varying in color to purplish. Attractive for the rock-garden on account of its masses of large dark leaves. Does well under trees. Prop. by division. Any soil. S. Eastern U. S. A.	6-12 in. *Shade*	Late May, early June
"White"	JAPANESE EVER-GREEN PACHY-SANDRA OR SPURGE	Pachysándra terminàlis	Low-growing evergreen plant. Flowers in short spikes. Dense foliage, bright green and glossy, makes it good for carpeting. Grows well under trees. Prop. by division. Any garden soil. Japan. *P. t. var. variegata,* foliage variegated. Hort.	6-12 in *Sun or shade*	Late May to mid. June
"White"	WHITE-FLOWERED PEONY	**Pæònia albiflòra *P. édulis*	Showy herbaceous Peony. Single flowers with a tuft of yellow stamens. For varieties see page 159. For cultivation see P. officinalis var. alba plena.	2-4 ft. *Sun or half shade*	"
"White" markings 26 dull	POPPY-FLOWERED TREE PEONY	Pæònia Moután var. papaveràcea *P. arbórea var. papaveràcea*	Beautiful shrub. Large flowers with delicate petals like a Poppy, red markings near the centre. Needs protected spot and winter covering. Handsome in the border, or on the edge of shrubbery. Prop. by grafting on roots of herbaceous species; it then does not flower until the third year. A gross feeder, it likes a rich rather moist loam enriched with cow manure.	3-6 ft. *Sun or half shade*	Mid. May to mid. June
"White"	STRIPED TREE PEONY	Pæònia Moután var. vittàta *P. arbòrea var. v.*	Single flowers shot with pink, sweet scented. See P. Moutan var. papaveracea.	3-6 ft. *Sun or half shade*	"

Color	English Name	Botanical Name and *Synonyms*	Description	Height and *Situation*	Time of Bloom
"White"	COMMON DOUBLE WHITE PEONY	Pæònia officinàlis var. álba plèna	Herbaceous Peony. Flowers double, white delicately tinted with crimson. Handsome divided foliage. Hardy and effective in the border, in clumps, or edging beds of shrubs. A half shady position is desirable. Prop. usually by division in early autumn. A gross feeder, it likes a deep rather moist loam enriched with cow manure. Hort.	2-3 ft. *Sun or half shade*	Mid. May to mid. June
"White"	ALPINE POPPY	**Papàver alpìnum	Dwarf Alpine species. Large fragrant flowers with yellow centres. Foliage finely divided. Stems leafless. Prop. by seed. Good rock plant. European Alps. Var. *album*, flowers spotted at the base. Also colored varieties. See Plate, page 87.	6 in. *Sun*	Mid. May to early June
"White"	WHITE ICELAND POPPY	**Papàver nudicaùle var. álbum	See page 19.		Late Apr. to mid. June, late Aug., Sept.
"White" 1	DOWNY PATRINIA	Patrínia villòsa	Valerian-like flowers in flat-topped clusters. Hairy leaves making a compact dwarf plant for margin of border or rock-garden. Light rich soil. Japan.	10-12 in. *Sun*	May, June
"White"	WHITE MOSS OR GROUND PINK	*Phlóx subulàta var. álba	See page 19.		Late Apr. to late May
"White"	MOSS OR GROUND PINK THE BRIDE	**Phlóx subulàta var. "The Bride"	See page 19.		"
"White"	NELSON'S MOSS OR GROUND PINK	*Phlóx subulàta var. Nelsoni	See page 19.		"
"White"	HIMALAYAN MAY APPLE	*Podophýllum Emòdi	Best species. Flowers cup-shaped 2 in. across, waxy and drooping beneath handsome large reddish leaves which die down in summer. Large bright red edible fruit. Wild garden plant. Prop. by seed and division. Deep soil, somewhat moist and peaty. Himalaya Region.	12-15 in. *Half shade*	Late May, June
"White"	MAY APPLE, WILD MANDRAKE OR LEMON	Podophýllum peltàtum	Resembles P. Emodi with dark green deeply divided foliage and large yellow fruit appearing in July. Wild garden plant. Prop. by seed and division. Deep soil, somewhat moist and peaty. N. Amer.	1-1½ ft. *Half shade*	May

ALPINE POPPY. *Papaver alpinum.*

BLOOD-ROOT. *Sanguinaria Canadensis.*

MAY

Color	English Name	Botanical Name and *Synonyms*	Description	Height and *Situation*	Time of Bloom
"White"	**WHITE JACOB'S LADDER OR CHARITY**	***Polemònium cærùleum var. álbum**	Dwarf bushy plant. Bell-shaped flowers nearly 1 in. across, in compact terminal panicles. Profusion of compound leaves. Good border plant. Prop. by seed and division. Rich well-drained loamy soil is best. Hort.	1-3 ft. *Half shade*	Mid. May to July
"Greenish white"	**GREAT OR SMOOTH SOLOMON'S SEAL**	**Polygonàtum gigantèum** *P. commucàtum*	Tubular flowers drooping from gracefully arched stems, leafy above. Very pretty for naturalization or wild garden. Prop. by division. Deep rich friable loam. Manitoba; U. S. A.	1-5 ft. *Half shade or shade*	Late May, June
"Greenish white"	**SOLOMON'S SEAL**	**Polygonàtum multiflòrum**	Tubular flowers droop from gracefully arched leafy stems. Very pretty for naturalization or wild garden. Prop. by division. Deep rich friable loam. Europe; Asia.	2-3 ft. *Half shade or shade*	Late May, early June
"White"	**THREE-TOOTHED CINQUE-FOIL**	**Potentílla tridentàta**	Spreading tufted plant. Flowers small but numerous. Dark strawberry-like foliage. Good for the rock-garden. Prop. by seed and division. Dry soil. Coast New England and north; Rocky Mts.; Scotland.	10 in. *Sun*	Mid. May to early June
"Bluish white"	**STRIPED SQUILL**	**Puschínia scilloìdes** *Adâmsia s.*	See page 20.		Late Apr., early May
"White" buds 23	**FLOWER-ING MOSS PYXIE**	**Pyxidanthèra barbulàta** *Diapénsia b.*	See page 20.		Apr., May
"White"	**WHITE ROSETTE MULLEIN**	***Ramónda Pyrenàica var. álba**	Pretty var. of this popular species. Flowers 1½ in. across with orange centres. Leaves in rosettes on the ground. Protect in winter. Plant in the rock-garden. Prop. by seed or by division. Deep peaty well-drained soil. Hort.	2–6 in. *Half shade*	Mid. May to July
"White"	**ACONITE-LEAVED BUTTER-CUP**	***Ranúnculus aconitifòlius**	Flowers in clusters. Foliage palmate, mostly in clumps at the base of the plant. Suitable for border or wild garden. Prop. by seed and division. Moist deep soil. Alps; Pyrenees. Var. *flore-pleno,* (*R. a. var. plenus*). Fair Maids of France, White Bachelor's Button. More effective than the species, having rosette-like double flowers. Pretty for the border.	½-2 ft. *Half shade*	"
"White"	**WHITE BUTTER-CUP OR CROWFOOT**	***Ranúnculus amplexi-caùlis**	Several flowers about 1 in. across, yellow centres, on slender stems. Foliage gray-green. Attractive for the rock-garden or border. Moist deep soil. S. Eastern Europe.	5-10 in. *Sun or half shade*	May, early June
"White"	**BLOOD-ROOT, RED PUCCOON**	***Sanguinària Canadénsis**	See page 20 and Plate, page 88.		Early Apr. to mid. May

Color	English Name	Botanical Name and *Synonyms*	Description	Height and *Situation*	Time of Bloom
"White"	ANDREW'S SAXIFRAGE	*Saxífraga Andrewsii	A good species. A hybrid between S. Geum and S. Aizoon. Numerous flowers dotted with red. Green spoon-shaped leaves. Plant in the rock-garden or border. Prop. by division. Sandy soil. Hort.	8-12 in. *Sun*	Late May, June
"Yellow-ish white"	LARGE STRAP-LEAVED SAXIFRAGE	*Saxífraga ligulàta	Small flowers in thick panicles on erect or wavy stalks. Leaves chiefly in rosette at base of plant. Prop. by division or offshoots. Ordinary garden soil. Alps; Apennines.	1-2 ft. *Half shade*	May
"White" or 25 pale	UMBRELLA PLANT	*Saxífraga peltàta	Flowers sometimes pale pink in clusters which appear before the foliage. Large shield-shaped leaves 1 ft. in diameter. Protect in winter. Border or margin of pond. Prop. by seed and division. Moist and peaty soil. Cal.	2-3 ft. *Sun*	Late May, early June
"Dull white"	EARLY SAXIFRAGE	Saxífraga Virginiénsis	See page 20.		Apr. to late June
"White"	WHITE WOOD HYACINTH	*Scílla festàlis var. álba *S. nùtans var. álba*	Fragrant drooping bell-shaped flowers in panicles on leafless stalks. Grass-like foliage in clumps. Rock-garden, border or shrubbery. Pretty when naturalized in woods. Bulbous. Soil enriched with manure. W. Europe.	8-12 in. *Half shade*	May, early June
"White"	WHITE SPANISH OR BELL-FLOWERED SQUILL, WHITE SPANISH JACINTH	**Scílla Hispánica var. álba	Most robust species. Strong pyramidal spikes of pendent bell-shaped flowers. Leaves spring from the ground. Attractive in wild garden, rock-garden border or when naturalized. Bulbous. Prop. by offsets in summer. Any soil. Top-dress yearly. Europe.	12-18 in. *Sun*	Late May, June
"White"	WHITE SIBERIAN SQUILL	**Scílla Sibírica var. álba	See page 4.		Mid. Mar. to early May
"White"	GALAX-LEAVED SHORTIA	Shórtia galacifòlia	Tufted plant, difficult to cultivate. Large nodding solitary flowers, white changing to rose. Evergreen foliage, tinged with bronze. Prop. by division or runners. Soil must contain "humus and leaf-mold." N. C. See Plate, page 91.	3-8 in. *Shade*	May, June
"White"	ALPINE CATCHFLY	Silène alpéstris	Dwarf compact Alpine plant. Flowers in tight clusters. Stems sticky in some forms. Excellent for the rock-garden. Prop. by seed or division. Any soil. Alps; Austria.	4-6 in. *Sun*	"

GREATER STITCHWORT. *Stellaria Holostea.*

GALAX-LEAVED SHORTIA. *Shortia galacifolia.*

LARGE-FLOWERED WAKEROBIN. *Trillium grandiflorum.*

Color	English Name	Botanical Name and *Synonyms*	Description	Height and *Situation*	Time of Bloom
"White"	CAUCASIAN CATCHFLY	*Silène Caucásica	Flowers usually solitary, rising from tufts of leaves. Good border or rock-garden plant. Thrives in sandy loamy soil. Caucasus.	4-5 in. *Sun*	Late May, June
"Cream white"	FALSE SOLOMON'S SEAL	*Smilacìna racemòsa	Profusion of flowers in irregular feathery panicles. Handsome foliage. Very ornamental in July and Aug. when in fruit. Wild garden or border. Prefers rich moist but well-drained soil. U. S. A.	1-3 ft. *Half shade or shade*	May, June
"White"	EASTER BELL, GREATER STITCH-WORT OR STARWORT, ADDER'S MEAT	*Stellària Holostèa *Alsìne Holostĕa*	A useful and pretty plant. Flowers numerous. Leaves insignificant. Good for border and for covering dry banks, where grass will not grow. Europe; N. Asia. See Plate, page 91.	6-18 in. *Sun or shade*	Mid. May to early June
"Cream white" often 37	COMMON COMFREY	Sýmphytum officinàle	Branching plant of somewhat coarse habit. Drooping flowers which vary to pink or pale purplish, in clusters. Wild garden. Any good soil. Europe; Asia.	2-3 ft. *Sun or half shade*	Late May to mid. July
"Greenish" turns to 37	FALSE ALUM ROOT	Tellìma grandiflòra	Tufted plant bearing one-sided spikes of bell-shaped flowers which turn red or pink. Pretty veined foliage. Wild spots or rock-garden. Prop. by division. Any good garden soil. Western N. Amer.	1½-2½ ft. *Shade*	May
"White"	FEATHERED COLUMBINE	*Thalíctrum aquilegifòlium	Fine erect plant. Large clusters of feathery flowers, purplish in centres. Handsome dark foliage resembling the Columbine. Border, rock-garden or naturalization. Prop. by seed or division. Well-drained loam. Europe; N. Asia.	1-3 ft. *Sun or half shade*	Late May to mid. July
"White"	PURPLISH MEADOW RUE	Thalíctrum purpuráscens *T. purpùreum* (Hort.)	Large lax leafy panicles of greenish flowers. Border or wild garden. Prop. in spring by seed and division. Well-drained loam. N. Amer.	4-7 ft. *Half shade*	Mid. May to July
"White"	MOUNTAIN BROOM-WORT	Thláspi alpéstre	Tufted plant with leafy clusters of flowers rising from rosettes of foliage. Neat for rockwork. Cool moist situation. Rocky Mts.	2-4 in. *Shade*	May, June
"White"	FOAM FLOWER	Tiarélla cordifòlia	Pretty starry flowers, borne in profusion in erect racemes. Buds faintly tinged with pink. Young leaves green; root leaves bronze. Divide every 2 years. Pretty in masses, Prop. by division. Prefers rich moist soil. Eastern N. Amer.	6-12 in. *Half shade or shade*	May
"White"	EASTER DAISY of Colorado.	Townséndia serícea	See page 20.		Apr., May

Color	English Name	Botanical Name and *Synonyms*	Description	Height and *Situation*	Time of Bloom
"White"	WHITE SPIDER-WORT	*Tradescántia Virginiàna var. álba	Bushy plant of free growth. Profusion of flowers in umbels. Long narrow leaves droop from the flower-stalks. Border or rock-garden. Prop. in spring by division. Ordinary garden soil. Eastern U. S. A.	1-3 ft. *Sun or half shade*	Late May to late Aug.
"White"	STAR FLOWER, CHICK-WEED-WINTER-GREEN	Trientàlis Americàna	Slender-stemmed erect plant bears starry flowers, singly or in pairs, above a whorl of leaves. Wild garden. Prop. by seed in spring or by division. Light peaty soil. Eastern N. Amer.	6-9 in. *Shade*	May
"White"	NODDING WAKE-ROBIN	Tríllium cérnuum	See page 20.		Late Apr. to early June
"White"	WHITE ILL-SCENTED WAKE-ROBIN	*Tríllium eréctum var. álbum	See page 20.		"
"White" turns to 36	LARGE-FLOWERED WAKE-ROBIN	**Tríllium grandiflòrum	Largest and most effective of the Trilliums. Solitary three-petaled flowers, changing to delicate rose. Wavy leaves 3 on a stem, near the flower. Excellent for shady situation. Prop. by seed and division. Rich moist soil. Eastern N. Amer. See Plate, page 92.	9-12 in. or more *Half shade*	May to early June
"Cream white"	EARLY WAKE-ROBIN	Tríllium nivàle	See page 20.		Late Apr., May
"White"	EGG-SHAPED TRILLIUM	Tríllium ovàtum	Resembles T. grandiflorum, but the petals are narrower. Sepals same length as petals. Protect in winter. See T. grandiflorum. Pacific States.	9-12 in. *Half shade*	May
"White"	PAINTED WAKE-ROBIN	Tríllium undulàtum *T. erythrocár-pum, T píctum*	See page 20.		Late Apr., May
"White"	TULIP	**Tùlipa vars.	**Single Early Bedding Tulips.** See page 64.		Late Apr. to late May
"White"	TULIP	**Tùlipa vars.	**Double Early Bedding Tulips.** See page 64.		"
"White"	AMERICAN BARREN-WORT	*Vancouvèria hexándra *Epimèdium hexándrum*	Hardy plant, with loose panicles of small flowers, valuable for its feathery delicate foliage. Grows well under trees. Plant in shady rock-garden or cultivate in masses. Prop. by division. Pacific Coast.	1 ft. *Shade*	May, June

MAY

Color	English Name	Botanical Name and *Synonyms*	Description	Height and *Situation*	Time of Bloom
"White"	WHITE GENTIAN-LEAVED SPEEDWELL	Verónica gentianoìdes var. álba	See page 23.		Late Apr. to late May
"Bluish white" tinged with 46	CREEPING SPEEDWELL	*Verónica rèpens	Creeping plant forming a mat of glossy green, covered with racemes of a few blue-tinged flowers. Good for rockwork and for covering bare spots. Prop. by seed and division. Prefers moist soil, but will grow where it is quite dry. Corsica.	Prostrate *Sun*	May
"White"	COMMON WHITE PERIWINKLE OR RUNNING MYRTLE	Vínca mìnor var. álba	See page 23.		Late Apr. to June
"White"	SWEET WHITE VIOLET	Vìola blánda	See page 23 and Plate, page 112.		Apr., May
"White" tinged with 43	CANADA VIOLET	Vìola Canadénsis	See page 23.		Late Apr. to mid. June
"White"	WHITE DOG VIOLET	Vìola canìna var. álba	See page 23.		Late Apr. to late May
"White"	WHITE HORNED VIOLET OR BEDDING PANSY	**Vìola cornùta var. álba	See page 23.		Late Apr. until frost
"White"	SWEET VIOLET	Vìola odoràta var. álba	See page 23.		Late Apr. to late May
"White"	WHITE BIRD'S-FOOT VIOLET	Vìola pedàta var. álba	A pretty species. Large flowers almost white. Flat petals pointed, and somewhat pansy-like. Leaves finely divided suggestive of a bird's foot. Wild garden or for naturalization. Prop. by seed and division. Dry sandy soil. U. S. A.	3-6 in. *Sun*	May, June
"Yellowish white" 2	TURKEY'S BEARD	Xerophýllum setifòlium *X. asphodeloìdes*	Stately plant which looks like an Asphodel. Flowers in long dense racemes terminating a tall stem which rises from a tuft of grassy leaves. Moist border or bog garden. Prop. by careful division in spring. Moist peaty soil. Eastern U. S. A.	1-4 ft. *Half shade*	"

Color	English Name	Botanical Name and *Synonyms*	Description	Height and *Situation*	Time of Bloom
"Yellow" 5	WOOLLY-LEAVED MILFOIL	*Achillèa tomentòsa	Downy plant of trim effect which forms a carpet. Yellow flowers in clusters. Good for cutting. Foliage feathery and evergreen. Pretty for the margin of border and rock-garden. Prop. by seed, division and cuttings. Any soil. N. Amer.; Europe; the East.	8-10 in. *Sun*	Late May to mid. Sept.
"Yellow" 5	SPRING ADONIS, OX EYE	**Adònis vernàlis *A. Apennìna, A. Davùrica*	See page 23.		Mid. Apr. to June
"Yellow" 4	GOLDEN GARLIC	**Állium Mòly	See page 24.		"
"Lemon yellow" 2	AUSTRIAN MADWORT	*Alýssum Gemonénse	See page 24.		Apr., May
"Golden yellow" 5	ROCK MADWORT, GOLDEN-TUFT	**Alýssum saxátile	See page 24.		Mid. Apr. to late May
"Yellow" 5	COMPACT ROCK MADWORT OR GOLDEN-TUFT	**Alýssum saxátile var. compàctum	See page 24		Mid. Apr. to June
"Golden yellow" bet. 1 & 2	YELLOW WOOD ANEMONE	*Anemòne ranunculoìdes	See page 24.		Apr., May
"Yellow" 2, centre 6	GOLDEN MARGUE-RITE, ROCK CAMOMILE	*Ánthemis tinctòria	Dense bushy plant vigorous and free flowering. Daisy-like flowers 1-2 in. across. Foliage deeply cut. Good to cut and for borders. Prop. by seed and division. Any ordinary soil. Europe.	2-3 ft. *Sun*	Mid. May to Oct.
"Yellow" 2	YELLOW CANADIAN COLUMBINE	*Aquilègia Canadénsis var. flaviflòra *A. C. var. flavêscens, A. cærùlea var. f.*	See page 24.		Late Apr. to early July
"Yellow" 3 & 2	GOLDEN SPURRED COLUMBINE	**Aquilègia chrysántha *A. leptocèras var. chrysántha*	A captivating species. Numerous fragrant flowers 2-3 in. across with long slender spurs. Graceful branching stems and handsome dark foliage. Hardier than many other Columbines. Plant in a sheltered place. Prop. by seed sown as early as possible and by division. Deep sandy loam. S. Western U. S. A. *Var. *flavescens* (*A. aurea, A. Canadensis var. aurea*); (color no. 2 marked with 32). Showy flowers marked with red, recurved spurs and pretty foliage. Mid. May to July. Hort.	3-4 ft. *Sun*	Late May to late Aug.

LARGE YELLOW LADY SLIPPER. *Cypripedium pubescens.*

LARGE YELLOW LADY SLIPPER. *Cypripedium pubescens.*

MAY

Color	English Name	Botanical Name and *Synonyms*	Description	Height and *Situation*	Time of Bloom
"Yellow" 5	PROPHET FLOWER	Arnèbia echioìdes *Macrotòmia e.*	See page 24.		Apr., May
"Yellow" 5	DOUBLE MARSH MARIGOLD	Cáltha palústris var. flòre-plèno *C. p. var. monstròsa plèna*	See page 24.		"
"Greenish yellow" 2 & green	BLUE COHOSH	Caulophýllum thalictroìdes	See page 27.		"
"Yellow" 1 & 21 light	LEMON-COLOR MOUNTAIN BLUET	Centaurèa montàna var. citrìna *C. m. var. sulphùrea*	Low running stems. Flowers resembling the Bachelor's Button, of good size and abundant, with brown centres and yellow petals. Foliage whitish. Border. Prop. by division. Ordinary garden soil. Armenia.	12-18 in. *Sun*	Late May to early July
"Greenish yellow" 2 greener	YELLOW CLINTONIA	Clintònia boreàlis *Smilacina boreàlis*	Flowers edged with yellow, in terminal clusters, succeeded by handsome blue berries and springing from a clump of dark shining leaves. Suitable for shady part of rock-garden or wild garden. Prop. by spring division. Moist sandy peaty soil. N. Amer.	1-2 ft. *Shade*	May, early June
"Golden yellow" 4	NOBLE FUMITORY	*Corýdalis nóbilis *Fumària nóbilis*	Bushy plant of erect habit. Large clusters of flowers white and golden yellow with chocolate spurs. Leaves finely divided. Good border plant. Prop. by seed and division. Siberia.	10-12 in. *Sun or half shade*	"
"Bright yellow" 2 deep & dull	SMALL YELLOW LADY'S SLIPPER	*Cypripèdium parviflòrum	Small flowers with flattened sac and slight perfume. Oval pointed leaves attached to flower stem. Protect in winter. Plant in wild or rock-garden. Prop. by division. Porous moist peaty soil. N. Amer.	1-2 ft. *Shade*	May, June
"Pale yellow" 2 deep & dull	LARGE YELLOW LADY'S SLIPPER	**Cypripèdium pubéscens *C. parviflòrum, C. hirsùtum*	One of the easiest to cultivate. Stems bear 1-3 greenish yellow flowers, spotted with brown, petals generally curled. For culture see C. parviflorum. Pa. to S. C. See Plates, pages 97 and 98.	1-2 ft. *Shade*	"
"Golden yellow" 5	AUSTRIAN LEOPARD'S BANE	*Dorónicum Austrìacum	Flowers numerous of the composite order and usually single. Foliage mostly in crown at base. Good for cutting. Plant in rough places or use as a border plant. Prop. by division. Any soil. S. Central Europe.	1½ ft. *Sun or half shade*	"
"Yellow" 5	CAUCASIAN LEOPARD'S BANE	*Dorónicum Caucásicum	Vigorous plant, with composite flowers, about 2 in. in diameter, generally solitary. Root leaves heart-shaped. Good for cutting. Prop. by division. Thrives in good loam. Woods of Caucasus; Sicily.	1-2 ft. *Sun*	May, early June

MAY

Color	English Name	Botanical Name and *Synonyms*	Description	Height and *Situation*	Time of Bloom
"Golden yellow" 5	CLUSIUS'S LEOPARD'S BANE	*Dorónicum Clùsii	Striking plant with large flowers. Foliage on flower stems, downy and toothed. Good for cutting. For cultivation see D. Caucasicum. Alps of Switzerland; Austria.	1½-2 ft. *Sun*	May, June
"Yellow" 5	CRAY-FISH LEOPARD'S BANE	*Dorónicum Pardaliánches	Flowers 1-5 on a stalk, rising high above a tuft of heart-shaped leaves. Good for cutting. Plant in wild spots. Prop. by division. Rich loamy soil. Europe.	1-2 ft. *Sun*	"
"Yellow" 5	PLANTAIN-LEAVED LEOPARD'S BANE	*Dorónicum plantagíneum	Composite flowers generally solitary, high above the clump of root leaves. Good for cutting. Prop. by division. Thrives in good loam. Sandy woods of Europe.	2 ft. *Sun*	"
"Orange yellow" 7	TALL PLANTAIN-LEAVED LEOPARD'S BANE "HARPUR CREWE"	*Dorónicum plantagíneum var. excelsum *D. excelsum "Harpur Crewe"*	Best of Doronica. Vigorous bushy plant with showy flowers and pointed leaves in a tuft at the base of plant. Prop. by division. Thrives in good loam. Europe.	1½-4 ft. *Sun*	"
"Yellow" 4	VITALE'S DOUGLASIA	Douglásia Vitaliàna *Arètia Vitaliàna, Gregòria Vitaliàna*	Rather delicate prostrate stems. Primrose-like flowers hardly projecting above the grayish foliage. Water and mulch in summer and protect in winter. Rock-garden. Prop. by seed and division. Moist peaty loam is best. Pyrenees; Alps.	2 in. *Half shade*	Late May, June
"Yellow" 4	AIZOON-LIKE WHITLOW GRASS	Dràba aizoìdes	Alpine plant with a profusion of small flowers above the tufts of dark foliage. Attractive in clumps for the border or rock-garden. Prop. by seed, but generally by division. Loose soil. Europe.	2-3 in. *Sun*	May, early June
"Yellow" 4	EVER-GREEN WHITLOW GRASS	Dràba Aìzoon	See page 27.		Late Apr., May
"Yellow" 4	ALPINE WHITLOW GRASS	Dràba alpìna	See page 27.		"
"Yellow" 1	BARREN-WORT	*Epimèdium pinnàtum	A good species. Strong growing attractive plant. Flowers in long clusters, with red spurs. Tufted foliage, beautiful and persistent, bronze-colored when young. Pleasing in rock-garden or border. Prop. by division. Any garden soil, preferably peaty. Persia; Caucasus.	1¼ ft. *Half shade*	May
"Yellow" 5 duller	LARGE-FLOWERED BARREN-WORT	*Epimèdium pinnàtum var. Cólchicum *E. Cólchicum*	Golden yellow flowers; spurs inconspicuous. See E. pinnatum. Persia.	9 in. *Sun*	"

Color	English Name	Botanical Name and *Synonyms*	Description	Height and *Situation*	Time of Bloom
"Sulphur yellow" 4	MOUNTAIN HEDGE MUSTARD	*Erýsimum alpínum	Fragrant flowers in profusion. Good for rock-garden or front of border. Frequent division advisable. Prop. by seed and division. Light soil. Norway.	6 in. *Sun*	May, June
"Sulphur yellow" 4	ALPINE WALL-FLOWER	Erýsimum ochroleùcum	See page 27.		Late Apr., May
"Yellow" 3	ROCK-LOVING HEDGE MUSTARD	*Erýsimum rupéstre *E. pulchéllum*	Plant resembling the Wallflower. Leaves narrow. Suitable for rock-garden. Prop. by seed or division. A well-drained light soil. Asia Minor.	6-8 in. *Sun*	May, early June
"Yellow" 1	COMMON ADDER'S TONGUE OR DOGTOOTH VIOLET	**Erythrò-nium Americànum	See page 27.		Late Apr. to late May
"Yellow" 15 yellower & lighter	LEMON-COLORED DOGTOOTH VIOLET	*Erythrò-nium citrìnum	See page 27.		Late Apr. to mid. May
"Yellow" 5 light	LARGE-FLOWERED DOGTOOTH VIOLET	**Erythrò-nium grandiflòrum *E. gigantèum*	See page 27.		"
"Yellow" 2 pale & greenish	MYRSIN-ITES-LIKE SPURGE	*Euphórbia Myrsinìtes	See page 28 and Plate, page 104.		Mid. Apr. to late May
"Bright yellow" 5	GOLDEN FRITILLARY	*Fritillària aùrea	See page 28.		Mid. Apr. to June
"Green-ish yel-low" 2 deep & greenish	PALE-FLOWERED FRITILLARY	*Fritillària pallidiflòra	See page 28.		"
"Yellow" 5	SHY FRITILLARY	*Fritillària pùdica	See page 28.		Mid. Apr. to mid. May
"Yellow" 6	YELLOW-FLOWERED MOUNTAIN AVENS	*Gèum montànum	Good species. A profusion of pretty cup-shaped flowers rising above leaves that lie close to the ground. Pleasing in rock-garden or border. Prop. by seed and division. Moist soil, preferably light. S. Europe.	9 12 in. *Sun*	Late May to mid. June
"Yellow" 6 lighter	PYRENEAN AVENS	Gèum Pyrenàicum	Drooping flowers in clusters of 2-4 on unbranching stems. Rock-garden or margin of border. Prop. by seed and division. Light moist soil. Pyrenees.	1½ ft. *Half shade*	Mid. May to mid. June

MAY

Color	English Name	Botanical Name and *Synonyms*	Description	Height and *Situation*	Time of Bloom
"Yellow" 5	**HOOPES'S SNEEZE-WEED**	****Helènium Hoòpesii**	Somewhat coarse habit. Large showy daisy-like flowers, good for cutting. Valuable border plant. Prop. by seed, division and cuttings. Thrives in rich moist soil. Western N. Amer.	1-3 ft. *Sun*	Late May to late June
"Yellow"	**DUTCH HYACINTH**	****Hyacínthus orientàlis vars.**	See page 59.		Late Apr., May
"Deep yellow" 5 deep	**YELLOW STAR GRASS**	**Hypóxis erécta** *H. hirsùta*	A low grass-like plant with small star-shaped flowers in clusters and grass-like leaves. Rock or wild garden. Prop. by division. Any soil. U. S. A. See Plate, page 103.	½ ft. *Half shade*	May, June
"Yellow"	**YELLOW-BANDED FLAG**	**Ìris orientàlis** *I. ochroleùca, I. gigantèa*	One of the largest species. Forms vigorous clumps. Two to three spikes of flowers. Petals fade into white at the margin. Glaucous leaves which twist gracefully. Prop. by division. Almost any soil. Asia Minor; Syria.	3-4 ft. *Sun or half shade*	Late May, June
"Yellow" 5 & 2	**YELLOW OR COMMON WATER FLAG**	***Ìris Pseudácorus** *I. Pseudácorus*	Forms luxuriant clumps having many stems which bear large broad-petaled flowers veined with brown. Long stiff gray-green leaves. Beautiful for the margin of water. Prop. by division. Europe.	1½-3 ft. *Sun*	Late May to late June
"Yellow" 5	**HOOP-PETTICOAT DAFFODIL**	***Narcíssus Bulbocòdium**	See page 28.		Late Apr. to late May
"Yellow mixed"	**STAR DAFFODIL**	****Narcíssus incomparàbilis & vars.**	See page 28, and Plate, page 103.		Mid. Apr. to mid. May
"Yellow" 3	**JONQUIL**	***Narcíssus Jonquílla & vars.**	See page 31.		Apr., early May
"Yellow mixed"	**COMMON OR TRUMPET DAFFODIL, LENT LILY**	****Narcíssus Pseùdo-Narcíssus vars.**	See page 31.		Late Apr., May
"Pale yellow" 1	**SESSILE-LEAVED BELLWORT**	**Oakèsia sessilifòlia** *Uvulària sessilifòlia*	Wild in woods. Graceful but inconspicuous plant; a few drooping flowers on slender stems. Pretty foliage. Naturalize under trees or in wild garden. Rich moist light soil. N. Amer.	1 ft. *Half shade*	May, June
"Yellow" 5 or 6 brilliant	**ICELAND POPPY**	****Papàver nudicaùle**	See page 35.		Late Apr. to July, late Aug. to Oct.

YELLOW STAR GRASS. *Hypoxis erecta.*

STAR DAFFODIL VAR. STELLA. *Narcissus incomparabilis Stella.*

MYRSINITES-LIKE SPURGE. *Euphorbia Myrsinites.*

MAY

Color	English Name	Botanical Name and *Synonyms*	Description	Height and *Situation*	Time of Bloom
"Yellow" 6	**SCABIOUS-LEAVED PATRINA**	**Patrínia scabiosæfòlia**	Flowers in flat-topped clusters. Foliage deeply divided. Margin of border or rock-garden. Prop. by seed. Light rich soil. N. Asia.	10-12 in. *Sun*	May, June
"Creamy yellow" 2	**WOOD BETONY**	**Pediculàris Canadénsis**	Valued for its fern-like foliage. Flowers sometimes purple tinged, rarely white, in short terminal spikes. Prop. by seed. Moist peaty soil. N. Amer.	5-12 in. *Sun*	"
"Yellow" 3	**SILVERY CINQUE-FOIL**	**Potentílla argéntea**	A tufted plant with flowers borne on leafy stems. Leaves deep green on top, whitish beneath. Useful for rock-work and can be grown where other plants will die. Prop. by seed and division. Sandy light soil. North Temperate Zone.	4-12 in. *Sun*	Early May to early July
"Lemon yellow" 3	**CALABRIAN CINQUE-FOIL**	**Potentílla Calábra**	A dwarf species with somewhat cup-shaped flowers 1 in. across and pretty silvery foliage. Rock-garden. Prop. by seed and division. Sandy light soil. Europe.	10-12 in. *Sun*	Late May to early July
"Golden yellow" 5	**PYRENEAN CINQUE-FOIL**	**Potentílla Pyrenàica**	Vigorous showy species with cup-shaped flowers. For rockwork or border. Prop. by seed and division. Light soil. Pyrenees.	6-15 in. *Sun*	May to Aug.
"Pale yellow" 2	**COMMON EUROPEAN OR TRUE PRIMROSE**	***Prímula acaùlis** *P. grandiflòra* *P. vulgàris*	See page 35.		Mid. Apr. to June
"Yellow" 3 light	**OXLIP**	**Prímula elàtior**	Resembles P. officinalis but with larger flowers. Blossoms broad and flat. Leaves wrinkled. Plant in the border or rock-garden. Protect in winter. Prop. by fresh seed, division or cuttings. Rich light soil, not dry. Mts., Europe.	8-12 in. *Half shade*	May
"Yellow" 3	**ENGLISH COWSLIP**	****Prímula officinàlis**	See page 35.		Late Apr. to late May
"Rich yellow" 3 light	**STUART'S PRIMROSE**	****Prímula Stùartii**	Fine vigorous species. Flowers droop in many-blossomed umbels. Leaves sometimes 1 ft. long. Sheltered, somewhat elevated position in rock-garden. Prop. generally by seed. Light deep soil, not dry. Mts., N. India.	9-15 in. *Half shade*	Late May, June
"Golden yellow" 5	**BACHE-LORS' BUTTON**	**Ranúnculus àcris var. flòre-plèno**	Downy plant of the Buttercup family. Unusually double globular flowers. Leaves deeply divided. Border. Prop. by seed and division. Any good loam, moist if possible. Hort.	½-3 ft. *Sun*	Mid. May to Sept.

MAY

Color	English Name	Botanical Name and *Synonyms*	Description	Height and *Situation*	Time of Bloom
"Yellow" 5	**DOUBLE ACONITE-LEAVED BUTTERCUP**	**Ranúnculus aconitifòlius var. lùteus-plènus**	Double rosette-shaped flowers in clusters. Foliage palmate mostly at base of plant. Suitable for the border or wild garden. Prop. by seed and division. Moist deep soil. Hort.	½-3 ft. *Half shade*	Mid. May to July
"Yellow" 5	**LESSER CELANDINE, FIGWORT**	***Ranúnculus Ficària** *Ficària Ficària*	Neat little plant covered with pretty flowers. Leaves form a dense mat of shining green. Plant under trees or in the rock-garden. Prop. by seed and division. Any garden soil. Europe; Western Asia. See Plate, page 107.	4-5 in. *Sun or half shade*	May
"Yellow" 5	**MOUNTAIN BUTTERCUP**	**Ranúnculus montànus**	Downy plant. Flowers solitary, about 1 in. wide. Plant in rock-garden. Prop. by seed and division. Sandy soil. Europe.	6-9 in. *Half shade*	May to early July
"Yellow" 5	**CREEPING DOUBLE-FLOWERED BUTTERCUP**	**Ranúnculus rèpens var. flòre-plèno**	Double form of common Buttercup. Flowers half spherical. Suitable for moist places. Prop. by seed and division. Loam deep and moist. Europe; Asia; N. Amer.	6-12 in. *Sun*	May to Aug.
"Yellow" 4	**STONECROP, WALL PEPPER, LOVE ENTANGLE**	***Sèdum àcre**	Useful dwarf plant, covered with small starry flowers on creeping stems. Leaves pulpy. Good on rocks and for carpeting sandy waste places. Prop. by division and cuttings. Prefers poor soil. Great Britain; Europe. Var. *aureum;* (color no. 4 greener), shoots have golden yellow tips in early spring.	2-3 in. *Sun or half shade*	Late May, June
"Yellow" 8 lighter	**CELANDINE POPPY**	**Stylóphorum diphýllum** *Chelidònium diphýllum Papàver Stylóphorum*	Resembles Chelidonium majus, but is a better plant. Grows in large tufts. Poppy-like flowers freely produced. Grayish foliage. Easily cultivated in rich rather loose moist soil. Central U. S. A.	1 ft. *Sun or shade*	May, early June
"Yellow" 4	**BEAN-LIKE THERMOPSIS**	***Thermópsis fabàcea** *T. montàna*	Effective plant. The blossoms, borne on long terminal spikes resemble the Lupine. Leaves large. Good border plant. Prop. by seed sown in fall or spring. Light soil. N. Amer.	1-1½ ft. *Sun*	May, June
"Yellow" 4	**ALLEGHANY THERMOPSIS**	***Thermópsis móllis**	Erect branching plant. Flowers in showy spikes. Good border plant. Prop. by division, but better by seed. Any soil, preferably light and well-drained. Southern Atlantic States.	2-3 ft. *Sun or shade*	Mid. May to Aug.
"Yellow"	**MOUNTAIN THERMOPSIS**	**Thermópsis montàna**	A graceful and distinct species. See T. fabacea. Western U. S A.	1½ ft. *Sun or shade*	May, June
"Orange yellow" 6	**ORANGE GLOBE FLOWER**	****Tróllius Asiáticus**	See page 35.		Late Apr. to late May, early Aug. to Oct.

LESSER CELANDINE. *Ranunculus Ficaria.*

SPRING FLOWERS

Color	English Name	Botanical Name and *Synonyms*	Description	Height and *Situation*	Time of Bloom
"Yellow" 5	**MOUNTAIN GLOBE FLOWER**	****Tróllius Europæus** *T. globòsus*	Branching plant. Round double flowers either solitary or in pairs. Beautifully shaped, deeply divided leaves. Good border plant. Prop. by seed and division; flowers the second year from seed. Flourishes in sandy peaty loam. N. Europe. Var. *Loddigesii*, flowers of deeper yellow. Hort.	6-15 in. *Half shade best*	Early May to early June
"Orange yellow" 7	**JAPANESE GLOBE FLOWER**	***Tróllius Japónicus**	Flowers deeper color than T. Europæus. Dense-growing plant with round double flowers like giant Buttercups. Handsome deeply divided leaves. Good border plant. Prop. by seed and division; flowers the second year from seed. Thrives in sandy peaty loam. Japan.	15 in. *Half shade*	May, June
"Yellow"	**TULIP**	****Tùlipa vars.**	**Single Early Bedding Tulips.** See page 64.		Late Apr. to late May
"Yellow"	**TULIP**	****Tùlipa vars.**	**Double Early Bedding Tulips.** See page 64.		"
"Yellow" 5 streaked with 18	**TURKISH TULIP**	***Tùlipa acuminàta**	Not effective for bedding. Petals very long and narrow and reflexed, streaked with red. Leaves have wavy margins. Plant in late Sept. or Oct. at a depth of 4 in. to the bottom of the bulb, allowing 4 in. between each bulb. It is wise to protect in winter and to give an occasional top-dressing. Prop. by offsets. Light soil. If drainage is not good put sand around the bulb. Hort.	1-1½ ft. *Sun*	May
"Yellow" 3 flushed with 18	**SOUTHERN TULIP**	***Tùlipa australis**	Not good for bedding. Similar to Tulipa sylvestris, but smaller in every way. Drooping buds. Bell-shaped flowers slightly funnel-form, tinged with red on the outside. For cultivation see T. acuminata. S. Europe; N. Africa.	1-1½ ft. *Sun*	"
"Lemon yellow" 3 lighter & warmer flushed with 29	**BATALIN'S TULIP**	***Tùlipa Batalìni**	Not good for bedding. One of the earliest Tulips. Flowers bell or funnel-shaped. Leaves lie on the ground. For cultivation see T. acuminata. Asia.	5 in. *Sun*	Early May
"Yellow" 3 lighter & warmer & 30	**BIEBERSTEIN'S TULIP**	***Tùlipa Biebersteiniàna**	Not good for bedding. Broad bell-shaped flowers tinted with reddish-pink on the edges, brownish at base. Buds slightly drooping. For cultivation see T. acuminata. Asia.	6 in. *Sun*	May
"Yellow" 5 light, 18 outside	**KAUFMANN'S TULIP**	***Tùlipa Kaufmanniàna**	Large bright flowers tinted with red near the top. Grayish leaves. See T. acuminata. Turkestan.	6-10 in. *Sun*	Mid. May

MAY

Color	English Name	Botanical Name and *Synonyms*	Description	Height and *Situation*	Time of Bloom
"Yellow" 6	PERSIAN TULIP	*Tùlipa Pérsica *T. pàtens*	Flowers about 3 in. wide, green-tinted outside, whitish within, yellow in centre. Plant in late Sept. or Oct. at a depth of 4 in. to the bottom of the bulb, allowing 4 in. between each. It is wise to protect in winter and to give an occasional top-dressing. Prop. by offsets. Light soil. If drainage is poor put sand around the bulb. Siberia.	3-9 in. *Sun*	Mid. May
"Yellow" 3 dull	REFLEXED TULIP	**Tùlipa retrofléxa	An elegant and graceful species. Flowers poised on slender stems, open widely, having very pointed and reflexed petals. More delicate in appearance than most hybrid Tulips. Plant in clumps or masses in the border of shrubbery beds, etc. There is also a red var. For cultivation, etc., see T. Persica. Hort. See Plate, page 111.	12-15 in. *Sun*	Early to late May
"Pale yellow" 3	WILD TULIP	*Tùlipa sylvéstris *T. Florentìna* *T. F. var. odoràta*	Fragrant flowers sometimes edged with red. Nodding buds and slender grayish leaves. For cultivation see T. Persica. England; Europe. See Plate, page 111.	10-15 in. *Sun*	Early May
"Sulphur yellow" 4	VITELLINE OR YELLOW TULIP	**Tùlipa vitellìna	Delicate yet showy in appearance, of sturdy habit with large shapely flowers of a soft color. One of the "Cottage Garden" Tulips. For location and cultivation see T. Persica. Hort.	1-2 ft. *Sun*	May
"Lemon yellow" 2	LARGE-FLOWERED BELLWORT	*Uvulària grandiflòra	This species has stouter stems and larger flowers than U. perfoliata. Eastern N. Amer.	10-18 in. *Shade*	May, June
"Cream yellow" 1 pale	PERFOLIATE BELLWORT, WILD OATS	Uvulària perfoliàta	A pretty but modest native. Long and narrow bell-shaped flowers droop gracefully amidst luxuriant soft green foliage. Wild garden plant. Prop. by division. Rich light soil. U. S. A. See Plate, page 112.	6-20 in. *Shade*	May
"Yellow" 3	LONG-LEAVED ITALIAN MULLEIN	*Verbàscum longifòlium *V. pannòsum*	A single erect stalk with flowers borne at the top in irregular branching racemes. Woolly leaves, lower ones very long. Coarse but imposing in the border; also looks well in front of shrubbery. Prop. by seed. Any soil. Italy.	3-4 ft. *Sun*	Late May to late June
"Yellow" 6	YELLOW HORNED VIOLET OR BEDDING PANSY	**Vìola cornùta var. lùtea major	See page 35.		Late Apr. until frost
"Yellow" 2	HALBERD-LEAVED YELLOW VIOLET	Vìola hastàta	Dwarf tufted Violet with halberd-shaped leaves. Showy when in bloom. Plant in shady part of rock-garden. Prop. by division. Southern States.	4-10 in. *Half shade*	May

WILD TULIP. *Tulipa sylvestris.*

REFLEXED TULIP. *Tulipa retroflexa.*

WILD OATS. *Uvularia perfoliata.* SWEET WHITE VIOLET. *Viola blanda.*

Color	English Name	Botanical Name and *Synonyms*	Description	Height and *Situation*	Time of Bloom
"Rich yellow" 5	**MOUNTAIN VIOLET**	***Vìola lùtea**	Dense growing Violet of neat habit, a profuse bloomer. Lower petals finely striped with black. Plant in margin of border or half shady spot in rock-garden. Prop. by division. Europe.	2-6 in. *Half shade*	Early to late May
"Yellow" effect 2 light	**HAIRY YELLOW VIOLET**	**Vìola pubéscens**	See page 35.		Late Apr. to late May
"Yellow" 4	**YELLOW OR BARREN STRAW-BERRY**	**Waldsteìnia fragarioìdes** *Dalibàrda fragarioìdes*	Strawberry-like plant of tufted habit. Several flowers on the stem. Glossy leaves. Good for covering dry banks or sunny position in rock-garden. East of the Mississippi.	4 in. *Sun*	May, June
"Yellow" 4	**EARLY OR GOLDEN MEADOW PARSNIP**	**Zízia aùrea**	See page 36.		Late Apr. to late May
"Orange yellow" 5	**GOLDEN BLOOM-ERIA**	**Bloomèria aùrea** *Nothoscor-dúm aùreum*	See page 36.		Apr., May
"Or-ange" 26 more orange	**DWARF CHILOE WATER FLOWER**	***Gèum Chiloénse var. miniàtum** *G. miniàtum*	Vigorous species. Color paler than G. Chiloense. Wide-open five-petaled flowers on stems which rise above the foliage. A group forms a brilliant and pretty effect in rock-garden or border. Prop. by seed and division. Easily cultivated. Moist soil preferable. Hort.	2-3 ft. *Sun*	May, June
"Pur-plish orange" near 3 & 41	**WATER AVENS OR WATER FLOWER**	***Gèum rivàle**	Few drooping flowers on unbranching stems. Leaves mostly at the base. Shady spot in rock-garden or border. Prop. by seed and division. Moist soil. N. Amer.	1-2 ft. *Half shade*	Late May, June
"Or-ange"	**DUTCH HYACINTH**	****Hyacínthus orientàlis vars.**	See page 59.		Late Apr., May
"Or-ange" 7 brighter	**HOARY PUCCOON, RED ROOT, INDIAN PAINT**	**Lithospèr-mum canéscens**	Erect stalks dividing near the summit into 2 elongated leafy arching clusters of tubular flowers. Rock-garden or border. Prop. by seed, division and cuttings. Well-drained soil, sandy or loamy. N. Amer.	9-15 in. *Sun or half shade*	Late May
"Deep orange" bet. 12 & 17	**ORANGE ICELAND POPPY**	***Papàver nudicaùle var. aurantià-cum**	See page 36.		Late Apr. to July, late Aug. to Oct.

MAY

Color	English Name	Botanical Name and *Synonyms*	Description	Height and *Situation*	Time of Bloom
"Deep orange" bet. 12 & 17	**SMALL ICELAND POPPY**	***Papàver nudicaùle var. miniàtum**	See page 36.		Late Apr. to July, mid. Aug. to Oct.
"Pale scarlet orange" 9	**HAIRY-STEMMED POPPY**	***Papàver pilòsum**	Covered with soft hairs. Flowers pale scarlet or intense orange, with white basal mark. Leaves pale green. Pleasing in elevated parts of rock-garden. Prop. by seed and division. Sandy loam. Greece.	12-18 in. *Sun*	Late May, June
"Red orange" 11	**ATLANTIC POPPY**	***Papàver rupífragum var. Atlánticum** *P. Atlánticum*	Flowers about 3 in. across. Foliage silvery and hairy. Border. Prop. by seed and division. Sandy loam. Morocco.	12-15 in. *Sun*	Late May to Aug.
"Scarlet" 19 & 4	**WILD COLUMBINE**	***Aquilègia Canadénsis**	See page 36.		Late Apr. to mid. June
"Orange red" 2 & 17	**HYBRID CALIFORNIAN COLUMBINE**	****Aquilègia formòsa var. hỳbrida** *A. Califórnica var. hỳbrida*	Showy drooping flowers, scarlet and yellow, spurs long and slender. Foliage gray-green. A good border plant for sheltered spot. Prop. by seed sown as early as possible. Deep sandy loam. Hort.	1-1¼ ft. *Sun*	Mid. May to July
"Orange red" 19 & 5	**MEXICAN COLUMBINE**	***Aquilègia Skínneri**	A large well-shaped plant with many-flowered stem. Sepals green; petals greenish orange; spurs bright red; stamens protruding beyond the petal. Good border plant. Light soil. Mexico.	1-2 ft. *Sun*	Late May, June
"Orange scarlet" 17	**CALIFORNIAN COLUMBINE**	***Aquilègia truncàta** *A. Califórnica, A. exímea*	See page 36.		Apr. to early June
"Scarlet" 18	**DWARF RED OR NORTHERN SCARLET LARKSPUR**	***Delphínium nudicaùle**	Pretty plant. Loose spikes of flowers ranging from light scarlet to crimson. Good for cutting. Protect in winter. Plant in the border. Prop. by seed. Rich sandy soil of good depth. N. Cal.	1-1½ ft. *Sun*	Late May, June
"Red" 26	**COMMON BARRENWORT, BISHOP'S HAT**	***Epimèdium alpìnum**	Not easy to obtain. Dainty plant with fanciful flowers. Outer sepals grayish, inner crimson, petals yellow. Foliage often evergreen. New leaves reddish tinted. Plant in rock-garden or edge of shady shrubbery. Prop. by division. Any garden soil, preferably peaty. Europe.	9 in. *Half shade*	Mid. May to early June

MAY

Color	English Name	Botanical Name and *Synonyms*	Description	Height and *Situation*	Time of Bloom
"Red" 26 darker	RED BARRENWORT	*Epimèdium rùbrum *E. alpìnum var. rùbrum*	Very similar in habit to E. alpinum. Red flowers with grayish outer petals. Foliage somewhat bronze-colored, good for cutting. Lasts all winter. Plant in rock-garden or on edge of shrubbery beds. Prop. by division. Any ordinary garden soil. Japan.	8-12 in. *Half shade*	May
"Crimson" 41 deeper	BLOOD-RED CRANES-BILL	**Gerànium sanguíneum	Good species. Large flowers "almost crimson," borne on branching stems. Pretty foliage. Good for wild garden, rock-garden or border. Prop. by seed and division. Any good soil. W. Asia; Europe.	1½-2 ft. *Sun or half shade*	Late May to mid. July
"Purplish red" 33 or 26 dull	APACHE PLUME, LONG-PLUMED PURPLE AVENS	*Gèum triflòrum *G. ciliàtum*	Low plant soft and hairy. Flowers followed by feathery red seed pods. Rock-garden or border. Prop. by seed and division. Light moist soil preferable. N. Amer.	6-18 in. *Half shade*	May, June
"Red"	DUTCH HYACINTH	**Hyacínthus orientàlis vars.	See page 59.		Late Apr., May
"Purple red" often 31 brilliant	RED OR MORNING CAMPION	Lýchnis diòica *L. diúrna*	Showy plant of somewhat rank habit. Flowers in loose clusters opening in the morning, borne on leafy stems. Color varies to pink and white. No fragrance. Blooms at intervals all summer. A good border plant. Prop. by seed and division. Any garden soil. Europe; Asia. Var. *flore-pleno*, a double form.	1-2 ft. *Sun*	Mid. May to late June
"Bright crimson"	ANOMALOUS PEONY	*Pæònia anómala & vars. *P. Fischeri, P. intermèdia*	Herbaceous Peony. Single species with brilliant blossoms of unusual size. Foliage finely divided. A half shady position is desirable for all herbaceous Peonies. Effective in the border but especially striking at a distance when edging beds of shrubs, walks, etc. Prop. by division usually in early autumn. Deep rather moist loam enriched with cow-manure. Europe; Asia. Var. *insignis*. The most popular variety, 1¼-2 ft. high. Var. *intermedia*. A variety with richer color than the type. 1¼-2 ft. high. Foliage more deeply cut. Europe.	1½-3 ft. *Sun or half shade*	May
"Dark red"	RAM'S HORN PEONY	*Pæònia arietìna *P. Crĕtica*	An effective herbaceous Peony with big solitary richly colored flowers and a bluish tinge to the foliage. Good for cutting. For location and cultivation see P. anomala. Many hort. vars. in different colors. S. Europe.	2-3 ft. *Sun or half shade*	"

MAY

Color	English Name	Botanical Name and *Synonyms*	Description	Height and *Situation*	Time of Bloom
"Pinkish red"	**SMALL-FRUITED PEONY**	***Pæònia microcarpa**	Herbaceous Peony. Very large single flowers of delicate color, rather low-growing. Half shade is desirable. Good in masses for distant effects, or for the border. Prop. usually by division in early autumn. Deep rather moist loam enriched with cow-manure. Spain.	1-1¼ ft. *Sun or half shade*	Late May
"Crimson" 33 redder, etc.	**COMMON GARDEN PEONY**	****Pæònia officinàlis & vars.** *P. fúlgida*	Herbaceous Peony. Common in old gardens. Large flowers and handsome divided foliage. Hardy and effective in the border or in masses with other Peonies. Half shade is desirable. The parent of many hort. vars., see page 159. Usually prop. by division in early autumn. Being gross feeders, they like a deep rather moist loam enriched with cow-manure. Europe. The following are good varieties: **Var. *anemonæflora;* (color no. 27), globular flowers with a quantity of crimson, twisted, yellow-edged stamens. **Var. *rosea-plena;* rich crimson double flowers. Var. *rubra-plena,* double crimson flowers. Hort.	2-3 ft. *Sun or half shade*	Mid. May to mid. June
"Deep crimson" bet. 27 & 28	**PARADOXICAL PEONY**	***Pæònia paradóxa**	Herbaceous Peony. Very compact and dwarf, with single flowers and somewhat glaucous foliage. Prop. usually by early autumn division. Deep moist loam enriched with cow-manure. S. Europe. Var. *fimbriata,* brilliant double flowers very handsome, but little known in U. S. A. Hort.	1-1¼ ft. *Sun or half shade*	Late May
"Crimson" 40 deeper	**STRAGGLING PEONY**	***Pæònia peregrìna**	Herbaceous Peony. Effective species resembling P. officinalis, crimson in the type, and having many hort. vars., see page 159. For cultivation see P. officinalis.	1½-2 ft. *Sun or half shade*	May
"Deep red" 20 richer	**FINE-LEAVED PEONY**	***Pæònia tenuifòlia**	Densely leafed herbaceous Peony. Large single flowers and fine feathery foliage. Good border plant. Half shade is desirable. Prop. generally by division in early autumn. Deep rather moist loam enriched with cow-manure. Caucasus. **Var. *flore-pleno,* a double-flowered variety.	1-1½ ft. *Sun or half shade*	Mid. May to mid. June
"Blood red" 17 redder	**BRACTEATE POPPY**	***Papàver bracteàtum** *P. orientàle var. bracteàtum*	One of the largest and most gorgeous of Poppies. Flowers 6-9 in. across, marked inside with purple-black spot, on erect stems. Handsome foliage in clump at base of plant. Effective in the border and against the green of shrubs. Prop. by seed, or division after flowering. Any soil. Caucasus, Persia.	3-4 ft. *Sun*	Late May to mid. June

MAY

Color	English Name	Botanical Name and *Synonyms*	Description	Height and *Situation*	Time of Bloom
"Light red" 16	**RUPIFRAGE POPPY**	***Papàver rupífragum**	Dwarf species of neat habit. Foliage grayish-green. Border. Prop. by seed and division. Sandy loam. Spain.	8-12 in. *Sun*	May to Aug.
"Crimson"	**CLEVELAND'S PENTSTEMON**	***Pentstèmon Clevelandi**	Tall narrow spikes of flowers. Foliage somewhat gray-green. Protect in winter. Prop. by seed and division. Any good garden soil, not too dry. S. Cal.	10-18 in. *Sun*	Late May to mid. June
"Bright red"	**RED SPIDERWORT**	**Tradescántia Virginiàna var. coccínea**	Bushy plant of free growth. Umbels of flowers borne in profusion. Long grass-like foliage. Border or rock-garden. Prop. by division in spring. Ordinary garden soil. S. Western U. S. A.	1-3 ft. *Sun or half shade*	Late May to late Aug.
"Brownish red" 28 neutral & lighter	**ILL-SCENTED WAKE-ROBIN**	***Tríllium eréctum** *T. pendùlum, T. purpùreum T. fœtidum*	See page 36.		Late Apr. to early June
"Red"	**TULIP**	****Tùlipa vars.**	**Single Early Bedding Tulips.** See page 64.		Late Apr. to late May
"Red"	**TULIP**	****Tùlipa vars.**	**Double Early Bedding Tulips.** See page 64.		"
"Deep scarlet"	**KEELED TULIP**	****Tùlipa carinàta**	Large open flowers with petals shading into bright yellow spot at base. Leaves slightly wavy. Plant in late Sept. or Oct. at a depth of 4 in. to the bottom of the bulb, allowing 4 in. between each. Best to protect in winter and to give an occasional top-dressing. Prop. by offsets. Light soil. If drainage is poor put sand around the bulb. Habitat unknown.	12-15 in. *Sun*	May
"Bright crimson"	**DIDIER'S TULIP**	***Tùlipa Didièri**	Large flowers 4½ in. across. Petals with deep purple spots at the base. Pointed wavy leaves. For location and cultivation see T. carinata. Alpine Region. See Plate, page 119.	1-1½ ft. *Sun*	"
"Bright red"	**ELEGANT-FLOWERED TULIP**	***Tùlipa élegans**	Graceful species with a yellow eye. Flowers about 3 in. long, with pointed slightly reflexed petals. For cultivation see T. carinata. Hort.	1-1½ ft. *Sun*	"
"Scarlet" 20	**BRILLIANT TULIP**	***Tùlipa fùlgens**	Said to be a garden form of T. Gesneriana. Brilliant flowers on tall stems. Leaves very wavy. For cultivation see T. carinata. Hort.	9-18 in. *Sun*	"

Color	English Name	Botanical Name and *Synonyms*	Description	Height and *Situation*	Time of Bloom
"Bright red" bet. 20 & 26	**COMMON GARDEN OR LATE TULIP**	****Tùlipa Gesnerià na**	One of the most gorgeous and effective of all the Tulips. Erect stem bears a large goblet-shaped scentless flower about 2 in. long, usually having a yellow or dark blotch at the base. Leaves often wavy. Beautiful in groups or masses or for edging shrubbery. There are vars. in different colors. Plant in late Sept. or Oct. 4 in. deep and 4 apart. Protect in winter and give an occasional top-dressing. Prop. by offsets. Light soil. If drainage is not good put sand around bulb. Origin unknown.	6-24 in. *Sun*	Mid. May to early June
"Red" 19	**GREIG'S TULIP**	****Tùlipa Greìgi**	See page 39.	3-8 in. *Sun*	Late Apr. to mid. May
"Scarlet"	**SUN'S-EYE TULIP**	***Tùlipa Óculus-sòlis**	Flowers about 4½ in. wide, with shining black spots edged with yellow clearly defined at the base of the petals. Plant in late Sept. or Oct. at a depth of 4 in. to the bottom of the bulb, allowing 4 in. between each. It is best to protect in winter and to give an occasional top-dressing. Prop. by offsets. Light soil. If drainage is not good put sand around the bulb. S. Europe.	12-18 in. *Sun*	May
"Scarlet" 18 tinged with yellow	**OSTROSK'S TULIP**	***Tùlipa Ostrowskiàna**	Allied to T. Oculus-solis. Scentless flowers spotted with black at the base, and sometimes edged at the top with yellow. For cultivation see T. Oculus-solis. Turkestan.	1 ft. *Sun*	Late May
"Scarlet" 18 marked with 3	**EARLY TULIP**	***Tùlipa præcox**	Very similar to T. Oculus-solis, but more vigorous and earlier blooming. See T. Oculus-solis. Europe.	12-18 in. *Sun*	May
"Rose pink" 29 dull	**ARETHUSA, INDIAN PINK**	**Arethùsa bulbòsa**	An orchid. Each plant bears one bright flower and one long and narrow leaf on an erect stem. Give northern exposure in rock or bog-garden. Prop. by pseudo-bulbs. Moist fibrous soil. Marshes U. S. A.	9 in. *Shade*	May, June
"Pink" 27 lighter	**CUSHION PINK, SEASIDE THRIFT, SEA TURF, CLIFF ROSE**	****Armèria marítima** *A. vulgàris*	Flowers in dense heads spring from tufts of evergreen foliage. Grows freely and blooms profusely. A trim plant excellent for rock-garden or edging. Prop. by seed and division. Europe; N. Amer.	3-6 in. *Sun*	Mid. May to mid. June
"Intense pink" 27 lighter	**LAUCHE'S THRIFT OR SEA PINK**	****Armèria marítima var. Laucheàna** *A. L.*	See page 39.		Late Apr. to mid. June

DIDIER'S TULIP. *Tulipa Didieri.*

MAIDEN PINK. *Dianthus deltoides.*

 GARLAND FLOWER. *Daphne Cneorum.*

Color	English Name	Botanical Name and *Synonyms*	Description	Height and *Situation*	Time of Bloom
"Pink" 31 deeper & redder	LEICHTLIN'S ROCK CRESS	*Aubriètia deltoìdea var. Leìchtlini	See page 39.		Late Apr., May
"Pink" 29 & white	ENGLISH DAISY	*Béllis perénnis	See page 39.		Mid. Apr. to mid. June
"Pink" 22	CALYPSO	Calýpso boreàlis *C. bulbòsa*	Pretty and rare Orchid, resembling the Lady's Slipper. Flowers solitary; large pink lip with brown spots and yellow crest within. One leaf on each plant. Suitable for bog garden but difficult of culture. Prop. by offsets. Light moist fibrous soil. Prop. by offsets. N. Amer.	3-6 in. *Half shade*	May, June
"Rose" 40 duller	ROSY MOUNTAIN BLUET	**Centaurèa montàna var. ròsea	Flowers resemble the Bachelor's Button. This var. is like the type. A pleasing border plant. Prop. by division. Ordinary garden soil. Hort.	12-20 in. *Sun*	Late May to early July
"Rose" 36 & 23	CAROLINA SPRING BEAUTY	Claytònia Caroliniàna	See page 39.		Late Apr., May
"Light pink" 36	SPRING BEAUTY	Claytònia Virgínica	See page 39.		Apr., May
"Pink" 29	SOLID-ROOTED FUMEWORT	Corýdalis bulbòsa *C. sólida, Fumària sólida*	Flowers borne in compact clusters rise from a mass of pretty gray-green foliage. Rock or wild garden. Prop. by seed or division. Any soil, not too dry. Europe; Asia.	6 in. *Sun or half shade*	May
"Purplish pink" 29	STEMLESS LADY'S SLIPPER	*Cypripèdium acaùle	Handsome solitary orchid-like flowers with pink sac and brownish sepals. Leaves like the Lily-of-the-valley. Protect in winter. Plant in shady rock-garden. Prop. by division. Moist porous peaty soil. Northern N. Amer.	8-12 in. *Shade*	Early May, June
"Deep pink" 29	GARLAND FLOWER	**Dáphne Cneòrum	See page 39 and Plate, page 120.		Late Apr., May
"Delicate pink" 27 light	CHEDDAR PINK	*Diánthus cæsius	Charming dwarf tufted species with pale fragrant flowers. Neat plant for rock-garden and border; prefers an elevated position. Prop. by seed and division. Rich well-drained soil. Europe.	8 in. *Sun*	Late May to early July
"Deep pink" 31	MAIDEN PINK	**Diánthus deltoìdes	A trim sturdy dwarf plant. Solitary crimson-centred flowers borne in abundance. Foliage forms a dense carpet. Very pretty for border or rock-garden. Prop. best by division. Rich soil. Japan; W. Europe. See Plate, page 120.	6-9 in. *Sun*	May, June

MAY

Color	English Name	Botanical Name and *Synonyms*	Description	Height and *Situation*	Time of Bloom
"Pink" 36 or 37	SCOTCH, COMMON GRASS, GARDEN OR PHEASANT'S EYE PINK	**Diánthus plumàrius	Often seen in old gardens. Fragrant flowers varying to white, purplish or parti-colored, with fringed or jagged petals. Small gray-green leaves form a dense tuft. Rock-garden or border. Prop. by seed and division. Rich soil. Siberia.	1 ft. *Sun*	Late May to late June
"Rose" 23	CALIFORNIA BLEEDING HEART	*Dicéntra formòsa	Heart-shaped flowers in drooping racemes hang above the pretty finely cut foliage. Protect slightly in winter. Attractive in old-fashioned gardens. Prop. by crown or root division. Good light soil. Cal.	1 ft. *Half shade best*	Late May, June
"Rose" 30	BLEEDING HEART	**Dicéntra spectábilis *Diélytra s.*	See page 40 and Plate, page 123.		Late Apr. to mid. July
"Rose" 29	JEFFREY'S SHOOTING STAR, AMERICAN COWSLIP	*Dodecàtheon Jéffreyi	Flowers in large umbels. Very large leaves springing from the root, and rising to the flowers. The foliage dies when the plant blossoms. Plant in sheltered places. Prop. slowly by seed, also by crown division. Moist well-drained soil. Western Coast of N. Amer.	1½-2 ft. *Half shade or shade*	May, June
"Pink-ish" 36 or 29	MAY-FLOWER, TRAILING ARBUTUS, GROUND LAUREL	*Epigæa rèpens	See page 40.		Apr., early May
"Rose" 38	ENDRES'S CRANES-BILL	Geránium Éndressi	Plentiful fringed blossoms with dark veining. Pretty foliage. Attractive in rough situations in the rock-garden. Prop. by seed and division. Any good soil. Pyrenees.	1-1½ ft. *Half shade*	Late May to late June
"Pink"	DUTCH HYACINTH	**Hyacínthus orientàlis vars.	See page 59.		Late Apr., May
"Pink" 29, spots 34	REDDISH LILY	*Lílium rubéllum	Showy clusters of funnel-form flowers about 3 in. across. Bright green foliage. Bulbous. Prop. by offsets or scales and very slowly by seed. Light well-drained rich soil. Avoid direct contact with manure. Japan.	3-4 ft. *Half shade*	Late May, early June
"Pink" 29 pale	RED ALPINE CAMPION	*Lýchnis alpìna *Viscària a.*	See page 59.		Late Apr., May
"Light pink" 29 pale	DOUBLE CUCKOO FLOWER OR RAGGED ROBIN	*Lýchnis Flós-cùculi var. pleníssima *L. pleníssima semperflòrens*	Pretty species, much grown in old-fashioned gardens. Double flowers in lax clusters, abundant and continuous in bloom, rise from a tuft of leaves. Prop. by seed, division and cuttings. Any soil. Hort.	1-2 ft. *Sun*	Late May to late June

BLEEDING HEART. *Dicentra spectabilis.*

ROSE-COLORED SPANISH SQUILL. *Scilia Hispanica var. rosea.*

MAY

Color	English Name	Botanical Name and *Synonyms*	Description	Height and *Situation*	Time of Bloom
"Rosy red" 31	GERMAN CATCHFLY	*Lýchnis Viscària	Tufted plant with a profusion of showy flowers in panicles. Grass-like leaves. Much used in old-fashioned gardens and borders. Prop. by seed, division or cuttings. Any soil. Europe; Siberia. Var. *elegans;* scarlet with stripes of white. Var. *flore-pleno;* (color no. 31), double, a good variety.	6-20 in. *Sun*	Mid. May to late June
"Rose" 29	LARGE ROSY TREE PEONY	*Pæònia Moután var. ròsea supérba *P. M. "Reine Elizabeth," "Triomphe de Grand"*	Beautiful shrub with large double flowers. Harder to grow than herbaceous Peony, needing protected spot and winter covering. Effective in isolated clumps, in the border or on edge of shrubbery. Prop. by grafting on roots of herbaceous species. It does not flower until the third year. Being a gross feeder, it likes a rich moist loam enriched with cow manure. Hort.	3-6 ft. *Sun or half shade*	Mid. May to mid. June
"Delicate rose" bet. 26 & 27	DOUBLE RED TREE PEONY	Pæònia Moután var. rùbra-plèna	Flowers somewhat double. See P. Moutan var. rosea superba. Hort.	3-6 ft. *Sun or half shade*	"
"Deep pink"	THREE-PART LEAVED PEONY	*Pæònia triternàta *P. daúrica*	Handsome herbaceous species with brilliant flowers and leaves of very regular shape. Effective in clumps on lawns and planted in masses in the garden. Usually prop. by autumn root-division. Deep moist loam, enriched with cow-manure. Siberia.	1½-2 ft. *Sun or half shade*	May
"Purple rose" 27 lighter & redder	CRAWLING PHLOX	*Phlóx réptans *P. stolonífera*	A creeping plant with many flower-bearing stems, forming pretty tufts in rock-garden or border. Prop. by seed, division or cuttings. Light rich soil. S. Eastern U. S. A.	4-8 in. *Sun*	May, June
"Magenta" bet. 44 & 38 dull	DARK PURPLE MOSS PINK	*Phlóx subulàta var. atropurpùrea	See page 40.		Late Apr., May
"Rose"	LEAFY MOSS PINK	*Phlóx subulàta var. frondòsa	See page 40.		"
"Purplish rose" 40 duller	FLOWERING WINTER-GREEN, GAY WINGS, FRINGED MILKWORT OR POLYGALA	Polýgala paucifòlia	Dainty trailing plant with fringed flowers, varying to white, on erect stems. Leaves purple-tinged. Wild garden. Prop. by seed and division. Moist peaty soil with sphagnum. N. Amer.	3-6 in. *Half shade*	May, June
"Deep rose" 26	NEPAL CINQUE-FOIL	*Potentílla Nepalénsis *P. formòsa, P. coccínea*	Good species. Somewhat furry cup-shaped flowers. Foliage compound. Border. Prop. by seed or division. Light rich soil. Himalayas.	1½-2 ft. *Sun*	"

Color	English Name	Botanical Name and *Synonyms*	Description	Height and *Situation*	Time of Bloom
"Rose" 26	CORTUSA-LEAVED PRIMROSE	*Prímula cortusoídes	Many flowers nearly 1 in. across in loose clusters terminate leafless stalks. Broad soft foliage in a rosette close to the ground. Charming in sheltered and slightly elevated position in rock-garden or border. Prop. by fresh seed, division or cuttings. Light rich soil. Siberia.	6-10 in. *Half shade*	May to early June
"Flesh color" 22	MISTASSINI OR DWARF CANADIAN PRIMROSE	*Prímula Mistassínica *P. farinòsa var. Mistassínica, P. pusílla*	Beautiful little plant bearing few flowers and small leaves. Moist position in the rock-garden. N. Amer.	3-6 in. *Half shade*	May, June
"Carmine pink" 30 bluer	ROSY HIMALAYAN PRIMROSE	*Prímula ròsea	Among the prettiest and best of alpine Primroses. Flowers with yellow eye, nearly 1 in. across, in drooping clusters. Smooth pale green leaves in tufts. Protect in winter. Rock-garden or border. Moist rich loam of good depth. Kashmir.	4-8 in. *Shade*	May
"Pink" 38 paler & more violet	ROCK SOAPWORT	*Saponària ocymoìdes	Trailing plant covered with flowers. Attractive in dry parts of rock-garden where it droops prettily over the rocks. Also used in the border. Prop. by seed or division. Any good soil. Central and S. Europe.	6-9 in. *Sun*	Late May to Aug.
"Pink" bet. 29 & 32	THICK-LEAVED SAXIFRAGE	*Saxífraga crassifòlia	See page 40.		Late Apr. to late May
"Purplish pink" often 46	NODDING WOOD HYACINTH	*Scílla festàlis var. cérnua *S. nùtans var. cérnua*	Fragrant nodding bell-shaped pinkish or bluish flowers in panicles on leafless stalks. Grass-like foliage in clumps. Pretty in rock-garden, border and edge of shrubbery, or naturalized in woods. Bulbous. Soil enriched with manure. Portugal.	8-12 in. *Half shade*	May, early June
"Pink" 36	PINK WOOD HYACINTH	Scílla festàlis var. ròsea *S. nùtans var. ròsea*	Differing from S. festalis var. cernua in its more rose-colored flowers. W. Europe.	8-12 in. *Half shade*	May to early June
"Flesh color" 22 dull	FLESH-COLORED SPANISH SQUILL	**Scílla Hispánica var. cárnea *S. campanulàta var. cárnea, S. pátula var. cárnea*	See S. Hispanica var. rosea. Spain; Portugal.		Late May, June
"Rose" 36	ROSE-COLORED SPANISH SQUILL	**Scílla Hispánica var. ròsea *S. campanulàta, S. pátula*	Most robust of the Scilla family. Strong pyramidal spikes of pendent bell-shaped flowers. Leaves narrow. Desirable in wild garden, rock-garden or edge of shrubbery. Bulbous. Enrich occasionally with top-dressing of manure. Spain; Portugal. See Plate, page 124.	1-1½ ft. *Sun or shade*	"

Color	English Name	Botanical Name and *Synonyms*	Description	Height and *Situation*	Time of Bloom
"Bright pink" 29	WILD PINK	*Silène Pennsylvánica	See page 40.		Late Apr., May
"Pale flesh color" white turns 36	TINY TRILLIUM	Tríllium pusíllum	See page 40.		Late Apr. to late May
"Pink"	TULIP	**Tùlipa vars.	**Single Early Bedding Tulips,** see page 64.		"
"Pink"	TULIP	**Tùlipa vars.	**Double Early Bedding Tulips,** see page 64.		"
"Rosy" bet. 36 & 37	DOUBLE PINK PERIWINKLE OR MYRTLE	Vínca mìnor var. ròsea plèna	See page 43.		Late Apr. to June
"Lilac" 26 purpler	LARGE-FLOWERED LEBANON CANDYTUFT	*Æthionèma grandiflòrum	Beautiful shrubby alpine plant of spreading habit, covered with rosy and lilac racemes of flowers. Good for cutting. Foliage light dull green. Well adapted for the rock-garden. Prop. by seed and cuttings. A light dry soil is necessary. Persia.	1½ ft. *Sun*	May to Aug.
"Purple" 47	BROAD-STEMMED ONIONWORT	Állium platycaùle *A. ánceps*	Flowers in umbels. Flourishes in sunny position in rock-garden. Bulbous. Prop. by offsets and seed. Light rich soil. Cal.	3-6 in. *Sun*	May, June
"Pale purple" white to 44 bluer	CAROLINA WIND-FLOWER	Anemòne Caroliniàna *A. decapètala*	See page 43.		Apr., May
"Lilac" 50	AMERICAN PASQUE FLOWER, WILD PATENS	*Anemòne pàtens var. Nuttalliàna *Pulsatílla hirsutíssima* (Brit.)	See page 43.		"
"Purple" 44 or 37	PASQUE FLOWER	*Anemòne Pulsatílla *A. acutipètala Pulsatílla vulgàris*	See page 43.		Early Apr. to late May
"Purple" 56 or 49	SHARP-SEPALED COLUMBINE	*Aquilègia oxysépala	Large showy flowers tinged with yellow and white. One of the earliest Aquilegias to bloom. Prop. by seed. Loose well-drained soil, moist and sandy. N. Asia.	2½ ft. *Sun*	May, June
"Purple" 56 or 49	EUROPEAN COLUMBINE	*Aquilègia vulgàris *A. stellàta, A. atràta*	Stems many-flowered. Petals ¾ in. long, spur stout and much incurved. Useful for wild garden or border. Prop. by seed. Light sandy soil, moist but well-drained. Numerous double and single vars. in odd colors not all of which are pleasing. Europe; N. Asia. Var. *Vervæneana* (var. *foliis-aureis,*	1½-2 ft. *Sun*	Mid. May to July

Color	English Name	Botanical Name and *Synonyms*	Description	Height and *Situation*	Time of Bloom
			var. *atroviolacea*), foliage marked with yellow. Hort. Var. *Olympica (A. Wittmaniana)*, has large short-spurred flowers, pale lilac, or mauve and white. Mt. Olympus.		
"Pinkish lavender" 37	ALPINE THRIFT	*Armèria alpìna	See page 43.		Late Apr. to mid. June
"Lavender" near 37	PLANTAIN-LIKE THRIFT	*Armèria plantagínea	Flowers in round compact heads spring from cushions of evergreen foliage. Free-growing and excellent for rock-garden and edging. Prop. by seed and division. Central and S. Europe.	10-12 in. *Sun*	Mid. May to late June
"Greenish purple"	ASARABACCA, HAZELWORT	Ásarum Europæ̀um	Hairy dwarf plant. Flowers lie on the ground under dark evergreen leaves resembling those of a Cyclamen. Useful for covering ground in shady places. A moist rich soil is necessary. Europe.	6-9 in. *Shade*	May, June
"Violet" 46	BLUE ALPINE ASTER	**Áster alpìnus	Large solitary single flowers with yellow centres, varying to pink and white. Leaves in clusters on the ground. Delightful for rock-garden and margin of border. Prop. by seed though generally by division. Easily cultivated in ordinary soil. Europe. See Plate, page 129. Var. *superbus*, a showy free-flowering form with large flowers. Hort.	3-10 in. *Sun or half shade*	Late May to late June
"Dark violet" 48 or 47	PURPLE ROCK CRESS	**Aubrìetia deltoìdea	See page 43.		Early Apr. to late May
"Dark violet" 48 lighter	EYRE'S PURPLE ROCK CRESS	Aubrìetia deltoìdea var. Eyrèi	See page 43.		Late Apr., May
"Dark violet" 46 & white	GRECIAN PURPLE ROCK CRESS	*Aubrìetia deltoìdea var. Græ̀ca	See page 43.		Mid. Apr. to late May
"Dark violet" 56	OLYMPIC PURPLE ROCK CRESS	*Aubrìetia deltoìdea var. Olýmpica	See page 44.		"
"Purple" 47 light	DEEP PURPLE ROCK CRESS	*Aubrìetia deltoìdea var. purpùrea	See page 44.		Mid. Apr. to late May
"Pale blue violet" 44 deep	GARGANO HAIRBELL	*Campánula Gargánica	Trailing tufted plant. Numerous flowers shading to white in the centre, in loose racemes on pendent stems. Pretty when hanging over rocky ledges. Prop. by cuttings in spring or by division. Rich well-drained loam. Italy.	3-6 in. *Sun*	May to Sept.

BLUE ALPINE ASTER. *Aster alpinus.*

COMMON SHOOTING STAR. *Dodecatheon Meadia.*

Color	English Name	Botanical Name and *Synonyms*	Description	Height and *Situation*	Time of Bloom
"Purple" 48	**WALL HAIRBELL**	**Campánula Porten-schlagiàna** *C. muràlis*	Plant generally erect. Bell-shaped flowers. Suitable for rock-garden. Prop. by seed, division or cuttings. Rich well-drained loam. Dalmatia.	6-8 in. *Sun*	May, June
"Rosy purple" 39	**BEAR'S EAR SANICLE**	**Cortùsa Matthioli**	See page 44.		Apr., May
"Pur-plish" effect 43	**RAM'S HEAD LADY'S SLIPPER**	***Cypripèdium arietìnum**	Solitary white flowers, veined with purple, on leafy stems. Winter protection necessary. Rock or wild garden plant. Prop. by division. Porous moist peaty soil. Northern N. Amer.	8-12 in. *Shade*	Late May to Aug.
"Rosy purple" 39	**COMMON OR EASTERN SHOOTING STAR, AMERICAN COWSLIP**	***Dodecàtheon Mèadia**	Umbels of drooping flowers of varying colors, with recurved petals and protruding stamens. Flower-stems naked. Leaves in an erect tuft. Plant in clumps in the rock-garden. Prop. by seed just gathered, or preferably by division. Rich well-drained moist loam. Southern U. S. A. See Plate, page 130.	9-24 in. *Half shade or shade*	May, June
"Pur-plish" 27 purpler	**BLACK-BERRIED HEATH, OR CROW-BERRY**	**Empètrum nìgrum**	Low heath-like bush. Insignificant flowers succeeded by black edible berries. Evergreen foliage. Associate with Heaths, etc., in rock-garden. Prop. by cuttings. Preferably moist soil of sand or peat. Northern Hemisphere.	9-12 in. *Sun*	May
"Pur-plish" 50	**BLUE OR BITTER FLEABANE**	**Erígeron ácris**	Aster-like flowers in clusters surmount branching stalks. Foliage rough. Mass in wild spots. Prop. by seed and division. Any ordinary soil. Europe; N. Asia; Western N. Amer.	1-1½ ft. *Sun*	Mid. May to early June
"Purple" 43	**THRIFT-LEAVED FLEABANE**	***Erígeron armeriæ-fòlius**	Tufted plant. Large yellow-centred aster-like flowers borne in racemes. Mass in wild garden or margin of border. Prop. by seed or division. Any ordinary soil. N. Asia; Western U. S. A.	1-1½ ft. *Half shade*	Late May, June
"Bluish purple" 43 deep & bluer	**ROBIN'S PLANTAIN, ROSE PETTY**	**Erígeron bellidifòlius** *E. pulchéllus*	Downy tufted plant. Aster-like yellow-centred flowers. Clump in wild garden. Prop. by seed and division. Eastern N. Amer. See Plate, page 133.	1 ft. *Half shade*	Mid. May to July
"Pur-plish" 32	**ROUGH ERIGERON**	***Erígeron glabéllus**	Aster-like flowers. Color varies to white. Plant in clumps in wild garden or border. Prop. by seed and division. Ordinary garden soil. Minn.; West.	6-15 in. *Half shade*	Late May, June
"Violet" 50 dull	**BEACH ASTER**	***Erígeron glaùcus**	Flowers resemble small single China Asters. Pale foliage in a tuft. A pretty plant for the border. Prop. by seed and division. Easily cultivated in any garden soil. Pacific Coast.	1 ft. *Half shade*	"

MAY

Color	English Name	Botanical Name and *Synonyms*	Description	Height and *Situation*	Time of Bloom
"Violet purple" 39 more violet	WALL ERINUS	*Erìnus alpìnus	Pretty alpine plant, with racemes of numerous flowers. Downy foliage. Plant on ledges of rock-garden. Prop. by seed and division. Thoroughly well-drained soil. Mts., W. Europe. There is a white var.	3-4 in. *Sun or half shade*	May, June
"Lilac" often bet. 43 & 45	COMMON DOGTOOTH VIOLET of Europe	**Erythròni-um Dens-Cànis	See page 44.		Late Apr., May
"Deep purple" 42 lighter	PURPLE FRITILLARY	*Fritillària atropurpùrea	See page 44.		Mid. Apr. to June
"Purple black" 56	TWO-FLOWERED FRITILLARY	*Fritillària biflòra	See page 44.		"
"Deep purple" 21 purpler	BLACK LILY	*Fritillària Camtschat-cénsis *Lìlium C.*	See page 44.		"
"Livid purple" 33	RUSSIAN FRITILLARY	*Fritillària Ruthénica	See page 47.		Mid. Apr. to mid. May
"Laven-der" 39	WILD GERANIUM, WILD OR SPOTTED CRANES-BILL	Gerànium maculàtum	Common wild species. Graceful plant with large flowers and pretty foliage. Wild garden. Prop. by seed and division. Any rather moist soil. See Plate, page 133. Var. *plenum,* double flowers of darker shade. N. Amer.	1-1½ ft. *Half shade*	Early May to July
"Pur-plish" 44 deeper	SWAMP OR STUD PINK	Helònias bullàta	See page 47.		Apr., May
"Pur-plish" 30 deeper to white	DOWNY HEUCHERA	Heùchera pubéscens *H. rubifòlia, H. pulver-ulénta*	Flowers, variegated with yellow, in panicles. Evergreen bronze-red foli-age, covered with powdery down. Bor-der or rock-garden. Prop. by seed and division. Ordinary soil. Mts. S. Eastern U. S. A.	1-3 ft. *Sun or half shade*	Late May, June
"Palest purple" 37 pale	LARGE HOUSTONIA	Houstònia purpùrea	Tufted plant with small flowers, sometimes varying to white, on slen-der stems which rise from a clump of tiny leaves. Rock or wild garden. Prop. by spring division. Dry soil. Eastern N. Amer	4-12 in. *Sun or half shade*	May, early June
"Purple & lilac"	DUTCH HYACINTH	**Hyacínthus orientàlis vars.	See page 59.		Late Apr., May
"Purple" often 50	APPEN-DAGED WATER-LEAF	Hydro-phýllum appendicu-làtum	Biennial plant. Erect bell-shaped flowers in loose clusters. Much di-vided leaves. Border. Prop. by seed and division. Fairly moist soil. N. Amer.	12-15 in. *Shade*	Mid. May to early June

ROBIN'S PLANTAIN. *Erigeron bellidifolius.*

WILD GERANIUM. *Geranium maculatum.*

LARGER BLUE FLAG. *Iris versicolor.*

DWARF FLAG. *Iris pumila.*

MAY

Color	English Name	Botanical Name and *Synonyms*	Description	Height and *Situation*	Time of Bloom
"Delicate lilac" 43 or 36	**GIBRALTAR CANDYTUFT**	***Ibèris Gibraltárica**	More showy and with larger flowers than other species of Iberis, but rather delicate, needing protection in winter. Prop. by seed and cuttings. Plant in rock-garden and do not disturb. Light soil. Gibraltar.	12-15 in. *Sun*	May, June
"Pale lilac" 44	**CRESTED DWARF IRIS**	****Ìris cristàta**	Tiny species. Delicate richly marked flowers borne close to the ground. Outer petals crested. Spreads rapidly. Plant in rock-garden or edge of border. Prop. by division. Light well-drained soil. Alleghany Mts.	4-9 in. *Sun*	Late May to July
"Purple & lavender"	**GERMAN IRIS, FLEUR-DE-LIS**	****Ìris Germánica vars.**	Large shapely fragrant flowers borne on stout stalks high above the broad sword-shaped leaves. Effective in isolated clumps, in masses, and along the edge of shrubbery. Prop. by division of rhizomes. To prevent crowding divide frequently. Any well-drained garden soil. **Albert Victor;* (color no. 46), standards soft blue, falls delicate lilac. **Darius;* (color no. 48 & 2), height 17 in., lilac flowers edged with white, beard rich orange. ***Purple King;* (color no. 55 warmer), purple. Hort. vars.	1½-3 ft. *Sun*	"
"Violet" 44 deep	**GREAT PURPLE FLAG, TURKEY FLAG**	****Ìris pállida** *I. Junònia, I. Asiática, I. sícula*	Flower-stems, much higher than foliage, bear 8-12 large fragrant flowers, rarely white. Beautiful species in border or in large groups. Prop. by division. Grows luxuriantly in any garden soil, though a gross feeder. S. Europe.	2-4 ft. *Sun or half shade*	"
"Lilac" 50	**PLAITED FLAG**	****Ìris plicàta** *I. aphýlla var. plicàta*	In habit resembles I. pallida. Flowers white veined and tinted with lilac at the edge of the petals. Inner petals much folded. Good border plant in masses or groups. Prop. by division. Origin unknown.	2-4 ft. *Sun or half shade*	"
"Blue violet" 49	**SLENDER BLUE FLAG**	***Ìris prismática** *I. Virgínica, I. grácilis*	Tall slender habit. Stems bear 1 or 2 flowers, yellow near the centre and veined with purple. Inner petals erect. Narrow leaves, shorter than flower-stem. Prop. by division. Rather moist soil advisable. New Brunswick to N. C.	1-2 ft. *Sun*	Mid. May to July
"Deep violet" 49	**DWARF FLAG**	***Ìris pùmila** *I. grácilis*	One of the best of dwarf Irises. Short-lived flowers over-large and close to the ground. Leaves sword-shaped. Spreads rapidly. Good for rock-garden or border. Prop. by division. Any garden soil. Europe. Var. *atroviolacea;* (color no. 48 deep), velvety purple flowers. There are other vars. in different colors. See Plate, page 134.	4-8 in. *Sun or half shade*	May

Color	English Name	Botanical Name and *Synonyms*	Description	Height and *Situation*	Time of Bloom
"Deep violet" 49	SIBERIAN FLAG	**Ìris Sibírica *I. acùta*	Attractive species distinguished by its tall slender stalks. Many flower-stems bear clusters of small but showy flowers, veined with white and bright violet, which rise above the dense tuft of grass-like foliage. There are vars. in other colors. Border. Prop. by division. Rich soil. Europe; E. Siberia.	2-3 ft. *Sun*	Late May to mid. June
"Blue violet" 55	SLENDER DWARF IRIS	*Ìris vérna	See page 47.		Late Apr., May
"Bright purple" 56 lighter	LARGER BLUE FLAG	*Ìris versícolor	Native Iris. Flowers marked with white, yellow and purple. Leaves slightly grayish. Good for margin of ponds and also for dry positions. Prop. by division. Canada; Northern U. S. A. See Plates, page 134.	1-3 ft. *Sun or half shade*	Late May, June
"Pur-plish" 45	VARIE-GATED NETTLE	Làmium maculàtum *L. purpù-reum* (Hort.)	Straggling plant. Flowers in whorls on leafy stalks. Heart-shaped foliage blotched with white. Common in old gardens. Useful for covering barren places. Prop. by division. Preferably sandy open soil. Europe; Asia; N. Africa.	6-8 in. *Sun*	Mid. May to late July
"Bluish violet" 39	SPRING BITTER VETCH	**Láthyrus vérnus *Orobus vêrnus*	See page 47.		Mid. Apr. to late May
"Lilac purple" 45	SLENDER BEARD-TONGUE	*Pentstèmon grácilis	Pretty species of slender growth. Tubular flowers sometimes whitish. Leaves springing from the root grow in pairs on the flower stalks. Prop. by seed and division. Good garden soil. Col. and North.	8-12 in. *Sun*	Late May to early July
"Rosy purple" 47	OVAL-LEAVED PENTSTE-MON	Pentstèmon ovàtus *P. Glaucus*	Vigorous plant of fine form. Small tubular flowers borne in great profusion above the bright foliage. Winter protection of leaves. Prop. by seed and division. Any good soil. Oregon; Western N. Amer.	2-3 ft. *Sun*	Late May to late June
"Pale violet" 47, edge white	DOWNY PENTSTE-MON	*Pentstèmon pubéscens	Not a showy species. Tubular flowers, sometimes flesh color, droop in long loose clusters. Mass in rock-garden. Border. Prop. by seed and division. Good rich soil. N. Amer.	2 ft. *Sun*	Late May to mid. July
"Bluish lilac" 44 or 50	WILD SWEET WILLIAM	**Phlóx divaricàta *P. Cana-dénsis*	Fragrant flowers in loose clusters which form a mass of color. Foliage good. Lovely plant for border or rock-garden. Prop. by seed, division or cuttings. Rich moist soil. N. Amer.	10-18 in. *Sun*	May
"Pinkish purple" 27 lighter & purpler	TRAILING PHLOX	*Phlóx procúmbens	See page 47 and Plate, page 137.		Late Apr. to late May

HORNED VIOLET AND TRAILING PHLOX. *Viola cornuta vars.* and *Phlox procumbens.*

 AMERICAN JACOB'S LADDER. *Polemonium cœruleum.*

MAY

Color	English Name	Botanical Name and *Synonyms*	Description	Height and *Situation*	Time of Bloom
"Pale lilac" 50	**LILAC MOSS PINK**	***Phlóx subulàta var. lilacìna**	See page 47.		Late Apr., May
"Bluish purple" bet. 44 & 50	**AMERICAN JACOB'S LADDER, CHARITY**	****Polemò-nium cærùleum**	Bushy bell-shaped flowers nearly 1 in. across, in compact terminal panicles. Profusion of compound leaves. Splendid border plant. Prop. by seed and division. Rich well-drained loamy soil is best. Vars. with white flowers and variegated foliage. N. Amer. See Plate, page 138.	1-3 ft. *Half shade*	Mid. May to July
"Pur-plish" 43 light	**EAR-LEAVED PRIMROSE**	***Prímula auriculàta** *P. longifólia*	Round head of flowers, each with a white eye. Pale green leaves in a rosette. Protect in winter. Sheltered position in rock-garden or border. Prop. generally by seed just gathered. Rich light soil. S. E. Europe.	4-6 in. *Half shade*	May, June
"Deep blue-purple" 46 deeper	**ROUND-HEADED HIMA-LAYAN PRIMROSE**	***Prímula capitàta**	Among the prettiest of Primroses resembling P. denticulata. Flowers in dense round heads. Pale green leaves in a tuft. Plant in slightly elevated position in rock-garden or margin of border. Light rich soil. Himalayas.	6-9 in. *Half shade*	May, early June
"Pale purple" 43 to white	**TOOTH-LEAVED PRIMROSE**	***Prímula denticulàta**	Flowers in dense clusters on stout erect stalks which rise high above the large rosettes of broad foliage. Pretty in rock-garden or border. Prop. by seed. A light rich loam is best. Himalayas. Var. *purpureum*, dark purple flowers.	6-12 in. *Half shade or shade*	"
"Dark purple" 44 deeper	**KASHMIR TOOTH-LEAVED PRIMROSE**	***Prímula denticulàta var. Cachemi-riàna** *P. Cachme-riàna*	Flowers with yellow centres in globe-shaped clusters. Foliage beautiful; pale green thickly dusted with mealy powder. Pretty in rock-garden or border. Prop. by seed. A moist loam is best. Western Himalayan Region.	8-12 in. *Half shade or shade*	"
"Lilac purple" 43 paler	**BIRD'S EYE PRIMROSE**	***Prímula farinòsa**	Flowers with yellow eye, in dense umbels, on stems that rise from tufts of silvery leaves. Somewhat elevated situation in rock-garden or border. Prop. generally by seed. Moist fibrous soil is best. Central and N. Europe.	3-9 in. *Half shade*	"
"Purple" near 30 deeper to white	**JAPANESE PRIMROSE**	****Prímula Japónica**	Fine and vigorous plant. Flowers 1 in. wide in whorls of twelve or more. Leaves oval, at base of plant. Excellent for border, rock-garden or wild garden. Prop. by seed. Soil moist and deep. Japan.	1-2 ft. *Half shade*	Late May to Aug.
"Violet purple" 52 duller & more violet	**ROSETTE MULLEIN**	***Ramónda Pyrenàica** *Ramóndia Pyrenàica*	Popular alpine plant. Flowers 1½ in. across, with orange centres. Leaves in rosettes on the ground. Protect in winter. Plant in chinks in the rock-garden. Prop. by seed and division. Deep peaty soil, well-drained. Pyrenees.	2-6 in. *Half shade*	Mid. May to July

Color	English Name	Botanical Name and *Synonyms*	Description	Height and *Situation*	Time of Bloom
"Bluish violet" 44	TWO-COLORED SAGE	**Sálvia bícolor**	Hardy biennial. Large hooded flowers; upper lip blue, lower white, sometimes fading reddish brown. Prop. by seed. Mediterranean Region.	2-3 ft. *Sun*	Late May, June
"Lilac" 39	HEART-LEAVED SAXIFRAGE	***Saxífraga cordifòlia**	See page 47.		Late Apr. to late May
"Purple" 53 brighter	PURPLE OR MOUNTAIN SAXIFRAGE	**Saxífraga oppositifòlia**	See page 48.		Late Apr., May
"Lilac" 43	LILAC WOOD HYACINTH	***Scílla festàlis var. lilacìna**	Fragrant drooping bell-shaped flowers in panicles on leafless stalks. Grass-like foliage in clumps. Rock-garden, border and shrubbery. Pretty naturalized in woods. Bulbous. Soil enriched with manure. W. Europe.	8-12 in. *Half shade*	May, early June
"Rich purple"	LARGE-FLOWERED BLUE-EYED GRASS	**Sisyrínchium grandiflòrum** *S. Doùglasii*	Pretty flowers drooping and bell-shaped. Foliage grass-like. Border, rock or wild garden. Prop. by division. Any garden soil. Western U. S. A.	1 ft. *Sun*	May, June
"Violet" 50 deep	BLUE MOONWORT	**Soldanélla alpìna**	Alpine plant. Flowers 2-4 on the stem, nodding, fringed, pale blue or violet. Leaves heart-shaped. Suitable for rock-garden. Prop. by seed and division. Alps, Europe.	2-3 in. *Half shade or shade*	May
"Purplish" 44, 48 or 49	COMMON SPIDER-WORT	***Tradescántia Virginiàna** *T. Virgínica*	Erect bushy plant of free and vigorous habit. Umbels of showy flowers in different shades of blue violet. Leaves grass-like. Useful for border or rock-garden. Prop. by division in spring. Ordinary garden soil. Eastern U. S. A. See Plate, page 84. Var. *cærulea;* (between color nos. 50 & 57), brighter blue flowers than the type.	1-3 ft. *Sun or half shade*	Late May to late Aug.
"Purple" 46	LONG-LEAF-STALKED TRILLIUM	**Tríllium petiòlatum**	Large flowers borne close to the ground. Pretty for the border or for naturalizing in moist soil. Prop. by seed. Moist rich soil. N. Western U. S. A.	6-8 in. *Half shade*	May to early June
"Pale violet" white, markings 53	SPRING STAR-FLOWER	**Triteleìa unifl̀ora** *Mílla unifl̀ora, Brodiæa unifl̀ora*	Hardy but not enduring. Solitary starry flowers, almost white with purple markings, on slender stems. Grass-like foliage. Protect in winter. Associate with Scilla, etc., in warm position in border or rock-garden. Bulbous. Prop. by seed or offsets. Rich soil, well-drained. Argentine Republic.	6-8 in. *Sun*	May
"Violet"	TULIP	****Tùlipa vars.**	**Single Early Bedding Tulips.** See page 64.		Late Apr. to late May

Color	English Name	Botanical Name and *Synonyms*	Description	Height and *Situation*	Time of Bloom
"Purplish" 46	HERBACEOUS PERIWINKLE OR MYRTLE	Vínca herbàcea	Trailing plant. Flowers more purple than V. minor with narrower petals. Leaves glossy and narrow but not evergreen like V. minor. Excellent for rockwork as it is not too bold. Prop. by division and cuttings. Moist loam is best. Eastern Europe.	Trailing *Half shade or shade*	Late May, June
"Violet" 49	COMMON PERIWINKLE, BLUE RUNNING OR TRAILING MYRTLE	*Vínca mìnor	See page 48.		Late Apr. to June
"Violet" 47 or 49	HORNED VIOLET, BEDDING PANSY	**Vìola cornùta	See page 48.		Late Apr. until frost
"Violet" 48 or 47	SWEET VIOLET	Vìola odoràta	See page 48.		Late Apr. to late May
"Violet" 53	EARLY BLUE VIOLET	Vìola palmàta *V. cucullàta var. palmàta*	See page 48.		Late Apr., May
"Violet" 46	COMMON BLUE OR LARGE AMERICAN VIOLET	Vìola palmàta var. cucullàta	See page 48.		Late Apr. to late May
"Lilac" usually 50	BIRD'S-FOOT VIOLET	*Vìola pedàta	A charming Violet. Flowers vary from deep lilac to white; petals flat and pansy-like. Leaves finely divided, resembling a bird's foot. Pretty for rock-garden, wild garden or naturalization. Prop. by seed and division. Dry sandy soil. U. S. A.	3-6 in. *Sun*	May, June
"Blue" 46	ERECT BUGLE	**Ajùga Genevénsis *A. rugòsa, A. alpìna*	Erect stem bears numerous flowers in whorls. Color varies. Forms a mat in border or edge of shrubbery. Prop. by seed and division. Thrives in any common soil. Europe.	6-8 in. *Sun or shade*	May
"Ultramarine blue" 54 pale & dull	CRISPED METALLIC BUGLE	*Ajùga metállica var. críspa	Shining flowers in twisted spikes. Adapted for carpeting and bedding. Prop. by seed and division. Germany.	4-5 in. *Sun*	May, June
"Blue" 46	BUGLE	**Ajùga réptans	Dense creeping plant; increases rapidly. Flowers numerous in erect spikes. Shiny leaves. Good for carpeting shady places. Prop. by seed and division. Any common soil. Europe. Var. *rubra;* Red-leaved Bugle; (color no. 46), more often cultivated than the type. Height 3-6 in. Valued for	3-4 in. *Sun or shade*	Early May to mid. June

Color	English Name	Botanical Name and *Synonyms*	Description	Height and *Situation*	Time of Bloom
			its dark purple foliage, rather than for its flowers. Europe. Var. *variegata,* variegated foliage. Hort.		
"Light blue" 58 lighter	**NARROW-LEAVED AMSONIA**	***Amsònia angustifòlia** *A. ciliàta*	Small and numerous flowers in panicles. Young leaves soft and downy. Valuable among shrubs and in the border because of its peculiar shade of color. Prop. by seed, division or cuttings in summer. Southern U. S. A.	1-3 ft. *Sun*	May, June
"Light blue" 58 pale	**AMSONIA**	****Amsònia Tabernæ-montàna** *A. latifòlia, A. salicifòlia Tabernæ-montàna Amsònia*	Small and numerous flowers in panicles succeeded by soft hairy pods. Smooth foliage resembling that of the Olive. Shrubbery and border. Prop. by seed, division and cuttings. N. C. to Tex. See Plate, page 143.	2-3 ft. *Sun*	Late May, early June
"Dark blue" 62	**BARRE-LIER'S ALKANET**	***Anchùsa Barrelièri**	Small flowers resembling Forget-me-nots in heliotrope-like clusters. Good for cutting. A valuable early bloomer for border or wild garden. Prop. generally by seed. Europe; Asia Minor.	2 ft. *Sun*	May, June
"Blue" 54	**CAPE ALKANET**	***Anchùsa Capénsis**	Attractive biennial plant. Flowers like Forget-me-nots with red and white markings. Long narrow leaves in clumps at base of plant. Often winter-killed but self perpetuating. Border. Prop. by seed. Well-drained soil. Cape of Good Hope.	1½ ft. *Sun*	Late May to mid. July
"Blue" 54, buds 41	**ITALIAN ALKANET**	***Anchùsa Itálica**	Best species. Attractive plant with trumpet-shaped flowers in panicles. Continuous in bloom if not allowed to seed. Large rough glossy foliage. Border. Prop. by seed. Any ordinary soil. S. Europe.	3-4 ft. *Sun*	"
"Sky-blue" 52	**APENNINE WIND-FLOWER**	***Anemòne apennìna**	See page 48.		Apr., May
"Sky-blue" 52 intense	**BLUE WINTER WIND-FLOWER OR ANEMONE**	***Anemòne blánda**	See page 51.		"
"Blue" 44 tinged blue	**ROBINSON'S WOOD ANEMONE**	***Anemòne nemoròsa var. Robinsoniàna** *A. n. var. cærùlea*	See page 51.		Mid. Apr. to June
"Blue" 46 & white	**ALPINE COLUMBINE**	***Aquilègia alpìna**	Showy flowers of delicate color, 2-5 on a stem. Fine foliage. Plant in a sheltered position in rock-garden or border. Prop. by seed started in earliest spring. A moist sandy loam is best. Switzerland. Var. *superba;* flowers with white centres. Siberia.	1 ft. or less *Sun*	May, June

AMSONIA. *Amsonia Tabernæmontana.* E

144 BLUE FALSE INDIGO. *Baptisia australis.*

MAY

Color	English Name	Botanical Name and *Synonyms*	Description	Height and *Situation*	Time of Bloom
"Blue" near 47	LONG-SPURRED COLUMBINE	**Aquilègia cærùlea *A. leptocèras, A. macrántha*	A beautiful plant. Large flowers varying to whitish, 2½-3 in. across with slender twisting green-tipped spurs. Leaves large and handsome. Best to treat as a biennial. Lovely in clumps or masses in rock-garden or border. Prop. by seed in spring. Light moist well-drained soil. Rocky Mts. Var. *hybrida;* sepals usually blue or pink and petals yellowish; good border plant. Hort.	1-1½ ft. *Sun*	Mid. May to July
"Deep blue" 63 & white	ALTAIAN COLUMBINE	**Aquilègia glandulòsa	Charming species. Large nodding flowers; petals tipped with cream; short incurving spurs. Treat as a biennial. Rock-garden or border. Prop. by seed in spring. Well-drained soil. Altai Mts.	1-1½ ft. *Sun*	May, June
"Lilac blue" 44 bluer & white	STUART'S COLUMBINE	*Aquilègia Stùarti	A fine plant. Flowers large, erect and beautiful. Hybrid of A. glandulosa and A. Olympica. Resembles A. glandulosa in color. Good border plant. Prop. by seed. Hort.	1½ ft. *Sun*	Mid. May to July
"Blue" 63	BLUE WILD OR FALSE INDIGO	**Baptísia austràlis *B. cærùlea, B. exaltàta*	The best species of Baptisia for the garden. Bushy plant resembling the Lupine. Large flowers in spikes. Foliage sea-green. Good border plant. Prop. by seed and division. Any soil. Pa.; south to Ga. and N. C. See Plate, page 144.	4-4½ ft. *Sun*	Late May to mid. June
"Pale blue" 44	CUSICK'S QUAMASH	*Camássia Cùsickii	Good species. Closely allied to the Squill. 30-50 flowers on a stalk. Good for cutting. Broad thick foliage 1 ft. long. Plant in early autumn and leave undisturbed. Bulbous. Any good garden soil. Oregon.	2-3 ft. *Sun or half shade*	May
"Purplish blue" 54	CAMASS	*Camássia esculénta	Nearly allied to the Squill. Flowers varying to white in racemes borne above the lance-shaped leaves. Plant in border or under trees. Bulbous. Good loamy soil. Western N. Amer.	1-2 ft. *Sun or half shade*	"
"Light blue" 57 deep	WILD HYACINTH	*Camássia Fràseri *Quamàsia hyacínthina, Scílla Fràseri*	Smaller in every way than C. esculenta. Flowers in dense racemes. Plant in border or naturalize in shady places. Bulbous. Good loamy soil. Eastern U. S. A.	12-18 in. *Half shade*	"
"Pale blue" 62 lighter	TUFTED HAIRBELL	*Campánula cæspitòsa	Vigorous tufted plant. Abundant light blue nodding bell-shaped flowers. Useful fôr edging; grows well in crevices of the rock-garden. Prop. by division. Rich loam and leaf mold preferable. Alps.	4-6 in. *Sun*	May to July
"Blue" 52	ALLEN'S GLORY OF THE SNOW	*Chionodóxa Álleni *Chionoscílla Álleni*	See page 4.		Mid. Mar. to early May

Color	English Name	Botanical Name and *Synonyms*	Description	Height and *Situation*	Time of Bloom
"Blue" 52 & white	GLORY OF THE SNOW	**Chiondóxa Lucíliæ	See page 7.		Mid. Mar. to early May
"Blue" 52 or 63	GIANT GLORY OF THE SNOW	**Chionodóxa Lucíliæ var. gigantèa	See page 7.		"
"Blue" 52	TMOLUS' GLORY OF THE SNOW	*Chionodóxa Lucíliæ var. Tmolùsi *C. T.*	See page 7.		"
"Gentian blue" 52 deeper	SARDIAN GLORY OF THE SNOW	*Chionodóxa Sardénsis	See page 7.		"
"Blue" 62 more violet	NODDING DRAGON'S-HEAD	*Dracocéphalum nùtans	Drooping racemes of quickly passing flowers which blossom well only in damp seasons. Rock-garden or border. Prop. by seed and division. Moist sandy loam. Russia; N. Asia. Var. *alpina* is less rare.	8-12 in. *Half shade*	May, June
"Dark blue" 63 lighter, markings 19 darker	TULIP-LEAVED FRITILLARY	*Fritillària tulipifòlia	Distinct and pretty species. Solitary dark tulip-shaped flowers with purple streaks, rusty purplish brown within. Bluish green foliage. Bulbous. Prop. by offsets; division every 3 or 4 years necessary. Deep sandy loam mixed with leaf mold. Caucasus.	2-8 in. *Shade*	Early May
"Blue" 62	GENTIANELLA, STEMLESS GENTIAN	*Gentiàna acaùlis	Handsome species. Very large bell-shaped flowers marked inside with yellow, rise from rosettes of glossy leaves. Useful for the rock-garden or border. Leave undisturbed. Prop. very slowly by freshly ripened seed and division. Good moist soil. Alps; Pyrenees.	2-4 in. *Half shade*	May, June
"Dark blue" 60	ALPINE GENTIAN	Gentiàna alpìna	A form of G. acaulis having small broad leaves. See G. acaulis. Alps.	3-4 in. *Half shade*	"
"Light blue" 60 deeper & brilliant	VERNAL GENTIAN	*Gentiàna vérna	See page 51.		Apr. to early June
"Light blue" 52 deep & dull	HAIR-FLOWERED GLOBE DAISY	*Globulària trichosántha	Rather large pale flowers in round heads. Foliage becomes blackish in autumn. Attractive in the rock-garden or border. Prop. by seed and division. Somewhat moist though well-drained soil. W. Asia.	6 in. *Half shade*	Late May to Aug.
"Bluish" near 44	SHARP-LOBED OR HEART LIVER LEAF	*Hepática acutíloba *H. tríloba var. acùta, Anemòne acutíloba*	See page 51.		Apr., early May

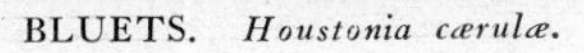

BLUETS. *Houstonia cærulæ.*

PERENNIAL FLAX. *Linum perenne.*

MAY

Color	English Name	Botanical Name and *Synonyms*	Description	Height and *Situation*	Time of Bloom
"Bluish" 53 lighter	FIVE-LOBED HEPATICA	*Hepática angulòsa *Anemòne angulòsa*	See page 51.		Late Apr., early May
"Bluish" near 44 or 46	ROUND-LOBED OR KIDNEY LIVER LEAF	*Hepática tríloba *H. Hepática, Anemòne H. A. tríloba*	See page 51.		"
"Pale blue" 57	BLUETS, INNOCENCE, QUAKER LADY	*Houstònia cærùlea	Dainty little native plant. Tufts of slender stems bear numerous star-shaped flowers. Pretty in rock or wild garden. Prop. by spring division. Moist soil preferable. N. Eastern U. S. A. See Plate, page 147.	3-6 in. *Sun or half shade*	May, early June
"Light blue" 61	AMETHYST HYACINTH	*Hyacínthus amethýstinus	Delicate and graceful plant. Drooping bell-shaped flowers borne along the flower-stem. Leaves lance-shaped springing from the root. Most effective when planted in masses. Bulbous. Prop. by offsets and seed. Any rich light soil. S. Europe.	½ ft. *Sun*	"
"Blue"	DUTCH HYACINTH	**Hyacínthus orientàlis vars.	See page 59.		Late Apr., May
"Pale blue" effect 44	WESTERN BLUE FLAG	*Ìris Missouriénsis *I. Tolmieàna*	Free bloomer. Flowers yellow near the centre, grow high above the foliage. Prop. by division. Good border plant in moist soil. Rocky Mts.	1-2 ft. *Sun*	Late May, June
"Blue" bet. 54 & 56	NEGLECTED FLAG	**Ìris neglécta	Stems many flowered. Blossoms 2½ in. across; falls much reflexed, deep blue veined with purplish red; beard bright yellow: standards pale blue; conspicuous yellow crest. Glaucous leaves. Prop. by division. Known only in cultivation.	1½-2 ft. *Sun*	Late May to early June
"Azure blue" 51	AZURE DWARF FLAG	*Ìris pùmila var. azùrea	One of the best dwarf Irises. Flowers close to the ground and short lived. Sword-shaped leaves. Spreads rapidly. Good for rock-garden or border. Prop. by division. Any garden soil. Europe.	4-8 in. *Sun or half shade*	May
"Deep blue" bet. 54 & 56	SPURIOUS IRIS	Ìris spùria	Not an attractive species. Stems rise above the glaucous grass-like leaves bearing several flower heads. Var. *notha* is a sturdier plant; flower stems 2-3 ft. Kashmir. Prop. by division. Europe.	15-20 in. *Sun*	Late May to late June
"Blue" 54 lighter	PERENNIAL FLAX	**Lìnum perénne	Light and feathery in effect. A free and continuous bloomer. Flowers on slender leafy stems. Foliage delicate. Very attractive in rock-garden or border. Prop. by seed and division. Rich light soil. Europe; Western U. S. A. See Plate, page 148.	1-1½ ft. *Sun or half shade*	Mid. May to Aug.

Color	English Name	Botanical Name and *Synonyms*	Description	Height and *Situation*	Time of Bloom
"Blue" 62	GENTIAN-BLUE CROMWELL	Lithospérmum prostràtum	A dwarf evergreen shrub of trailing habit. Low leafy clusters of gentian-blue flowers. Protection needed. Pretty rock or border plant. Prop. by "cuttings of previous year's wood." Well-drained sandy or loamy soil. S. Europe.	4 in. *Sun or half shade*	May, June
"Blue" 62 greener	NOOTKA LUPINE	*Lupìnus Nootkaténsis	A showy plant with large spikes of flowers. Deep green digitate foliage. Prop. by seed. Border plant doing well in ordinary garden soil. Nootka Sound.	12-15 in. *Sun*	Late May to early July
"Blue" 57 deeper, turns to 29	VIRGINIAN COWSLIP, BLUE BELLS	*Merténsia pulmonarioìdes *M. Virgínica*	See page 52.		Late Apr. to late May
"Blue" 51	SIBERIAN LUNGWORT	*Merténsia Sibírica	Rare but excellent species. Similar to Virginian Cowslip. Funnel-formed flowers in nodding clusters. Buds pinkish. Broad grayish leaves. Do not disturb. Good for rock-garden or border. Prop. preferably by seed just ripe. Sheltered spots in good loam. Rocky Mts.; Siberia.	1-5 ft. *Sun*	May, early June
"Pale blue" 63 or paler	COMMON GRAPE HYACINTH	**Muscàri botryoìdes *Hyacínthus botryoìdes*	See page 52.		Apr., May
"Dark blue" 49 redder	DARK PURPLE GRAPE HYACINTH	*Muscàri commutàtum	See page 52.		"
"Blue" 63 duller	TUFTED GRAPE HYACINTH, PURSE TASSELS, TUZZY-MUZZY	*Muscàri comòsum *Hyacínthus Comòsus*	See page 52.		"
"Blue" 63 duller	FEATHERED FAIR-HAIRED OR TASSELED HYACINTH	**Muscàri comòsum var. monstròsum *M. plumòsum* *M. p. var. m.*	See page 52.		"
"Dark blue" 61	STARCH GRAPE HYACINTH	*Muscàri racemòsum *Hyacínthus r.*	See page 52.		"
"Deep sky-blue" 58 brighter	EARLY FORGET-ME-NOT	*Myosòtis dissitiflòra	See page 52.		Late Apr. to July
"Bright blue" 58 brighter	TRUE FORGET-ME-NOT	*Myosòtis palústris	Dwarf Alpine plant with many small flowers in lax racemes. Foliage forms a low lying tuft. Good for carpeting and edging. Prop. by seed or cuttings. Any rather moist soil. Europe; Asia.	6-18 in. *Half shade or shade*	May, June

EVER-FLOWERING FORGET-ME-NOT. *Myosotis palustris var. semperflorens.*

GENTIAN-LEAVED SPEEDWELL. *Veronica gentianoides.*

 GREEK VALERIAN. *Polemonium reptans.*

Color	English Name	Botanical Name and *Synonyms*	Description	Height and *Situation*	Time of Bloom
"Blue" 57	EVER-FLOWERING FORGET-ME-NOT	**Myosòtis palústris var. sempérflorens	Dwarf plant of spreading habit, called semperflorens from its long season of bloom. Flowers in loose clusters. Good for damp shady spots of rock-garden. Prop. by seed and cuttings. Moist soil. Hort. See Plate, page 151.	8 in. *Shade*	May to Sept.
"Blue" 58 darker	WOOD FORGET-ME-NOT	*Myosòtis sylvática	One of the best species for spring bedding. Loose clusters of small flowers with yellow eyes. Attractive for fringing walks or in garden beds and pretty when naturalized in the grass. Prop. by seed. Any good soil. Europe; Asia.	1-2 ft. *Sun or shade*	May, early June
"Blue" 58 brilliant	ALPINE WOOD FORGET-ME-NOT	*Myosòtis sylvática var. alpéstris *M. alpéstris*	Dwarf variety with more compact flower clusters than the type. Very attractive when planted in masses. Good for fringing walks. Prop. by seed. Any good soil. Europe.	3-8 in. *Sun or shade*	"
"Blue" 58 deeper	CREEPING FORGET-ME-NOT	*Omphalòdes vérna	See page 55.		Apr., May
"Bright blue" 46	BLUE SMOOTH BEARD-TONGUE	*Pentstèmon glàber var. cyanánthus *P. cyánthus*	Improvement on the type; taller with brighter flowers in close clusters, broader and greener foliage. Very pretty for the garden. Prop. by seed and division. Any good rich soil. Rocky Mts.	12-18 in. *Sun*	Late May, June
"Pale blue" 50	CHICK-WEED PHLOX	*Phlóx stellària	See page 55.		Late Apr., May
"Light blue" bet. 46 & 52	GREEK VALERIAN	*Polemònium réptans	See page 55 and Plate, page 152.		Late Apr. to early June
"Azure blue" 46 bluer	HIMALAYAN VALERIAN	*Polemònium réptans var. Himalayànum *P. grandiflòrum, P. cærùleum var. grandiflòrum*	Alpine plant. Bell-shaped flowers 1½ in. in diameter, in large panicles. Foliage fern-like. Good border plant. Prop. by seed and division. Good garden soil. Himalayas.	1-2 ft. *Half shade*	Late May to mid. July
"Blue" 46 bluer	BLUE COWSLIP	*Pulmonària angustifòlia	Compact plant with funnel-shaped flowers and thick tufts of foliage. Easily cultivated. Divide every few years. Grows best in half shade. Wild garden or border. Prop. in early spring by division. Light fairly moist soil. Europe.	6-12 in. *Sun or half shade*	May
"Rose, turns blue" turns 54	LUNGWORT	*Pulmonària officinàlis *P. maculàta*	Tufted plant. Flowers in upright clusters turning blue. Foliage coarsely hairy, handsomely spotted. Plant in semi-wild places or border. Prop. by division. Light soil not too dry advisable. Europe.	6-12 in. *Sun or half shade*	"

Color	English Name	Botanical Name and *Synonyms*	Description	Height and *Situation*	Time of Bloom
"Blue" 51	BETHLE-HEM SAGE	Pulmonària saccharàta	See page 55.		Late Apr. to late May
"Blue" 61	STAR HYACINTH	*Scílla amœna	See page 7.		Mid. Mar. to early May
"Dark blue" 62	EARLY SQUILL	*Scílla bifòlia	See page 7.		"
"Blue" 61 deeper	SIBERIAN SQUILL	**Scílla Sibírica *S. amœna var. præcox*	See page 7.		"
"Blue" 32 turns 61	PRICKLY COMFREY	Sýmphytum aspérrimum	Somewhat coarse plant. Tubular drooping flowers reddish turning to blue, in scattered clusters. Prickly foliage. Mass in wild garden. Any garden soil. Caucasus.	3-5 ft. *Sun or half shade*	Late May to mid. July
"Blue" bet. 60 & 61	ANGEL'S OR BIRD'S EYES, GERMAN-DER SPEED-WELL	**Verónica Chamædrys	Compact plant with many slender branches which ascend from a creeping rootstock. Large flowers in loose racemes 3-6 in. long. Excellent border plant. Prop. by division. Any good garden soil. Europe.	1-1½ ft. *Sun*	Late May, June
"Blue" 46 or 53	GENTIAN-LEAVED SPEED-WELL	**Verónica gentianoìdes	See page 55 and Plate, page 152.		Late Apr. to late May
"Light blue" 50	COMMON SPEED-WELL	Verónica officinàlis	Native. Rapidly trailing plant with thick-flowered racemes. Good for cutting. Of easiest culture and continuous bloom. Useful for carpeting under trees. Prop. by cuttings. Any soil. Europe; N. Amer.	4-6 in. *Shade*	May, July
"Deep blue" near 47	SCAL-LOPED-LEAVED SPEED-WELL	*Verónica pectinàta	Pretty trailing plant with long many-flowered racemes. Suitable for a dry position in rock-garden or edge of border. Prop. by seed and division. Almost any soil. Asia Minor.	Prostrate *Sun or shade*	May, June
"Pur-plish blue" 54	ROCK SPEED-WELL	**Verónica rupéstris *V. fruticulòsa*	Neat closely trailing plant with dense erect spike-like racemes of flowers borne in profusion. A splendid border plant forming a thick mat. Prop. by seed or division. Any good garden soil.	4-5 in. *Sun*	Mid. May to late June
"Pale blue" 53 pale	BASTARD SPEED-WELL	*Verónica spùria *V. paniculàta V. amethýstina*	Produces a wealth of long racemes crowded with small blossoms, but grows weedy after flowering. Prop. by seed and division. Forests of S. Europe and Russia. Var. *elegans;* more graceful and widely branching than the type.	1-1½ ft. *Sun*	Mid. May to June

MAY

Color	English Name	Botanical Name and *Synonyms*	Description	Height and *Situation*	Time of Bloom
"Blue" bet. 60 & 61	HUNGARIAN SPEED-WELL	****Verónica Teùcrium**	Rapidly spreading, of dense growth, covered with thickly-flowered racemes. Excellent for border or rock-garden. Prop. by seed and division. Any garden soil. Central and S. Europe; Central Asia.	½-1 ft. *Sun*	Late May to early June
"Purplish blue" 47 or 48	HAIRY VIOLET	***Vìola hìrta**	Similar to V. odorata, though almost scentless and with narrower leaves. Plant in shady spot in rock-garden. Europe.	4-6 in. *Half shade*	May
"Purplish blue" 53	ARROW-LEAVED VIOLET	**Vìola sagittàta** *V. dentàta*	Large flowers with short spurs, and bearded petals. Leaves varying from heart-shaped to spear-shaped. Border or rock-garden. Prop. by seed, division or runners. N. Amer.	2-3 in. *Half shade*	May, early June
Parti-colored 26, 43, & white	LARGE-FLOWERED BARREN-WORT	***Epimèdium macránthum**	See page 56.		Late Apr. to late May
1, markings 17	TWISTED-LEAVED FRITILLARY	***Fritillària oblìqua**	See page 56.		Mid. Apr. to mid. May
1, markings 56	SLENDER FRITILLARY	***Fritillària tenélla** *F. montàna*	Tubular yellowish flowers, checkered with purplish brown, on an almost erect stem. Border and rock-garden. Bulbous. Prop. by offsets. Rich soil well-drained. Maritime Alps.	1 ft. *Shade*	May
White & 46	ELDER-SCENTED FLAG	***Ìris sambucìna**	Resembles I. squalens but less vigorous and with smaller flowers. Prop. by division. Rather moist soil. Europe.	1½-2½ ft. *Sun*	Late May
Often 46 deep with 1 & 35	BROWN-FLOWERED IRIS	***Ìris squàlens**	Several flowered species. Flowers much smaller than I. Germanica, inner petals purplish buff or brownish and yellow; outer petals dull violet, beard yellow. Grayish green leaves shorter than flower-stems. Prop. by division. Central Europe to Caucasus.	3 ft. *Sun*	Late May, June
Parti-colored	TULIP	****Tùlipa vars.**	**Single Early Bedding Tulips.** See page 64.		Late Apr. to late May
"White & rose"	LADY TULIP	**Tùlipa Clusiàna**	Dainty very fragrant flower about 2 in. across, flushed with rose and purplish black at base. Leaves long, narrow and folded. Pretty for border or rock-garden. Plant in late Sept. or in Oct. at a depth of 4 in. to the bottom of the bulb, allowing 4 in. between each. It is best to protect in winter and to give an occasional top-dressing. Bulbous. Prop. by offsets. If drainage is not good put sand around the bulb. S. Europe.	12-18 in. *Sun*	May

Color	English Name	Botanical Name and *Synonyms*	Description	Height and *Situation*	Time of Bloom
Various white to 48 duller	DOUBLE-FLOWERED EUROPEAN COLUMBINE	*Aquilègia vulgàris var. flòre-plèno	Flowers much doubled. Color varies from white to deep blue. Useful for wild garden or border. Prop. by seed. Light sandy soil, moist but well-drained. Hort.	1-1½ ft. *Sun*	Mid. May to July
Often 7	WALL-FLOWER	*Cheiránthus Cheiri	Bushy plant. Yellow flowers varying to brown and purple, ¾ in. across, grow in clusters. Foliage good. Should be renewed every year from seed. Flowers the second year. Pretty in rock-garden or border. Prop. by seed or cuttings. Any good garden soil. S. Europe	1-2½ ft. *Sun*	May
Often near 22	FLESH-COLORED HEATH	Erìca cárnea	See page 56.		Apr., May
4, 13 or 19 more orange	CROWN IMPERIAL	**Fritillària Imperiàlis	See page 56.		Mid. Apr. to mid. May
Various	GUINEA-HEN FLOWER, SNAKE'S HEAD, CHECK-ERED LILY	**Fritillària Meleàgris	See page 59.		Late Apr. to late May
Various	DUTCH HYACINTH	**Hyacínthus orientàlis vars.	See page 59.		Late Apr., May
Some-times 43 dull	VIRGINIA WATER-LEAF	Hydrophýl-lum Virgínicum	White or violet erect bell-shaped flowers in clusters. Useful for shady places in rock-garden or border. N. Amer.	1-2 ft. *Sun or shade*	Mid. May to July
Various	GERMAN FLAG, FLEUR-DE-LIS	**Ìris Germánica vars.	The name is given to a large group of varieties. Large and beautiful flowers with erect standards and yellow beard. The blossoms vary in color from pure white to violet and purple, sometimes being yellow or variegated with yellow. They appear on stout, usually branching stalks high above the broad sword-shaped leaves. Effective in isolated clumps and in masses in the border or edge of shrubbery. Prop. by division of rhizomes. Divide when crowded, but disturb as little as possible. They are gross feeders, but will flourish in any well-drained garden soil. Europe Among the best vars. are: **Albert Victor;* (color no. 46 deep), standards soft blue, falls a delicate lilac. ***Bronze Beauty;* standards dull yellow, falls rose lavender. **Darius;* (color nos. 48 & 2), lilac flowers edged with white, beard rich orange. ***Duc de Nemours*	2-3 ft. *Sun*	Late May, June

FLEUR-DE-LIS. *Iris Germanica.*

TREE PEONY. *Pæonia Moutan.*

Color	English Name	Botanical Name and *Synonyms*	Description	Height and *Situation*	Time of Bloom
			(color no. 55 and white), purple margined with white. **Gracchus;* (color nos. 2 & 41), early bloomer. Standards pale yellow, falls yellow suffused with deep red. Height 1½ ft. ***Hector;* (color nos. 5 light & 13 deep), standards delicate yellow, falls very deep crimson. ***Mme. Chereau;* (tinged with color no. 44), white beautifully feathered and margined with blue. ***Mrs. H. Darwin;* (tinged with color no. 48), white netted with violet; a free bloomer. ***Purple King;* (color no. 55 warmer), purple. ***Victorine;* (white and color no. 55), blue and white. See Plate, page 157.		
Various	WHITE-FLOWERED PEONY	**Pæònia albiflòra vars. *P. édulis*	Showy herbaceous Peony, white in the type, but ranging in the many varieties through shades of pink and red, also parti-colored. For location and cultivation see P. officinalis. Siberia.	2-4 ft. *Sun or half shade*	Late May to mid. June
Various	TREE PEONY	Pæònia Moután *P. arbòrea*	Beautiful shrub with very large flowers. Many double or single varieties, in colors from white to crimson. Harder to grow than other species, needing protected spot and winter covering. Plant in isolated clumps or in border or on the edge of shrubbery. Prop. by grafting on roots of herbaceous species. Does not flower until third year. Being a gross feeder, it requires a rich moist loam enriched with cow manure. China. See Plate, page 158.	3-6 ft. *Sun or half shade*	Mid. May to mid. June
Often 33 redder	COMMON GARDEN PEONY	**Pæònia officinàlis vars. *P. fúlgida*	Herbaceous Peonies, typically crimson, but the parents of many colored horticultural vars. Large single or double flowers varying from white to deep crimson and parti-colored. Handsome divided foliage. Hardy and effective in clumps or masses in large border or for edging beds of shrubs. Half shade is desirable. Prop. usually by division in early autumn. Being gross feeders, they require a deep rather moist loam enriched with cow manure.	2-3 ft. *Sun or half shade*	"
Often 40 deeper	STRAGGLING PEONY	*Pæònia peregrìna vars.	Herbaceous Peony. Effective species resembling P. officinalis; the flowers of the many vars. range from pale rose to deep crimson. Dark glossy foliage. For location and cultivation see P. officinalis. S. Europe.	1½-2 ft. *Sun or half shade*	May
Often 31 duller	HAIRY PHLOX	**Phlóx amœna *P. procúmbens* (Gray)	See page 63.		Late Apr., May

Color	English Name	Botanical Name and *Synonyms*	Description	Height and *Situation*	Time of Bloom
Various	**DOWNY PHLOX**	***Phlóx pilòsa** *P. aristàta*	Beautiful species with a profusion of white, purplish or pink flowers in flat clusters. Resembles P. Drummondi. Rock-garden or border. Prop. by seed, division or cuttings. Good light garden soil. N. Amer.	1-1½ ft. *Sun*	May, June
40, 36, 43, white, etc.	**MOSS OR GROUND PINK**	****Phlóx subulàta** *P. nivàlis*	See page 63.		Late Apr., May
34 & others	**AURICULA**	***Prímula Aurícula** *Aurícula*	See page 63.		"
Various	**POLYANTHUS**	***Prímula Polyántha**	See page 63.		"
Mixed white, 48, etc.	**VON SIEBOLD'S PRIMROSE**	****Prímula Sièboldi** *P. cortusoìdes var. amœna, var. grandiflòra & var. S.*	See page 63.		"
Various	**TONGUE-LEAVED SAXIFRAGE**	***Saxífraga ligulàta** *S. Schmídtii*	See page 64.		"
Various	**COMMON BLUEBELL of England, WOOD HYACINTH**	***Scílla festàlis** *S. nùtans, S. nonscrípta, S. cèrnua*	Drooping bell-shaped fragrant flowers, blue, purple, white or pink, on tall stems. Leaves long and grass-like. Charming in rock-garden or margin of shrubbery. Bulbous. Prop. by offsets. Enrich occasionally with top-dressing of manure. Europe.	8-12 in. *Half shade*	May, early June
Various	**SPANISH SQUILL OR JACINTH, BELL-FLOWERED SQUILL**	****Scílla Hispánica** *S. campanulàta, S. pátula*	The most robust species. Bell-shaped flowers about 1 in. across, blue, white or rose, hang from the flower-stalks. Leaves long and narrow. Charming in masses for rock-garden, wild garden, border and naturalization. Bulbous. Prop. by offsets in summer. Any soil with a yearly top-dressing. Spain; Portugal.	12-18 in. *Sun*	Late May, June
34 or green	**SESSILE-FLOWERED WAKE-ROBIN**	**Tríllium séssile**	See page 64.		Late Apr., May
Various	**LARGE CALIFORNIAN WAKE-ROBIN**	**Tríllium séssile var. gigantèum** *T. s. var. Califórnicum*	See page 64.		"
Various	**TULIP**	****Tùlipa vars.**	**Single Early Bedding Tulips.** See page 64.		Late Apr. to late May
Various	**TULIP**	****Tùlipa vars.**	**Double Early Bedding Tulips.** See page 64.		"

LATE TULIPS

DARWIN TULIP. *Tulipa "Darwin,"* AND ALPINE WOOD FORGET-ME-NOT. *Myosotis alpestris.*

Color	English Name	Botanical Name and *Synonyms*	Description	Height and *Situation*	Time of Bloom
Various	**TULIP**	****Tùlipa vars.**	The following is a carefully selected list of some of the best single and double late Tulips of garden origin. They have longer and more graceful flower-stalks than the early species and are good for cutting. There are many other good vars. in the trade. Plant in late Sept. or in Oct. at a depth of 4 in. to the bottom of the bulb, allowing 4 in. between each. Protect in winter and give an occasional top-dressing. Bulbous. Prop. by offsets. Light soil. If drainage is poor, put sand around the bulb. See Plate, page 161.		
			Single Late Bedding Tulips. Vars. *Bizarres;* (marked with color no. 35); yellow ground, striped or feathered with crimson, purple or white. *Byblœmens;* white ground, striped or marked with purple, lilac or black *Roses;* white ground, marked with scarlet, pink or red. *Bouton d'or;* (color no. 5), bright yellow globular flowers; very pretty. *Golden Crown;* (effect color no. 6), yellow edged with red. *Picotée;* white edged with deep pink; charmingly shaped flowers on graceful stems.	12-18 in. *Sun*	Mid. to late May
			Double Late Bedding Tulips. Vars.: *Belle Alliance;* (color no. bet. 19 & 20), red, striped with white. *Blue Flag;* (color no. 46), violet blue; height 16 in. *Duke of York;* (color no. 32), dark rose with white border. *La Candeur;* very large and pure white. *Mariage de ma Fille;* (color no. 22), pure white striped with rose. *Peony Gold;* (color no. 5 marked with 19), golden yellow, striped with red. *Rose Blanche;* (color no. 22 lighter), snow white and pink; height 9 in.	9-16 in. *Sun*	"
Often mixed or 39, 25, or 3 & green, etc.	**DARWIN TULIP**	****Tùlipa "Darwin"**	Single late-flowering Tulip of robust growth, producing very large finely-formed flowers, on tall graceful stems, and in unusual colors, ranging from palest lilac to purplish black, including shades of rose, slate, mahogany, and crimson. Excellent for cutting. Beautiful in mixed colors for beds or for grouping on the edge of shrubbery, or in border. For cultivation see T. Gesneriana vars. Garden origin. See Plate, page 162. Among the most desirable are:— *Alabama;* (color no. 39), deep rosy lilac. *Glory;* salmon scarlet. *Grande Duchesse;* (color no. 19), carmine red. *Gustave Doré;* vivid rose. *Herold;* (color no. 43), pale violet. *Hippolyte;*	1½-2 ft. *Sun or half shade*	Late May to early June

Color	English Name	Botanical Name and *Synonyms*	Description	Height and *Situation*	Time of Bloom
			(color no. 41 brighter), bluish violet. *Liberia;* (color no. 35), deep brown, shaded black. *Longfellow;* (color no. 37 or 38), deep rose. *Olga;* (tinged with color no. 43), white and lilac. *Pales;* (color no. 43), pale lilac. *Richelieu;* (color no. 35), deep red. *Scylla;* (color no. 29), pink.		
Various	**DUC VAN THOL TULIP**	****Tùlipa "Duc van Thol"**	See page 68.		Late Apr., early May
Various	**COMMON GARDEN OR LATE TULIP**	****Tùlipa Gesneriàna vars.**	The type of this effective flower is described under color "red." There are many varieties in shades of orange, yellow, carmine, etc. Beautiful in groups and masses. Plant in late Sept. or in Oct. at a depth of 4 in. to the bottom of the bulb, allowing 4 in. between each. It is best to protect in winter and to give an occasional top-dressing. Bulbous. Prop. by offsets. Light soil. If drainage is not good put sand around the bulb.	6-24 in. *Sun*	Mid. May to early June
Often mixed 18 & 3 or 3 & green	**PARROT OR DRAGON TULIP**	****Tùlipa Gesneriàna var. Dracóntia**	Single late flowering Tulip, odd, showy and interesting. Large flowers with petals curiously cut or feathered, in brilliant colors, sometimes combining shades of green. Good for cutting. Excellent in beds or borders. For cultivation see T. Gesneriana vars. Of garden origin. Among the most desirable are:— *Admiral of Constantinople;* (color no. 20), dark red; large flowers. *Café Brun;* (effect color no. 14), coffee color and yellow. *Lutea Major;* (color no. 5), bright yellow, sometimes streaked with red. *Markgraaf;* (color nos. 4 & 20 striped with green), yellow, scarlet and green striped; large. *Monstre Rouge;* (color no. 20 more brilliant), beautiful deep scarlet, large flower, very handsome. *Perfecta;* (color nos. 4 & 17), yellow, spotted with red. See Plate, page 165.	12-18 in. *Half shade*	Mid. May to June
46 etc.	**DOG VIOLET**	**Vìola canìna**	See page 68.		Late Apr. to late May
Often 48 darker	**MUNBY'S VIOLET**	**Vìola Munbyana**	Very pretty species. Vigorous and free-flowering. Large flowers, deep purple or yellow. Resembles V. cornuta. Mass in the rock-garden. Algeria.	4-5 in. *Half shade*	Early to late May
Various	**PANSY, HEARTS-EASE**	****Vìola trícolor**	See page 68.		Mid. Apr. to mid. Sept.

PARROT TULIP. *Tulipa Gesneriana var. Dracontia.*

A JUNE BORDER

JUNE

WHITE TO GREENISH

Color	English Name	Botanical Name and *Synonyms*	Description	Height and *Situation*	Time of Bloom
"White"	SNEEZE-WORT	Achillèa Ptármica	Straggling plant. Flowers in loose flat clusters. Border or wild garden. Prop. by seed, division or cuttings. Any ordinary soil. N. Temp. Zone.	1-2 ft. *Sun*	June to mid. Sept.
"White"	DOUBLE SNEEZE-WORT	**Achillèa Ptármica var. "The Pearl"	Large and fine var. Double flowers in small clusters. Foliage insignificant. Excellent for cutting. Border. Prop. by division and cuttings. Rich soil. N. Temp. Zone. See Plate, page 276.	1½-2½ ft. *Sun*	June to Oct.
"Cream white"	WHITE BANE-BERRY	Actæa álba *A. rùbra*	See page 11.		Early Apr. to July
"White"	COHOSH, HERB CHRISTO-PHER	Actæa spicàta	See page 71.		May, June
"White"	WHITE BUGLE	*Ajùga réptans var. álba	See page 71.		May to mid. June
"White" 39 under petals	NARCISSUS-FLOWERED ANEMONE	Anemòne narcissiflòra *A. umbellàta*	See page 71.		May to Aug.
"White" often 36	WOOD ANEMONE	Anemòne nemoròsa	See page 11.		Late Apr. to early June
"White"	CANADA ANEMONE	*Anemòne Pennsyl-vánica *A. Canadénsis, A. dichótoma*	See page 71.		Mid. May to early July
"Cream white" tinged with 39	SNOWDROP WIND-FLOWER	**Anemòne sylvéstris	See page 11.		Late Apr. to mid. July
"Green-ish white"	TALL OR VIRGINIAN ANEMONE	Anemòne Virginiàna	Hairy species, sturdy and branching. Flowers 1-1½ in. wide. Good for naturalizing in wild garden. N. Amer.	2-3 ft. *Sun or shade*	June, July
"White"	ST. BRUNO'S LILY	*Anthéricum Liliástrum *Paradísea Liliástrum*	See page 72.		Late May to early July

Color	English Name	Botanical Name and *Synonyms*	Description	Height and *Situation*	Time of Bloom
"White"	BRANCHING ANTHERICUM	*Anthéricum racemòsum *A. Graminifòlium*	Small lily-like flowers in racemes on branching stems. Tufted grass-like foliage. Winter protection. Pretty in border. Prop. by seed, division and stolons. S. Europe.	1½-2 ft. *Sun*	June
"Pure white"	DWARF WHITE-FLOWERED COLUMBINE	*Aquilègia flabellàta var. nàna-álba *A. f. var. flòre-álba*	Dwarf var. of A. flabellata. Flowers have short spurs. Prop. by seed and division. Should have a light well-drained but not dry soil. Hort.	8-12 in. *Sun*	"
"White"	WHITE COLUMBINE	**Aquilègia vulgàris var. nívea	See page 72.		Mid. May to July
"White"	WHITE PLANTAIN-LIKE THRIFT	*Armèria plantagínea var. leucántha *A. dianthoìdes*	See page 72.		Late May to late June
"White"	GALIUM-LIKE WOODRUFF	Aspérula Galioìdes	Erect plant with pleasing flowers. Useful in rock-garden in rather dry position. Europe.	1-1¼ ft. *Half shade*	June, July
"White"	SWEET WOODRUFF, HAY PLANT	Aspérula odoràta	See page 75.		Early May to mid. June
"White"	BRANCHING ASPHODEL	*Asphódelus álbus	Small lily-like flowers borne in spikes which rise high above the tufts of long narrow leaves. Good border plant. Prop. by division. Any good soil. S. Europe.	2-3 ft. *Sun or half shade*	June, July
"White"	WHITE ALPINE ASTER	**Áster alpìnus var. álbus	See page 75.		Late May to late June
"White"	SMALL WHITE ASTER	*Áster vimíneus	Pretty plant with numerous showy white flowers. Wild garden. U. S. A.	2-4 ft. *Sun*	June, July
"Creamy white"	FALSE GOAT'S BEARD	**Astílbe decándra *A. biternàta*	Flowers in terminal panicles. Foliage somewhat heart-shaped. Effective in masses. Border. Prop. by division. Any good garden soil. U. S. A.	3-6 ft. *Half shade*	Early June to early July
"White"	JAPANESE FALSE GOAT'S BEARD	**Astílbe Japónica *Hoteìa Japónica, H. barbàta, Spiræa Japónica*	Plant of compact habit. Flowers form feathery panicles which rise above the handsome dark foliage. Effective in masses and in the border. Prop. by division. Any soil. Japan. See Plate, page 169. Var. *grandiflora;* (color no. 22 pale), has larger flowers and more compact racemes. Var. *multiflora;* many flowered var., flowers larger and more compact than in the type. Var. *variegata,* bears pretty variegated foliage.	1-3 ft. *Half shade*	Mid. June to mid. July

JAPANESE FALSE GOAT'S BEARD. *Astilbe Japonica.*

WHITE PEACH-LEAVED BELLFLOWER. *Campanula Persicifolia var. alba.*

Color	English Name	Botanical Name and *Synonyms*	Description	Height and *Situation*	Time of Bloom
"White"	**LARGE WHITE WILD INDIGO**	***Baptísia leucántha**	Stately plant. Flowers an in. long, in terminal racemes. Foliage clover-like. Prop. by seed and division. Any good garden soil. Eastern U.S.A.	2-4 ft. *Sun*	Early June to mid. July
"White"	**WHITE CARPATHIAN HAIRBELL**	****Campánula Carpática var. álba**	Vigorous compact plant. The large erect cup-shaped flowers are borne in continuous profusion above the pretty foliage. Invaluable for rock-garden or border. Prop. by seed, division or cuttings. Rich well-drained loam. Transylvania.	9 in. *Sun*	Late June to late Aug.
"White"	**WHITE PEACH-LEAVED BELL-FLOWER**	****Campánula persicifòlia var. álba**	A very charming plant. Many large cup-shaped flowers 2 in. across, ranging along tall stems are borne above a tuft of pretty foliage. Good for cutting, and a graceful border plant. A charming effect is given by massing against a dark background. Prop. by seed and division. Easily cultivated in rich soil. See Plate, page 170. Var. *flore-pleno* has double camellia-like flowers. Good for cutting. Central Europe. Var. *alba grandiflora* is a favorite Campanula. The very large cup-shaped flowers make it an imposing plant for the border. Prop. by division. Hort.	1½-3 ft. *Sun*	Early June to early July
"White"	**BACKHOUSE'S PEACH-LEAVED BELL-FLOWER**	****Campánula persicifòlia var. Báckhousei**	A popular form of C. persicifolia. Attractive as a rock-garden plant, showy in the border, and good for cutting. See C. persicifolia var. alba. Hort.	1½-3 ft. *Sun*	June, July
"White"	**MOERHEIM'S PEACH-LEAVED BELL-FLOWER**	****Campánula persicifòlia var. Moerheimi**	Recently introduced. A good and very hardy form of this popular Bell-flower. Spikes of camellia-like flowers, 2½-3½ in. in diameter, are produced in profusion. Good for cutting. See C. persicifolia var. alba. Hort.	1½-2 ft. *Sun*	"
"White"	**SPOTTED BELL-FLOWER**	***Campánula punctàta**	An interesting species. Two or three large drooping flowers on erect stems, throats spotted with red. Not good for cutting. Rough and hairy foliage mostly at base of plant. Attractive border plant. Prop. by division. Rich loam well-drained. Siberia; Japan.	1-1½ ft. *Sun*	June to late July
"White"	**WHITE BLUE BELLS OF SCOTLAND**	***Campánula rotundifòlia var. álba**	A lovely little plant with drooping bell-flowers on slender stems. Not so leafy as the type. Pretty for border, though especially suited for the rock-garden or for naturalization in wild spots. Prop. by seed, division or cuttings. Rich well-drained loam. N. Temp. Regions.	6-12 in. *Sun*	June to Sept.

Color	English Name	Botanical Name and *Synonyms*	Description	Height and *Situation*	Time of Bloom
"White"	**WHITE COVENTRY BELLS**	**Campánula Trachèlium var. álba**	A white form of a common and very hardy species. Flowers bearded inside in loose racemes 1 ft. or more long. Good border plant. Prop. by seed, division or cuttings. Rich well-drained loam. Europe. Var. *alba plena,* a double form.	2-3 ft. *Sun*	June
"White"	**WHITE MOUNTAIN BLUET OR KNAPWEED**	****Centaurèa montàna var. álba**	See page 75.		Late May to early July
"White"	**WHITE JUPITER'S BEARD**	**Centránthus rùber var. álbus**	Compact bushy plant covered with bold clusters of flowers which terminate leafy stems. Gray-green foliage. Often found in old-fashioned gardens. Prop. by seed or division. Mediterranean Region.	1-3 ft. *Sun*	June, July
"White"	**BIEBERSTEIN'S MOUSE-EAR CHICKWEED**	**Cerástium Bièbersteinii**	Evergreen creeping plant. Small flowers. Broad glossy silvery leaves forming a dense mat. Used chiefly for edging. Prop. by division or cuttings. Any garden soil. Asia Minor.	6 in. *Sun or half shade*	"
"White"	**BOISSIER'S MOUSE-EAR CHICKWEED**	**Cerástium Boissierii**	Spreading plant. Large flowered var. Silvery foliage. Stems weak. Good for covering dry banks. Any ordinary garden soil. Spain.	10-12 in. *Sun or half shade*	June
"White"	**LARGE-FLOWERED MOUSE-EAR CHICKWEED**	***Cerástium grandiflòrum**	See page 75.		May, June
"White"	**COTTONY MOUSE-EAR CHICKWEED**	**Cerástium tomentòsum**	Spreading plant of weak stem and small flowers. Foliage silvery. Used principally for edging. Europe.	3-6 in. *Sun*	June
"White"	**PRINCE'S PINE**	**Chimáphila umbellàta** *C. corymbòsa, Pýrola umbellàta*	Small shrubby evergreen plant. Flowers sometimes reddish in few-flowered clusters. Dark green glossy leaves. Suitable for wild garden or rock-garden. Plant among Pyrolas, Wintergreen, etc. Prop. by division. Sandy peaty soil with leaf mold. N. Amer.; Europe; Japan.	3-6 in. *Half shade*	"
"White"	**DR. JAMES'S SNOW-FLOWER**	**Chionophila Jamesii**	Tufted alpine plant. Creamy flowers in dense spikes. Leaves rather thick, mostly about the root. Desirable in rock-garden. Prop. by seed or division. Well-drained soil. Rocky Mts. of Col.	1-3 in. *Sun*	June, July
"White"	**LARGE-FLOWERED WHITE-WEED**	****Chrysánthemum máximum**	Single flowers like large Daisies. Long leaves narrow at the base, growing on the lower part of flower stalks. Admirable border plant. Prop. by seed, division, cuttings or suckers.	1 ft. *Sun*	"

JUNE

Color	English Name	Botanical Name and *Synonyms*	Description	Height and *Situation*	Time of Bloom
			Rich soil. Mulch and water well. Pyrenees. Var. *filiformis* has thread-like petals. Hort. Var. *"Triumph"* has larger flowers about 4 in. wide, with broader overlapping petals. Hort.		
"White"	SHASTA DAISY	*Chrysánthemum "Shasta Daisy"	Very large daisy-like flowers about 4 in. across. Good for cutting and as a border plant. For cultivation see C. maximum. Hort.	1-2 ft. *Sun*	June to Sept.
"White"	TURFING DAISY	*Chrysánthemum Tchihátchewii	A dwarf species of compact tufted habit, bearing a profusion of yellow-centred flowers and finely-divided dark leaves. Excellent for covering dry barren places. Prop. by seed, division or cuttings. Any soil. Siberia.	2-9 in. *Sun*	June, July
"White"	WHITE HERBACEOUS VIRGIN'S BOWER	**Clématis récta *C. erécta*	Splendid erect tufted plant. One of the best herbaceous kinds. Profusion of fragrant flowers in broad terminal clusters. Prop. by seed or cuttings. Rich deep soil. S. Europe. Var. *lathyrifolia* has Pea-leaves. Var. *umbellata* has flowers in umbels.	2-3 ft. *Sun*	Early June to mid. July
"White"	RAMONDIA-LIKE CONANDRON	Conándron ramondioìdes	Similar to Dodecatheon. Numerous nodding flowers, sometimes lilac or purple, on erect leafless stems, dark green wrinkled leaves in flat tufts. Plant in shady part of rock garden. Tuberous plant. Moist soil. Japan.	6 in. *Half shade*	June, July
"White"	LILY-OF-THE-VALLEY	**Convallària majàlis	See page 75.		Mid. May to mid. June
"Greenish white"	DWARF CORNEL, BUNCHBERRY	Córnus Canadénsis	See page 75 and Plate, page 175.		May, June
"White"	HEART-LEAVED COLEWORT	*Crámbe cordifòlia	Effective plant with dense sprays of small fragrant flowers rising above the clump of broad heart-shaped leaves. Suitable for bold effects. Blooms third year from seed and then dies. Rich soil. Caucasus.	5-7 ft. *Sun*	June, July
"White" 36 in effect	SMALL WHITE LADY'S SLIPPER	*Cypripèdium cándidum	See page 76.		May to early June
"White"	MOUNTAIN LADY'S SLIPPER OR MOCCASIN FLOWER	*Cypripèdium montànum	See page 76.		"

Color	English Name	Botanical Name and *Synonyms*	Description	Height and *Situation*	Time of Bloom
"White"	**SHOWY LADY'S SLIPPER**	****Cypripèdium spectábile** *C. regìnæ*	One of the handsomest species. Large flowers variegated with purple stripes. Oval leaves. Protect in winter. Plant in the rock or wild garden. Needs to be watered during flowering season. Prop. by division. Porous moist vegetable or peaty soil. N. Eastern U. S. A.	1-2½ ft. *Half shade or shade*	June
"White"	**SAND PINK**	***Diánthus arenàrius**	Fragrant flowers purple at centres with fringed petals. Excellent for cutting. Good for rock-garden or front of border. Prop. by seed, division or cuttings. Rich light soil. Europe.	1-1½ ft. *Sun*	Early June to early July
"White" sometimes 36	**TWO-COLORED PINK**	***Diánthus bicolor**	Solitary flowers white above and dull gray beneath, on leafy stems. Border plant, also desirable for rock-garden. Prop. by seed, division or cuttings. Rich light soil. S. Russia.	1-2 ft. *Sun*	June
"White"	**PINK MISS SIMKINS**	****Diánthus "Miss Simkins"**	See page 76.		Late May to late June
"White"	**DOUBLE WHITE GARDEN PINKS**	****Diánthus plumàrius vars. álba plèna & "White Witch"**	These vars. are double; otherwise they resemble the type, being fragrant and excellent for cutting. Good border plants. Prop. by cuttings or layers. Rich light soil. Hort.	8-12 in. *Sun*	June
"White"	**SPREADING PINK**	***Diánthus squarròsus**	Fragrant flowers with toothed petals. Curved grass-like leaves. For border or rock-garden. Prop. best by seed, division or cuttings. Rich light soil. S. Russia.	6 in. *Sun*	June, July
"White"	**WHITE BLEEDING HEART**	***Dicéntra spectábilis var. álba** *Diélytra s. var. a.*	See page 76.		Late May, June
"White"	**GAS PLANT, BURNING BUSH, FRAXINELLA, DITTANY**	****Dictámnus álbus** *D. Fraxinélla*	Erect bushy plant. Fragrant flowers in racemes surmount the glossy foliage. Should not be crowded or often disturbed. Flowers last well when cut. Effective on margin of shrubbery or in the border. Prop. by seed planted when ripe, or by division. Re-sows itself. Moderately rich heavy soil. Europe; N. Asia.	2-3 ft. *Sun or half shade*	June, July
"White"	**WHITE FOXGLOVE**	****Digitàlis purpùrea var. álba** *D. tomentòsa var. álba*	Usually biennial. The prettiest var. of D. purpurea. Tubular flowers droop from tall flower stalks. Foliage in clumps at the base. Stately and effective whether grouped among shrubs, in the wild garden, in the border, or naturalized in masses. If cut	2-3 ft. *Sun or half shade*	June, early July

BUNCH BERRY. *Cornus Canadensis.*

FOX GLOVE VARS. *Digitalis purpurea vars.*

Color	English Name	Botanical Name and *Synonyms*	Description	Height and *Situation*	Time of Bloom
			down it will keep on flowering somewhat through the summer. Prop. by seed and division. Rich light soil preferable. See Plate, page 176.		
"Purplish white"	**FULLER'S TEASEL**	**Dípsacus Fullònum**	Biennial or perennial. Coarse-growing plant of curious habit with prickly heads. Handsome foliage. Wild garden. Prop. by seed. Any ordinary soil. Europe.	4-6 ft. *Sun*	June, July
"White"	**MOUNTAIN AVENS**	***Drŷas octopétala**	See page 76.		May, June
"White"	**WHITE GREAT WILLOW HERB OR FIRE WEED**	***Epilòbium angustifòlium var. álbum** *E. spicàtum var. álbum*	Dense bushy plant. Flowers in spikes. Good for cutting. Foliage like that of the Willow. Plant with rough shrubbery or beside water. Prop. by seed or division. Any ordinary soil. N. Amer.	3-5 ft. *Sun*	June to early Aug.
"Greenish white"	**MOTTLED SWAMP-ORCHIS, FALSE LADY'S SLIPPER**	**Epipáctis Royleàna** *E. gigantèa*	Vigorous Orchid. Five to ten flowers, with deep purple veins; lip tinged with orange or yellow. Sword-shaped leaves. Moist soil in sheltered position. Western U. S. A.; Himalayas.	1-4 ft. *Half shade*	June, July
"White"	**HIMALAYAN EREMURUS**	***Eremùrus Himalaìcus**	Very striking plant. Star-shaped flowers in dense spikes 2 ft. long, on stems which rise above the clump of long leaves. Protect crowns in winter and in spring. Group in sheltered situation. Prop. by division. Deep rich sandy loam. Himalayas.	4-8 ft. *Sun*	"
"White"	**HORSE-WEED, BUTTER-WEED**	**Erígeron Canadénsis**	Aster-like flowers in panicles borne in profusion. Mass in the wild garden. Prop. by seed or division. Any garden soil. Central N. Amer.	1-4 ft. *Sun*	June to Sept.
"White"	**LARGE WHITE MOUNTAIN DAISY**	***Erígeron Coúlteri**	See page 77.		Late May to late June
"Yellowish white"	**RUNNING FLEABANE**	***Erígeron flagellàris**	See page 77.		Mid. May to late June
"White"	**LAVENDER MOUNTAIN DAISY**	**Erígeron salsuginòsus** *Áster salsuginòsus*	Aster-like flowers, usually solitary, on somewhat sticky stems. Color varies to lilac. Plant in masses. Prop. by seed or division. Any garden soil. Sierras.	1-1½ ft. *Sun*	June
"White"	**RATTLESNAKE-MASTER, BUTTON SNAKE-ROOT**	**Erýngium aquáticum** *E. yuccæfòlium*	Curious and striking plant. Thistle-like flowers on bright steel-blue stems. Long and narrow bristly leaves. Winter protection necessary. Effective in borders. Prop. best by seed. Either dry or moist soil. U. S. A.	2-6 ft. *Sun*	June to Oct.

Color	English Name	Botanical Name and *Synonyms*	Description	Height and *Situation*	Time of Bloom
"White"	WHITE GOAT'S RUE	*Galèga officinàlis var. álba *G. Pêrsica**	Bushy plant with pea-shaped flowers in compact racemes, with graceful luxuriant foliage. Good for cutting. Plant groups in border or wild garden. Prop. by seed and division. Any good soil. Europe.	2-3 ft. *Sun or half shade*	June, July
"White"	NORTHERN BEDSTRAW	Gàlium boreàle *G. septentrionàle*	Tiny flowers in small clusters. Showy leaves in regular whorls. Rock-garden or border. Prop. by division. U. S. A.	6-18 in. *Half shade*	Early June to mid. July
"White"	WHITE OR GREAT HEDGE BEDSTRAW, WILD MADDER	Gàlium Mollùgo	Small clusters of tiny blossoms resembling Gypsophila. Good for cutting. Leaves in regular whorls of six to eight. Rock-garden. Prop. by seed and division. Any good soil. Europe.	1-3 ft. *Half shade*	Early June to late Aug.
"White"	WHITE BLOOD CRANES-BILL	*Gerànium sanguíneum var. álbum	See page 77.		Late May to mid. July
"White" often near 41	WHITE WATER AVENS OR WATER FLOWER	*Gèum rivàle var. álbum	See page 77.		Late May, June
"White"	AMERICAN IPECAC	*Gillènia stipulàcea *Porteránthus stipulàtus*	Charming graceful plant resembling Spiræa. Flowers small in clusters. Foliage deeply divided. Plant in border, shrubbery or wild garden, grouped with Lilies or Irises. Prop. by seed and division. Any garden soil. N. Amer.	2-4 ft. *Sun*	June, early July
"White"	BOWMAN'S ROOT, INDIAN PHYSIC	*Gillènia trifoliàta *Porteránthus trifoliàtus*	Grows wild in rich woods. Closely resembles G. stipulacea, but is rather taller and less compact in habit. For cultivation and location see G. stipulacea. N. Amer.	2-4 ft. *Half shade*	"
"White"	ELEGANT CHALK-PLANT	*Gypsóphila élegans	A dainty though ineffective plant. Flowers sometimes pink, grow in dense panicles. Foliage inconspicuous. Sprays good for cutting. Rock-garden. Prop. by seed, division or cuttings. Dry soil. Asia Minor.	1 ft. *Sun*	June, July
"White"	STEVEN'S CHALK-PLANT	*Gypsóphila Stèveni *G. glaùca*	Flowers larger and in smaller panicles than G. paniculata. Foliage grayish green. Border or rock-garden. Prop. by seed, division or cuttings. Fairly dry garden soil. Caucasus.	1-2 ft. *Sun*	June, early July
"White"	UMBEL-FLOWERED SUN ROSE	Heliánthemum umbellàtum	Shrubby evergreen plant. Flowers in whorls or umbels. Foliage somewhat sticky, when young, forming a spreading mat. Protect in winter. Rock-garden or border. Prop. by seed or cuttings, but generally by division. Any ordinary soil. Europe.	1-2 ft. *Sun*	June, July

JUNE

Color	English Name	Botanical Name and *Synonyms*	Description	Height and *Situation*	Time of Bloom
"White"	DOUBLE WHITE SWEET ROCKET	***Hésperis matronàlis var. álba plèna**	Showy spikes of double fragrant flowers, borne in profusion. Good to cut. Border or rock-garden. Prop. by seed or division. Rich rather moist soil. Europe.	2-3 ft. *Sun*	June, July
"White"	WHITE CORAL BELLS	**Heùchera sangúinea var. álba** *H. álba*	Not so noticeable as the type. Rather inconspicuous flower-panicles rise from a clump of handsome pale foliage. Good for cutting. Mass in wild garden, rock-garden or among shrubs. Prop. by seed and division. Any soil. Hort.	1-1½ ft. *Sun or half shade*	June to late Sept.
"Whitish" faintly 32	HAIRY ALUM ROOT	**Heùchera villòsa** *H. caulèscens*	Small inconspicuous flowers in loose panicles on an almost leafless stalk. Worth growing for its foliage, which lasts long when cut. Border or rock-garden. Prop. by seed and division. Ordinary garden soil. Md. to Ga. and West.	1-3 ft. *Sun*	Late June to Sept.
"Yellowish white"	ALPINE HUTCHINSIA	***Hutchinsia alpìna**	See page 77.		May, June
"Cream white"	BROAD-LEAVED WATER-LEAF	**Hydrophýllum Canadénse**	Grows wild in rich woods. Bell-shaped flowers, sometimes slightly greenish or purplish, in dense clusters. Leaves palmately divided. Good under shrubbery, succeeding remarkably well in shade. Separate occasionally. Prop. by seed and division. Fairly moist soil. N. Amer.	1 ft. *Half shade or shade*	June, July
"White"	GARREX'S CANDYTUFT	***Ibèris Garrexiàna**	See page 78.		May, June
"White"	ROCK CANDYTUFT	***Ibèris saxátilis**	See page 78.		May, early June
"White"	LEATHERY-LEAVED ROCK CANDYTUFT	***Ibèris saxátilis var. corifòlia** *I. corifòlia*	See page 78.		May, June
"White"	EVERGREEN CANDYTUFT	****Ibèris sempérvirens**	See page 78.		May, early June
"White"	TENORE'S CANDYTUFT	****Ibèris Tenoreàna**	See page 78.		"
"White"	FLORENTINE FLAG, ORRIS ROOT	****Ìris Florentìna**	See page 78.		"
"White"	WHITE SIBERIAN FLAG	****Ìris Sibírica var. álba**	See page 81.		Late May to mid. June
"White"	WHITE VARIEGATED NETTLE	**Làmium maculàtum var. álbum** *L. álbum*	See page 81.		Mid. May to late July

Color	English Name	Botanical Name and *Synonyms*	Description	Height and *Situation*	Time of Bloom
"White"	**MADONNA LILY**	****Lílium cándidum**	One of the most familiar and deservedly popular species, though subject to disease. Lovely fragrant flowers in clusters crown an erect leafy stalk. Very effective in small clumps against a dark background of shrubs, in borders, or when edging paths and beds of shrubs, Rhododendrons in particular. Bulbous. Prop. by offsets, scales or very slowly by seed. Rich well-drained soil. Avoid contact with manure. S. Europe. See Plate, page 181.	3-5 ft. *Sun or half shade*	June, July
"White"	**NEVADA OR WASHINGTON LILY**	***Lílium Washingtoniànum**	Lovely fragrant flowers. dotted with purple and often tinged with pink or lilac, grow in elongated racemes. Excellent in clumps against background in border. Bulbous. Prop. by offsets, scales or very slowly by seed. Light well-drained soil. Avoid contact with manure. Cal.	2-5 ft. *Sun or half shade*	June
"White"	**WHITE PERENNIAL FLAX**	***Lìnum perénne var. álbum**	See page 81.		Mid. May to Aug.
"White"	**WHITE MANY-LEAVED LUPINE**	****Lupìnus polyphýllus var. albiflòrus** *L. p. var. álbus, L. grandiflòrus var. álbus*	Pea-shaped flowers in long spikes rise above the handsome clump of satiny palmate leaves. Effective for border and naturalizing. Prop. by seed and division. Any garden soil. Cal. See Plate, page 182.	2-5 ft. *Sun*	June, July
"White"	**SINGLE & DOUBLE WHITE MALTESE CROSS**	****Lýchnis Chalcedónica vars. álba & álba plèna**	Var. *alba plena* much superior to the single var. Clusters of double flowers on leafy stalks. Excellent in the border. Prop. in spring by seed and division. Rich sandy loam. Var. *alba,* a less effective species. Russia.	2-3 ft. *Sun*	June to early Aug.
"White"	**WHITE MULLEIN PINK OR DUSTY MILLER**	***Lýchnis coronària var. álba**	Woolly plant with striking mullein-like foliage. Solitary flowers terminate spikes of pale silvery leaves. Border. Prop. in spring by seed and division. Any good soil. Europe.	1-2½ ft. *Sun*	June, July
"White"	**SIEBOLD'S LYCHNIS**	***Lýchnis coronàta var. Sièboldii** *L. Sièboldii, L. fúlgens, var. Sièboldii*	Tender. Very large flowers generally borne in loose panicles. Foliage downy. Showy in the border. Protect slightly in winter. Prop. in spring by seed and division. Light rich soil. Japan; China.	1-1½ ft. *Sun or half shade*	"
"White"	**JAPANESE LOOSESTRIFE**	****Lysimàchia clethroìdes**	Graceful plant with long nodding spikes of starry flowers. Good for cutting. Foliage when dying varies in tint. Good in the border. Prop. by seed and division. Any ordinary soil. Japan.	2-3 ft. *Sun*	Mid. June to late July

MADONNA LILY. *Lilium candidum.*

MANY-LEAVED LUPINE VARS. *Lupinus polyphyllus vars.*

Color	English Name	Botanical Name and *Synonyms*	Description	Height and *Situation*	Time of Bloom
"White"	DOUBLE SCENTLESS CAMOMILE	*Matricària inodòra var. pleníssima *M. i. var. ligulòsa, var. múltiplex, M. grandiflòra, Chrysánthemum inodòrum var. flòre-plèno*	Creeping plant growing freely in dense masses. Chrysanthemum-like flowers, numerous and good for cutting. Finely divided feathery foliage. Good carpeting for border. Prop. by division and cuttings. Any garden soil. Hort.	12-18 in. *Sun*	June to Sept.
"Yellowish white"	BALM	Melíssa officinàlis	Grown for its fragrance. Flowers in clusters. Good for naturalization. Var. *variegata* is used for edging. Prop. by seed and division. Warm soil. S. Europe.	1-2 ft. *Sun*	June to early Aug.
"Whitish"	VARIEGATED ROUND-LEAVED MINT	Méntha rotundifòlia var. variegàta	Leaves variegated with pale yellow. Flowers unimportant. Ornamental foliage its chief attraction. Useful for edging and for covering waste places. Prop. by division. Moist peaty soil. S. Europe.	1-2 ft. *Sun*	June, July
"White"	PARTRIDGE BERRY, SQUAW-BERRY	Mitchélla rèpens	Familiar little trailing evergreen. The tiny funnel-shaped flowers, borne in pairs, are less conspicuous than the succeeding bright red berries. Foliage glossy and pretty. Charming carpeting for rock-garden. Prop. by division. Sandy peaty soil. Eastern N. Amer. See Plates, page 185.	2-3 in. *Half shade*	June
"White"	WHITE WOOD FOR-GET-ME-NOT	*Myosòtis sylvática var. álba	See page 81.		May, early June
"White" turns to 36	STEMLESS EVENING PRIMROSE	*Œnothèra acaùlis *Œ. Taraxacifòlia*	See page 82.		Late May to Aug.
"White"	TUFTED EVENING PRIMROSE	*Œnothèra cæspitòsa *Œ. eximea, Œ. marginàta*	See page 82.		Late May, June
"White"	COMMON STAR-OF-BETHLEHEM, TEN-O'CLOCK	Ornithógalum umbellàtum	See page 85.		May, June
"White"	ALLEGHANY MOUNTAIN SPURGE	Pachysándra procúmbens	See page 85.		Late May, early June

Color	English Name	Botanical Name and *Synonyms*	Description	Height and *Situation*	Time of Bloom
"White"	JAPANESE EVER-GREEN PACHY-SANDRA OR SPURGE	Pachysándra terminàlis	See page 85.		Late May to mid. June
"White"	HERBA-CEOUS PEONY	**Pæònia vars.	See color "various," page 268.		June
"White"	WHITE-FLOWERED PEONY	**Pæònia albiflòra *P. édulis*	See page 85.		Late May, June
"White"	POPPY-FLOWERED TREE PEONY	Pæònia Moután var. papaveràcea *P. arbòrea var. p.*	See page 85.		Mid. May to mid. June
"White"	STRIPED TREE PEONY	Pæònia Moután var. vittàta *P. arbòrea var. v.*	See page 85.		"
"White"	COMMON DOUBLE WHITE PEONY	Pæònia officinàlis var. álba plèna	See page 86.		"
"White"	ALPINE POPPY	**Papàver alpìnum	See page 86.		Mid. May to early June
"White"	WHITE ICELAND POPPY	**Papàver nudicaùle var. álbum	See page 19.		Late Apr. to mid. June, late Aug., Sept.
"Silvery white"	ORIENTAL POPPY SILVER QUEEN	**Papàver orientàle "Silver Queen"	Low-growing var., with pale silvery pink flowers and handsome foliage, which dies in middle of summer. Excellent for border if supplemented by other later blooming plants. Prop. by seed or division after flowering. Hort.	3-3½ ft. *Sun*	Early June to early July
"White"	AMERICAN FEVERFEW, PRAIRIE DOCK	Parthènium integrifòlium	Numerous heads of flowers which last a long time in good condition. Large foliage at base of plant. Good for margin of shrubbery. N. Amer.	3-4 ft. *Sun*	June to Sept.
"White" 1	DOWNY PATRINIA	Patrínia villòsa	See page 86.		May, June
"White"	HARMALA RUE	Peganum Harmala	Uncommon and low-branching. Flowers fairly large, veined with green. Rock-garden. Prop. by root division. Light soil. Europe.	6-8 in. *Sun*	June, July

PARTRIDGE BERRY. *Mitchella repens.*

SOLOMON'S SEAL. *Polygonatum multiflorum.*

Color	English Name	Botanical Name and *Synonyms*	Description	Height and *Situation*	Time of Bloom
"White"	FOXGLOVE BEARD-TONGUE	**Pentstèmon lævigàtus var. Digitàlis *P. Digitàlis*	Erect free-growing and slender plant. Tube-shaped flowers marked with purple, about 1 in. long, grow in long loose clusters. Protect in winter. Good border plant. Prop. by seed and division. Any good garden soil. U. S. A.	3-4 ft. *Sun*	Early June to mid. July
"White"	WHITE BALLOON FLOWER	**Platycòdon grandiflòrum var. álbum *Campánula g. var. álba, Wahlenbérgia g. var. á.*	Sturdy bushy plant. Showy open bell-shaped flowers 2-3 in. across, very numerous at the summit of erect leafy stalks. Very useful border plant. Prop. in the spring by seed or division. Well-drained loamy soil. China; Japan.	1-3 ft. *Sun or half shade*	June to Oct.
"White"	HIMA-LAYAN MAY APPLE	*Podophýl-lum Emòdi	See page 86.		Late May, June
"White"	WHITE JACOB'S LADDER OR CHARITY	*Polemònium cærùleum var. álbum	See page 89.		Mid. May to July
"Green-ish white"	GREAT OR SMOOTH SOLOMON'S SEAL	Polygonàtum gigantèum *P. commutà-tum*	See page 89.		Late May, June
"White"	SOLOMON'S SEAL	Polygonàtum multiflòrum	See page 89 and Plate, page 186.		Late May, early June
"White"	THREE-TOOTHED CINQUE-FOIL	Potentílla tridentàta	See page 89.		"
"White"	WHITE JAPANESE PRIMROSE	*Prímula Japónica var. álba	White var. very similar to the type. Excellent for border, wild or rock-garden. Prop. generally by seed. Soil moist and deep. Hort.	1-2 ft. *Shade*	June, July
"White"	ROUND-LEAVED WINTER-GREEN, INDIAN LETTUCE	Pýrola rotundifòlia	Wild evergreen, common in damp or sandy woods, difficult to cultivate. Resembles Lily-of-the-Valley. Fragrant flowers, sometimes pink, borne in racemes, rise on erect stems from tufts of round leathery leaves. Bog or rock-garden or under shade of evergreens. Prop. by division. Peaty soil. Eastern U. S. A.	6-12 in. *Shade*	"
"White"	WHITE ROSETTE MULLEIN	*Ramónda Pyrenaìca var. álba	See page 89.		Mid. May to July
"White"	ACONITE-LEAVED BUTTERCUP	*Ranúnculus aconitifòlius	See page 89.		"
"White"	WHITE BUTTERCUP OR CROWFOOT	*Ranúnculus amplexicaùlis	See page 89.		May, early June

JUNE

Color	English Name	Botanical Name and *Synonyms*	Description	Height and *Situation*	Time of Bloom
"Green"	PYRENEAN MIGNON-ETTE	Resèda glaùca	Spreading plant with spikes of small flowers and somewhat shiny leaves. Margin of border or elevated position in rock-garden. Flourishes in dry soil. Pyrenees.	10-12 in. *Sun*	June, July
"White"	RODGER'S BRONZE-LEAF	Rodgérsia podophýlla	Rather inconspicuous plumy panicles of small flowers. Strikingly large leaves somewhat bronze-colored cut into five divisions. Prop. by cuttings. Peaty soil. Japan.	3-4 ft. *Sun*	June, early July
"White"	SILVERY CLARY, SILVER-LEAVED SAGE	*Sálvia argéntea	Noteworthy for its silvery foliage. Irregular downy tubular flowers. Leaves woolly. Prop. by seed in spring. Deep rich well-drained soil. S. Europe.	2-4 ft. *Sun*	"
"White"	WHITE MEADOW SAGE	*Sálvia praténsis var. álba	Flowers 1 in. long clustered in whorls. Leaves heart-shaped and wrinkled. Border or wild garden. Prop. by seed. Good garden soil. Europe.	2-3 ft. *Sun*	"
"Whit-ish"	CANADIAN OR WILD BURNET	Sanguisórba Canadénsis	Wild herb, seldom cultivated. Flowers in round heads about 4 in. wide. Small leaves in profusion. Border. Prop. by seed. Any soil, preferably poor. Western U. S. A.	5-6 ft. *Sun*	Mid. June to late July
"Green-ish"	BURNET	Sanguisórba mìnor *Potèrium Sanguisórba*	Round or elongated flower heads terminate tall stems. Border. Prop. by seed. Any soil, preferably poor. Europe; Asia.	1-2½ ft. *Sun or half shade*	Late June to mid. July
"Cream white"	LIVELONG SAXIFRAGE OR ROCKFOIL	*Saxífraga Aizòon *S. rosulàris, S. rècta*	Small flowers in clusters surmount a tall stalk rising from a rosette of silvery leaves. Foliage striking. Rock-garden and edge of border. Prop. by seed and division. Light well-drained soil. Europe; N. Amer.	6-15 in. *Sun or shade*	June
"White"	ANDREW'S SAXIFRAGE	*Saxífraga Andrewsii	See page 90.		Late May, June
"White"	PYRAMIDAL COTYLEDON SAXIFRAGE	*Saxífraga Cotylèdon var. pyramidàlis	A vigorous var.: numerous flowers in pyramidal panicles rise on a tall stalk far above a pretty rosette of silvery leaves. Rock-garden. Protect with leaves in winter. Prop. by seed and division. Mts., Europe.	1-2 ft. *Sun or shade*	June, July
"White" or 25 pale	UMBRELLA PLANT	*Saxífraga peltàta	See page 90.		Late May, early June
"Dull white"	EARLY SAXIFRAGE	Saxífraga Virginiénsis	See page 20.		Apr. to late June

Color	English Name	Botanical Name and *Synonyms*	Description	Height and *Situation*	Time of Bloom
"White"	**WHITE CAUCASIAN SCABIOUS**	***Scabiòsa Caucásica var. álba**	Free blooming vigorous plant. Flowers in flat heads on long slender stems. Good for cutting. Sparse grayish foliage. Border or picking garden. Protect in winter. Prop. by seed and division. Fairly good garden soil. Hort.	1½-2 ft. *Sun*	June, July
"White"	**WHITE WOODLAND SCABIOUS**	***Scabiòsa sylvática var. albiflòra**	Thick bushy plant with large leaves. The flowers are produced singly on long stems and are useful for cutting. Prop. by seed. Grows well in any ordinary garden soil. Europe.	1-2 ft. *Sun*	Early June to late Sept.
"White"	**WHITE WOOD HYACINTH**	***Scílla festàlis var. álba** *S. nùtans var. álba*	See page 90.		May, early June
"White"	**WHITE SPANISH OR BELL-FLOWERED SQUILL, WHITE SPANISH JACINTH**	****Scílla Hispánica var. álba**	See page 90.		Late May, June
"White"	**MONREGALENSIS STONECROP**	**Sèdum Monregalénse** *S. cruciàtum*	Small flowers. Leaves cross-shaped. Good for rock-garden or front of border. Corsica; N. Italy.	6 in. *Sun*	June, July
"Greenish"	**ROSEROOT, ROSEWORT**	**Sèdum ròseum** *S. Rhodìola*	Flowers, sometimes tinged with red, in small flat clusters. Foliage sparse. Root aromatic. Rock-garden or edging. Prop. by seed and offsets. Sandy soil. Europe; Asia; N. Amer.	6-8 in. *Sun*	"
"White"	**AMERICAN ORPINE**	***Sèdum telephoìdes**	Excellent foliage. Flowers in close clusters. Glossy green leaves in dense rosettes. Good to naturalize in dry rocky places. Prop. by seed and offsets. S. Alleghanies.	6-12 in. *Sun*	June
"White"	**GALAX-LEAVED SHORTIA**	**Shórtia galacifòlia**	See page 90.		May, June
"White"	**WHITE SIDALCEA**	***Sidálcea cándida**	Erect somewhat branching plant. Flowers 1 in. wide in tall spikes, like racemes. Blooms profusely. Border. Prop. by seed and division. Any garden soil. Rocky Mts.	2-3 ft. *Sun*	June
"White"	**ALPINE CATCHFLY**	**Silène alpéstris**	See page 90.		May, June
"White"	**CAUCASIAN CATCHFLY**	***Silène Caucásica**	See page 93.		Late May, June
"White"	**DOUBLE SEASIDE CATCHFLY**	***Silène marítima var. flòre-plèno**	Rambling trailer. Double flowers larger than the type, but few on a stem. Border, rock-garden or edging. Prop. by cuttings.	6-8 in. *Sun*	June, July

Color	English Name	Botanical Name and *Synonyms*	Description	Height and *Situation*	Time of Bloom
"Cream white"	**FALSE SOLOMON'S SEAL**	***Smilacìna racemòsa**	See page 93 and Plate, page 191.		May, June
"White"	**GOAT'S BEARD**	****Spiræ̀a Arúncus** *Arúncus sylvéster*	Erect branching herb growing wild in rich woods. Abundant flowers in plumy panicles. Handsome compound foliage, invaluable for rough places and grouped with foliage plants. Prop. by division. Any soil. N. Europe; Asia; N. Amer.	3-5 ft. *Sun or shade*	June, early July
"Creamy white"	**ASTILBE-LIKE MEADOW SWEET**	****Spiræ̀a astilboìdes** *S. Arúncus var. a., Arúncus a., Astîlbe a., A. Japònica*	Not so tall as S. Aruncus and of neater habit. Flowers in graceful plumes. Foliage compound, dark, glossy and exceedingly decorative Effective in border or as a high edging. Prop. by seed and division. Garden soil preferably moist. Japan. Var. *floribunda;* vigorous; flowers profusely.	2 ft. *Sun or half shade*	"
"Creamy white"	**FINGERED SPIRÆA**	***Spiræ̀a digitàta**	Closely related to S. palmata. Flowers in broad feathery clusters on erect stems. Tufted root leaves, much divided. Charming for border or water-side. Prop. by seed and division. Fairly rich moist soil. E. Siberia.	1-3 ft. *Half shade*	Early June to early July
"Yellowish white"	**DROPWORT**	***Spiræ̀a Filipéndula** *Filipéndula hexapétala, Ulmària Filipéndula*	Tufted plant. Tiny fragrant flowers often tipped with red, grow in loose irregular clusters. Fern-like foliage. Border or for edging beds of shrubs. Prop. by division. Rather dry soil. Europe; Asia. Var. *flore-pleno,* flowers double.	1-2½ ft. *Sun*	June, early July
"Muddy white"	**WHITE PALMATE-LEAVED MEADOW SWEET**	***Spiræ̀a palmàta var. álba**	Broad clusters of feathery flowers borne on erect stems. Tufted root-leaves, lighter green than the type, palmately divided. Charming for border or water side. Prop. by seed and division. Fairly rich moist soil. Japan.	1-3 ft. *Half shade*	Late June, July
"Cream white"	**ENGLISH MEADOW SWEET, MEADOW QUEEN, HONEY SWEET**	***Spiræ̀a Ulmària** *Filipéndula Ulmària, Ulmària pentapétala U. palùstris*	Less beautiful than many other Spiræas. Fragrant flowers in thick clusters. Leaves sometimes silvery and hairy beneath. Good for rough places. Prop. by seed and division. Moist rich soil. Europe; W. Asia. Var. *aurea variegata;* (color no. 2), foliage variegated with creamy yellow; decorative. Var. *flore-pleno,* double flowers.	2-4 ft. *Half shade*	June, July
"White"	**EASTER BELL, GREATER STITCH-WORT OR STARWORT, ADDER'S MEAT**	***Stellària Holostèa** *Alsìne Holostèa*	See page 93.		Mid. May to mid. June

FALSE SOLOMON'S SEAL. *Smilacina racemosa.*

WHITE BIRD'S-FOOT VIOLET. *Viola pedaea var. alba.*

CANADA VIOLET. *Viola Canadensis.*

ADAM'S NEEDLE. *Yucca filamentosa.*

JUNE

Color	English Name	Botanical Name and *Synonyms*	Description	Height and *Situation*	Time of Bloom
"Creamy white" often 37	**COMMON COMFREY**	**Sýmphytum officinàle**	See page 93.		Late May to mid. July
"White"	**FEATHERED COLUMBINE**	***Thalíctrum aquilegifòlium**	See page 93.		"
"White"	**PURPLISH MEADOW RUE**	**Thalíctrum purpuráscens** *T. purpùreum* (Hort.)	See page 93.		Mid. May to July
"White"	**MOUNTAIN BROOMWORT**	**Thláspi alpéstre**	See page 93.		May, June
"White"	**MOUNTAIN WILD THYME**	**Thŷmus Serpýllum var. montànus** *T. montànus, T. Chamædrys*	Creeping evergreen with longer branches than the type. Tiny flowers in clusters. Leaves small and rather woolly. Useful for covering bare places and in rock-garden. Prop. by division, occasionally by seed. Europe.	1-2 in. *Sun*	Early June to mid. Aug.
"White"	**WHITE SPIDERWORT**	***Tradescántia Virginiàna var. álba**	See page 94.		Late May to late Aug.
"Yellowish white" 2	**HUNGARIAN CLOVER**	**Trifòlium Pannónicum**	Effective herb. Terminal flowers on unbranching stems. Clover-like foliage. Border. Prop. by seed and division. Any soil. Europe; Asia.	1-2 ft. *Sun*	Late June to mid. July
"White"	**NODDING WAKE-ROBIN**	**Tríllium cérnuum**	See page 20.		Late Apr. to early June
"White"	**WHITE ILL-SCENTED WAKE-ROBIN**	***Tríllium eréctum var. álbum**	See page 20.		"
"White" turns to 36	**LARGE-FLOWERED WAKE-ROBIN**	****Tríllium grandiflòrum**	See page 94.		May to early June
"White"	**AMERICAN BARRENWORT**	***Vancouvèria hexándra** *Epimèdium h.*	See page 94.		May, June
"White"	**WHITE SPIKE-FLOWERED SPEEDWELL**	****Verónica spicàta var. álba**	Dense spikes of free-blooming long-continuing flowers. Excellent in border. Prop. by division. Garden soil. Europe; Asia.	2-2½ ft. *Sun*	Early June, July
"White" tinged with 43	**CANADA VIOLET**	**Vìola Canadénsis**	See page 23 and Plate, page 191.		Late Apr. to mid. June

Color	English Name	Botanical Name and *Synonyms*	Description	Height and *Situation*	Time of Bloom
"White"	**WHITE HORNED VIOLET OR BEDDING PANSY**	****Vìola cornùta var. álba**	See page 23.		Late Apr. until frost
"White"	**WHITE BIRD'S-FOOT VIOLET**	**Vìola pedàta var. álba**	See page 95 and Plate, page 191.		May, June
"Yellow-ish white" 2	**TURKEY'S BEARD**	**Xerophýllum setifòlium** *X. asphode-loìdes*	See page 95.		"
"Creamy white" 2 greener	**ADAM'S NEEDLE, BEAR OR SILK GRASS, THREADY YUCCA**	****Yúcca filamentòsa**	Of tropical appearance. Large pendent bell-shaped flowers are clustered thickly on a branching stalk which rises high above a large clump of sword-like foliage. Effective and beautiful grouped against a dark background in the border or shrubbery. Resists drought. Prop. by division of root-stock. Light well-drained soil preferred. U. S. A. See Plate, page 192. Var. *filamentosa bicolor*, foliage striped with white.	6 ft. *Sun*	June, July
"Yellow" 5	**WOOLLY-LEAVED MILFOIL**	***Achillèa tomentòsa**	See page 96.		Late May to mid. Sept.
"Yellow" 4 pale	**PYRENEAN MONKS-HOOD**	**Aconìtum Anthòra** *A. Pyrenaì-cum*	Helmet-shaped flowers, fairly large in racemes. Leaves deeply divided, smooth above, hairy beneath. Root poisonous. Border plant. Prop. by seed or division. Rich soil preferable. S. Europe.	1-2 ft. *Sun or half shade*	June, July
"Yellow" 4	**PIGMY SUN-FLOWER**	**Actinélla grandiflòra**	Very downy plant somewhat branching. Flower heads 3 in. across. Pretty for rock-garden. Prop. by seed or division. Light soil. Col.	6-12 in. *Sun*	June, early July
"Yellow" 6	**SILVERY MADWORT**	**Alýssum argéntium** *A. alpêstre*	Compact dwarf plant. Flowers in dense clusters. Leaves small and white beneath. Rock-garden and border. Prop. by seed, division and cuttings. Any well-drained garden soil. Europe.	12-15 in. *Sun*	June to early Aug.
"Yellow" 4	**BEAKED MADWORT**	**Alýssum rostràtum** *A. Wièrz-bickii*	Alpine plant of compact habit. Flowers in dense heads. Foliage hairy. Good rock-plant. Prop. by seed, division and cuttings. Any well-drained garden soil. Asia Minor.	12-15 in. *Sun*	Early June to early Aug.
"Yellow" 5	**GOLDEN MARGUE-RITE**	***Ánthemis Kélwayi** *A. tinctòria var. Kélwayi*	Dense bushy plant with a very long blooming season. Daisy-like flowers of a richer yellow than A. tinctoria with more finely cut foliage. Good for cutting and for the border. Prop. by seed or division. Any soil. Hort.	2-3 ft. *Sun*	Mid. June to Oct.

Color	English Name	Botanical Name and *Synonyms*	Description	Height and *Situation*	Time of Bloom
"Yellow" 2, centre 6	GOLDEN MARGUE-RITE, ROCK CAMOMILE	*Ánthemis tinctòria	See page 96.		Mid. May to Oct.
"Yellow" 2	YELLOW CANADIAN COLUMBINE	*Aquilègia Canadénsis var. flaviflòra *A. C. var. flavéscens, A. cærùlea var. f.*	See page 24.		Late Apr. to early July
"Yellow" 3 & 2	GOLDEN-SPURRED COLUMBINE	**Aquilègia chrysántha *A. leptocèras var. c.*	See page 96.		Late May to late Aug.
"Yellow" 4	TRUE ASPHODEL, KING'S SPEAR	*Asphodelìne lùtea *Asphódelus lùteus*	The Asphodel of the ancients. Flowers in tall spikes 1 ft. long. Foliage grass-like and glossy, attached to the flower stalk. Attractive in clumps and good for cutting. Prop. by seed and division. Any ordinary garden soil. Mediterranean Region.	2-4 ft. *Sun or half shade*	June, July
"Yellow" 4	YELLOW MILK VETCH	Astrágalus alopecuroìdes	Leguminous plant. Flowers in pretty spikes. Foliage slightly hairy. Appropriate for the rock-garden or border. Prop. by seed. Rather light soil. Siberia.	2-5 ft. *Sun*	"
"Yellow"	CHINESE MILK VETCH	Astrágalus Chinénsis	Creeping plant with pinnate leaves. Thrives in the rock-garden and covers the rocks well. China.	4-5 ft. *Sun*	"
"Yellow"	GALEGA-LIKE MILK VETCH	Astrágalus galegifòrmis	Erect plant. Profusion of pale drooping flowers in long clusters. Compound leaves. Plant in wild garden. Prop. by seed just ripe, also by division. Light soil, rather dry. Siberia.	3-5 ft. *Half shade*	Early June to early July
"Yellow" 6	WILLOW-LEAVED OX-EYE	*Bupthálmum salicifòlium *B. grandiflòrum*	Showy plant of compact habit. Solitary flowers with long rays, numerous and large. Leaves scattered along the flower stalks. Good for cutting. Prop. by division. Any soil. Europe; W. Asia.	18 in. *Sun or shade*	June, July
"Yellow" 6	SHOWIEST OX-EYE	*Bupthálmum speciosíssimum	Terminal solitary flowers with long rays. Pointed heart-shaped leaves. Good for shrubbery and border. Prop. by division. Any ordinary garden soil. Europe; W. Asia.	2-5 ft. *Sun or shade*	"
"Yellow" 5	SHOWY OX-EYE	*Bupthálmum speciòsum *B. cordifòlium*	Vigorous showy plant. Large daisy-like flowers. Large heart-shaped leaves, mostly at the base of the plant. Good for shrubbery and border. Prop. by division. Any soil. Central Europe.	3-4 ft. *Sun or shade*	"
"Yellow" 1 & 21 light	LEMON-COLOR MOUNTAIN BLUET	Centaurèa montàna var. citrìna *C. m. var. sulphùrea*	See page 99.		Late May to early July

Color	English Name	Botanical Name and *Synonyms*	Description	Height and *Situation*	Time of Bloom
"Pale yellow" 4 very pale	**ALPINE CEPHALARIA**	***Cephalària alpìna**	Appearance rather coarse but an excellent border plant, with showy scabious-like flowers. Prop. by seed. Any good soil. Greece; the Alps.	5 ft. *Sun*	Late June to late July
"Greenish yellow" 2 greener	**YELLOW CLINTONIA**	**Clintònia boreàlis** *Smilacìna b.*	See page 99.		May, early June
"Yellow" 3	**GOLD JOINT**	**Chrysógonum Virginiànum**	A profusely blooming dwarf plant suitable for sunny borders. The flo ers are produced singly. Prop. y division in spring and runners. Loamy soil mixed with leaf mold and peat is desirable. Penn. and South.	1 ft. *Sun*	June, July
"Yellow" 5	**LARGE-FLOWERED TICKSEED**	****Coreópsis grandiflòra**	Daisy-like showy flowers, larger than other garden species, few on a stem and lasting a long time. Foliage small. Excellent for cutting. A group forms a bright mass of color in the border. Prop. by seed or division. Any soil. Southern U. S. A.	1-2 ft. *Sun*	June to Sept.
"Yellow" 6	**LANCE-LEAVED TICKSEED**	****Coreópsis lanceolàta**	Resembles C. grandiflora with smaller flowers; yellow or brown centres and broad rays, flowers solitary on the stalks which rise high above the dense tufts of glossy foliage. Good for cutting. Border. Prop. by seed or division. Any ordinary garden soil. Eastern States.	1-2 ft. *Sun*	"
"Yellow" 5	**STIFF TICKSEED**	***Coreópsis palmàta** *C. prǽcox*	Flowers, 1½-2½ in. across, have brown or yellow centres with lemon rays. Border plant. Prop. by seed or division. Ordinary garden soil. Middle U. S. A.	1½-3 ft. *Sun*	June, July
"Golden yellow" 4	**NOBLE FUMITORY**	***Corýdalis nóbilis** *Fumària n.*	See page 99.		May, early June
"Bright yellow" 2 deep & dull	**SMALL YELLOW LADY'S SLIPPER**	***Cypripèdium parviflòrum**	See page 99.		May, June
"Pale yellow" 2 deep & dull	**LARGE YELLOW LADY'S SLIPPER**	****Cypripèdium pubéscens** *C. parviflòrum, C. hirsùtum*	See page 99.		"
"Yellow" 2	**ZALIL'S LARKSPUR**	***Delphínium Zàlil** *D. sulphùreum, D. hýbridum var. sulphùreum*	Large flowers on tall racemes. Dark green leaves. Suitable for the border. Prop. by seed, division and cuttings. Deep sandy soil, well enriched. Persia.	1-2 ft. *Sun*	June, July

Color	English Name	Botanical Name and *Synonyms*	Description	Height and *Situation*	Time of Bloom
"Yellow" 2	YELLOW FOXGLOVE	*Digitàlis ambígua *D. grandiflòra* *D. ochroleùca*	Perennial. Large tubular flowers with brown markings, which droop from long stalks rising high above the clump of broad downy leaves. Effective planted in groups in the border or wild garden. Prop. by seed and division. Rather moist light loam. Europe; Asia.	2-3 ft. *Sun or half shade*	Early June to mid. July
"Cream color" 2	WOOLLY FOXGLOVE	*Digitàlis lanàta	A perennial Foxglove with long dense spikes of small flowers, sometimes grayish or slightly purplish. See D. ambigua. S. Eastern Europe.	2-3 ft. *Sun or half shade*	Early June to late July
"Golden yellow" 5	AUSTRIAN LEOPARD'S BANE	*Dorónicum Austrìacum	See page 99.		May, June
"Yellow" 5	CAUCASIAN LEOPARD'S BANE	*Dorónicum Caucásicum	See page 99.		May, early June
"Golden yellow" 5	CLUSIUS'S LEOPARD'S BANE	*Dorónicum Clùsii	See page 100.		May, June
"Yellow" 5	CRAY-FISH LEOPARD'S BANE	*Dorónicum Pardaliánches	See page 100.		"
"Yellow" 5	PLANTAIN-LEAVED LEOPARD'S BANE	*Dorónicum plantagíneum	See page 100.		"
"Orange yellow" 7	TALL PLANTAIN-LEAVED LEOPARD'S BANE "HARPUR CREWE"	*Dorónicum plantagíneum var. excélsum *D. e. "Harpur Crewe"*	See page 100.		"
"Yellow" 4	VITALE'S DOUGLASIA	Douglásia Vitaliàna *Arètia V.* *Gregòria V.*	See page 100.		Late May, June
"Yellow" 4	AIZOON-LIKE WHITLOW GRASS	Dràba aizoìdes	See page 100.		May, early June
"Yellow" 2 deeper	OLYMPIAN WHITLOW GRASS	*Dràba Olýmpica *D. bruniæfòlia*	Dainty tufted alpine plant bearing a profusion of small flowers which rise from the dense spreading tufts of foliage. Useful for rockwork in elevated positions. Prop. by seed, though generally by division. Well-drained soil. Caucasus.	4 in. *Sun*	June
"Bright yellow" 4, anthers 13	BUNGE'S EREMURUS	Eremùrus Búngei	Pretty dwarf species. Bell-shaped flowers with long stamens in spikes 5 in. long. Flower stems slender. Leaves grass-like in a clump at base. Plant in groups. Prop. by division. Deep rich sandy loam. Persia.	1-3 ft. *Sun*	June, July

Color	English Name	Botanical Name and *Synonyms*	Description	Height and *Situation*	Time of Bloom
"Yellow" 6 lighter	WOOLLY BAHIA	Eriophýllum cæspitòsum *Bàhia lanàta, Actinélla lanàta*	Branching grayish plant. Numerous fairly large rayed flowers on leafy stalks. Suitable for the wild garden. Prop. by seed and division. Light well-drained sandy soil. N. Amer.	1-1½ ft. *Sun*	June, July
"Sulphur yellow" 4	MOUNTAIN HEDGE MUSTARD	*Erýsimum alpìnum	See page 101.		May, June
"Yellow" 3	ROCK-LOVING HEDGE MUSTARD	*Erýsimum rupéstre *E. pulchéllum*	See page 101.		May, early June
"Yellow" 3	INDIAN STRAW-BERRY	*Fragària Índica	Trailer with small solitary flowers and many red berries. Leaves divided into 3 lobes. Protect slightly in winter. Suitable for rock-gardens and garden vases. Prop. by division. Any soil not too damp. India.	6-8 in. *Sun*	June
"Yellow" 4	LADY'S BEDSTRAW	Gàlium vèrum	Of low-growing spreading habit. Numbers of tiny flowers in panicles. Charming feathery foliage. Good for naturalization on sterile ground. Prop. by seed and division. Any soil. Europe.	1-3 ft. *Sun*	June to Sept.
"Yellow" bet. 5 & 6	DYER'S GREEN-WEED, BASE BROOM	*Genísta tinctòria	Effective sub-shrub, to be seen on the Salem hills in Mass. Pea-shaped flowers in racemes which form a mass of golden bloom. Foliage inconspicuous. Valuable for covering dry banks and also for rock-gardens. Prop. by seed and green cuttings. Any soil. Europe; Asia; Naturalized in Mass., Me., and N. Y. Var. *plena* has double flowers.	3 ft. *Sun*	Late June, July
"Yellow" 6	YELLOW-FLOWERED MOUNTAIN AVENS	*Gèum montànum	See page 101.		Late May to mid. June
"Yellow" 6 lighter	PYRENEAN AVENS	Gèum Pyrenàicum	See page 101.		Mid. May to mid. June
"Yellow" 4	BROAD-LEAVED GUM-PLANT	Grindèlia squarròsa	An unimportant sub-shrub of coarse habit. Rayed flowers terminate branching stalks. Foliage stiff. Wild garden or border. Prop. by seed, spring sown in green-house, division or cuttings. Any soil. Western U. S. A.	1-2 ft. *Sun*	Late June to Sept.
"Yellow" 6	BOLAND-ER'S SNEEZE-WEED OR SNEEZE-WORT	*Helènium Bolánderi	Wild in low ground. Robust, free-blooming plant resembling Sunflower. Showy flowers 2-3 in. wide with long daisy-like drooping petals surrounding a dark brown centre. Border. Prop. by seed, division or cuttings. Cal.	1-2 ft. *Sun*	June to Sept.

Color	English Name	Botanical Name and *Synonyms*	Description	Height and *Situation*	Time of Bloom
"Yellow" 5	HOOPES'S SNEEZE-WEED	**Helènium Hoòpesii	See page 102.		Late May to late June
"Yellow" 2 to 5	ROCK OR SUN ROSE	*Heliánthe-mum vulgàre	Pretty shrubby evergreen plant. Flowers like small single Roses in lax racemes. Foliage silvery beneath, in a low thick carpet. Desirable for the rock-garden. Needs slight protection in winter. Prop. by seed, division or cuttings. Sandy loam. Many different-colored vars. Europe; N. Africa; W. Asia.	8-15 in. *Sun*	Early June, July
"Yellow" 6 deep	LEMON LILY	**Hemerocál-lis flàva	Very hard and vigorous. Fragrant lily-like flowers in clusters surmount a mound of graceful grass-like leaves. Good for cutting. Spreads rapidly. Charming in masses on the banks of streams, in clumps in the border or rock-garden, and for edging walks or beds of shrubs. Prop. by division. Any moist rich soil. Europe; N. Asia. See Plate, page 201.	3 ft. *Half shade*	June, early July
"Orange yellow" 8	MIDDEN-DORF'S YELLOW DAY LILY	**Hemero-cállis Middendorfii	The prettiest species. Very fragrant. Resembles H. Dumortieri, with paler flowers borne in threes or fours, on long stems and broader bright green curving leaves in clumps. For planting and cultivation see H. flava. Amurland.	1-3 ft. *Half shade*	Late June, July
"Yellow" 6 lighter	THUN-BERG'S YELLOW DAY LILY	**Hemero-cállis Thúnbergii	Closely resembles H. flava though the flowers are not so large. See H. flava. Japan.	3-4 ft. *Sun or half shade*	"
"Golden yellow" 5	SHAGGY HAWK-WEED	Hieràcium villòsum	Plant covered with white silkiness, effective for its large flowers and silvery foliage. Naturalize on barren ground. More easily kept from spreading than other species of this genus. Prop. by seed and division. Any soil. Europe.	1-2 ft. *Sun*	June to mid. Aug.
"Deep yellow" 5 deep	YELLOW STAR GRASS	Hypóxis eréćta *H. hirsùta*	See page 102.		May, June
"Yellow" 5	LARGE-FLOWERING ELECAM-PANE	*Ínula grandiflòra	Vigorous plant with large flowers in profusion. Foliage coarse. Border. Resists drought. Prop. by seed and division. Any good soil. Asia.	2-3 ft. *Sun*	June
"Yellow"	YELLOW-BANDED FLAG	Ìris orientàlis *I. ochroleùca, I. gigantèa*	See page 102.		Late May, June
"Yellow" 5 & 2	YELLOW OR COMMON WATER FLAG	*Ìris Pseudácorus *I. Pseud-ácorus*	See page 102.		Late May to late June

Color	English Name	Botanical Name and *Synonyms*	Description	Height and *Situation*	Time of Bloom
"Yellow" effect 11	**TUCK'S FLAME FLOWER**	***Kniphòfia Túckii**	Brilliantly striking plant. Flowers, reddish when first in bloom, rise in dense slender spikes high above tufts of sword-shaped leaves. Group among tropical-looking plants in sheltered border with good background. Leave undisturbed and protect slightly in winter. Prop. by division. Any well-drained soil. Central Africa.	4-5 ft. *Half shade*	June, to Sept.
"Yellow"	**EDELWEISS**	***Leontopòdium alpìnum** *Gnaphàlium Leontopòdium*	Prized by Alpine climbers. Downy star-like flowers rise well above a tuft of narrow hoary leaves. Pretty for rock-garden. Prop. by seed. Rather light soil. Switzerland.	4-12 in. *Sun*	June, July
"Yellow" 5	**GRAY-HEADED CONE-FLOWER**	***Lépachys pinnàta** *Ratìbida pinnàta Rudběckia pinnàta*	Flowers cone-shaped; purple centres and abruptly drooping rays. Resembles Rudbeckia. Good for back of border or in shrubbery. Prop. by seed or division. N. Y.; W. and S.	3-5 ft. *Sun*	June to mid. Sept.
"Yellow" 7 dull	**YELLOW CANADA LILY**	***Lílium Canadénse var. flàvum**	Picturesque native, wild in moist meadows and bogs. Flowers, with dark spots, drooping in clusters. Lovely in masses or scattered among shrubs, or naturalized. Bulbous. Prop. by offsets or scales. Light well-drained soil. Avoid contact with manure. Eastern N. Amer. See Plate, page 202.	1-4 ft. *Sun or half shade*	June, July
"Yellow" 3	**CAUCASIAN LILY**	***Lílium monadélphum** *L. Szovitziànum, L. Cólchicum*	Conspicuous species. Fragrant turban-shaped flowers, tinged with purple and spotted with black, crown a vigorous leafy stalk. Mass or scatter against dense background in border. Bulbous. Prop. by offsets, scales or very slowly by seed. Any light soil. Avoid contact with manure. Persia.	2-5 ft. *Half shade*	"
"Yellow" 5	**PARRY'S LILY**	****Lílium Párryi**	Best yellow species. Flowers with reddish brown markings and spreading reflexed tips. Foliage in whorls. Effective massed in border against background. Bulbous. Prop. by offsets, scales, or very slowly by seed. Light well-drained soil. Avoid contact with manure. Cal.	2-6 ft. *Shade*	"
"Golden yellow" 6 more orange	**HUMBOLDT'S LILY**	***Lílium pubérulum** *L. Califórnicum, L. Humboldtii, L. Bloomeriànum*	Graceful and stately. Drooping flowers 6-8 in. across, dotted with pale purple turning to red or brown, in panicles. Bright green leaves in large whorls. Margin of shrubbery. Bulbous. Prop. by offsets, scales or very slowly by seed. Any light well-drained soil. Avoid direct contact with manure. Cal.	3-5 ft. *Half shade*	"

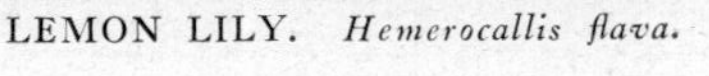

LEMON LILY. *Hemerocallis flava.*

YELLOW CANADA LILY. *Lilium Canadense var. flavum.*

Color	English Name	Botanical Name and *Synonyms*	Description	Height and *Situation*	Time of Bloom
"Buff" 1	NANKEEN LILY	**Lìlium testàceum *L. excèlsum, L. Isabellìnum*	Very choice and stately species, widely cultivated. Clusters of fragrant flowers with conspicuous orange stamens, hang nodding from the top of tall leafy stalks. Group in border. Bulbous Prop. by offsets or scales. Any light soil. Avoid contact with manure. Supposed hybrid.	2-6 ft. *Sun or half shade*	Mid. June to mid. July
"Yellow" 3	DALMATIAN TOAD-FLAX	*Linària Dalmática	Showy, graceful and free-growing. Large snapdragon-like flowers in profusion. Foliage dense. Border. Prop. by seed and division. Any light soil. S. Eastern Europe.	1-3 ft. *Sun or shade*	Early June to early July, late July to late Aug.
"Yellow" 3 & 7	MACEDONIAN TOAD-FLAX	Linària Macedónica	Distinct species, with broader leaves than L. Dalmatica. Snapdragon-like flowers in tall terminal racemes bloom continuously. Inside of lower lip orange. Prop. by seed or division. Any ordinary soil. Macedonica.	2-3 ft. *Sun or shade*	June, July
"Yellow" 5	YELLOW FLAX	*Lìnum flàvum	Gay flowers in loose clusters. Foliage pretty but inconspicuous. Invaluable plant forming neat bush. Border or rock-garden. Prop. by seed and division. Light rich soil. Europe.	1-2 ft. *Sun*	"
"Yellow" 3	BIRD'S-FOOT TREFOIL, BABIES' SLIPPERS	Lòtus corniculàtus	Hardy spreading trailer. Pea-like flowers often marked with red, in umbels of five to ten. Foliage in tufts. Rock-garden. Any soil. Var. *flore-pleno.* Double and more showy variety; flowers last longer, preferred to the type. N. Hemisphere; Britain.	4-18 in. *Sun*	June to Oct.
"Yellow" 5	MONEY-WORT, CREEPING JENNY	*Lysimàchia nummulària	Rapid creeper. Masses of pretty cup-shaped flowers on prostrate branchlets. Good for carpeting where grass will not grow, for rock-gardens and for drooping from garden vases. Prop. by seed and division. Any soil. Europe. Var. *aurea,* leaves wholly or partly yellow.	1-2 in. *Sun or half shade*	Early June to late July
"Yellow" 5	SPOTTED LOOSE-STRIFE	*Lysimàchia punctàta *L. verticulàta*	Of sturdy erect habit. Flowers in whorls around leafy stalks. Good in border. Prop. by seed and division. Good garden soil. Europe; W. Asia. See Plate, page 205.	1-3 ft. *Sun*	June, July
"Yellow" bet. 5 & 6	BULB-BEARING LOOSE-STRIFE	Lysimàchia strícta *L. terrêstris*	Wild in low or wet ground. Flowers in thick terminal racemes above long narrow leaves, grayish underneath. Wild garden Prop. by seed and division. Any soil. Eastern U. S. A.	8-24 in. *Sun*	"

Color	English Name	Botanical Name and *Synonyms*	Description	Height and *Situation*	Time of Bloom
"Yellow" 4	COMMON YELLOW LOOSE-STRIFE	*Lysimàchia vulgàris	Flowers, often with red margins, in spikes. Pointed leaves in whorls of 3 or 4 about the flower stalks. Quite showy in clumps on banks of streams. Prop. by seed and division. Any moist soil. Europe; Asia.	2-3 ft. *Sun*	June, July
"Yellow" 5	WELSH POPPY	Meconópsis Cámbrica	Rather large poppy-like flowers are borne erect on slender stems. Foliage light green much divided. Rock-garden. Prop. by seed and division. Any good soil. W. Europe.	1 ft. *Sun*	"
"Pale yellow" 1	SESSILE-LEAVED BELLWORT	Oakèsia sessilifòlia *Uvulària s.*	See page 102.		May, June
"Yellow" 3	LARGE-FLOWERED BIENNIAL EVENING PRIMROSE	*Œnothèra biénnis var. grandiflòra *Œ. Lamarckiàna*	Biennial. Vigorous and sometimes branching. A large variety of the common Evening Primrose. Showy flowers 3-4 in. across. Effective in masses in the wild garden and border. Prop. by seed and cuttings. Any soil. S. W. U. S. A.	4-5 ft. *Sun*	June to Sept.
"Yellow" 5	SUNDROPS	**Œnothèra fruticòsa	One of the best species. Shrubby plant with red stems and clusters of broad showy flowers continuously covering the plant with golden bloom. Excellent border plant. Prop. by seed and cuttings. Any light soil. N. Amer. Var. *Youngii.* Flowers lemon-yellow. Branches slender. Vigorous and a free bloomer. Var. *major;* (color no. 3), is thick and bushy. See Plate, page 206.	1-3 ft. *Half shade*	June, July
"Yellow" 5	FRASER'S EVENING PRIMROSE	**Œnothèra glaùca var. Fràseri *Œ. Fràseri*	A good variety. Upright plant becoming bushy and leafy at the top, covered with clusters of flowers 2 in. across. Continuous in bloom. Effective in masses for rock-garden or border. Easy cultivation. Prop. by seed and cuttings. Any good garden soil. Southern U. S. A.	2-3 ft. *Sun*	June to Sept.
"Lemon yellow" 3	LINEAR-LEAVED EVENING PRIMROSE	Œnothèra lineàris *Œ. fruticòsa var. lineàris, Œ. ripària*	Bushy plant. Flowers small but numerous, forming a mass of bloom. Leaves very narrow. Border plant. Prop. by division or cuttings in the spring. Rich sandy soil. Eastern U. S. A.	1½ ft. *Sun*	June to early Aug.
"Yellow" 3	MISSOURI PRIMROSE	**Œnothèra Missouriénsis *Œ. macrocárpa, Megaptèrium Missouriênse*	A low species with prostrate ascending branches. A profuse bloomer. Solitary flowers, often 5 in. across, cover the plant. Attractive in open border or rock-garden. Prop. by division or cuttings. Any light soil. S. Western U. S. A.	10 in. *Half shade*	"

TUBEROUS JERUSALEM SAGE. *Phlomis tuberosa.*

SPOTTED LOOSE-STRIFE. *Lysimachia punctata.*

SUNDROPS. *Œnothera fruticosa.*

ICELAND POPPY. *Papaver nudicaule.*

Color	English Name	Botanical Name and *Synonyms*	Description	Height and *Situation*	Time of Bloom
"Dull yellow" 3 duller	BARBERRY FIG, COMMON PRICKLY PEAR	Opúntia vulgàris *O. Opúntia*	Native of our shores. Straggling light green Cactus plant with grayish spines. Flowers 2 in. across. Rock-garden. Prop. by cuttings more easily than by seed. Any well-drained soil. Atlantic States.	10-12 in. *Sun*	June to Sept.
"Yellow"	HERBACEOUS PEONY	**Pæònia vars.	See color "various," page 268.		June
"Yellow" 5 or 6 brilliant	ICELAND POPPY	**Papàver nudicaùle	See page 35 and Plate, page 206.		Late Apr. to July, late Aug. to Oct.
"Yellow" 6	SCABIOUS-LEAVED PATRINIA	Patrínia scabiosæfòlia	See page 105.		May, June
"Creamy yellow" 2	WOOD BETONY	Pediculàris Canadénsis	See page 105.		"
"Pale yellow" 2	CROWDED PENTSTEMON OR BEARD-TONGUE	*Pentstèmon confértus	Numerous tube-shaped flowers varying to cream white, in spikes. Foliage rather long and narrow. Protection of leaves advisable. Rock-garden and border. Prop. by seed and division. Any good garden soil. Rocky Mts. Var. *cæruleo-purpureus*, purple flowers. Colo. W. & N.	1-2 ft. *Sun*	June, early July
"Yellow" 6 shading to 13	JERUSALEM SAGE	Phlòmis fruticòsa	Tender branching shrub. Showy flowers in dense whorls. Branches and wrinkled leaves covered with yellowish down. Sheltered position in border, shrubbery or wild garden. Protect in winter. Prop. by seed or cuttings. Any soil. Mediterranean Region.	2-4 ft. *Sun*	June, July
"Yellow" 3	SILVERY CINQUEFOIL	Potentílla argéntea	See page 105.		Early May to early July
"Yellow" 1	SILVERY-LEAVED CINQUEFOIL OR FIVE-FINGER	Potentílla argyrophýlla *P. insígnis*	Small strawberry-like flowers borne on silky leafy stems. Leaves whitish beneath. Border or rock-garden. Prop. by seed and division. Dry sandy soil. Himalaya.	2-3 ft. *Sun*	June, July
"Lemon yellow" 3	CALABRIAN CINQUEFOIL	Potentílla Calabra	See page 105.		Late May to early July
"Yellow" 3	SHRUBBY CINQUEFOIL OR FIVE-FINGER	Potentílla fruticòsa	Dwarf shrub bearing bright strawberry-like flowers in profusion. Foliage silky. Border, rock-garden or margin of shrubbery. Prop. by seed, division or green cuttings. Moist soil. Europe; Asia; N. Amer.	½-4 ft. *Sun*	June to Sept.

JUNE

Color	English Name	Botanical Name and *Synonyms*	Description	Height and *Situation*	Time of Bloom
"Golden yellow" effect 7	**HYBRID CINQUE-FOIL OR FIVE-FINGER**	***Potentílla "Gloire de Nancy"**	Rather tender double hybrid with a profusion of buttercup-like flowers. Foliage excellent. Border. Protect in winter. Prop. by root division in spring. Dry sandy soil. Hort.	2 ft. *Sun*	June to Sept.
"Yellow" 3	**LARGE-FLOWERED CINQUE-FOIL OR FIVE-FINGER**	**Potentílla grandiflòra**	Flowers borne on leafy branching stems. Silky compound leaves. Border or rock-garden. Prop. by seed and division. Dry sandy soil. Europe; Asia.	10-20 in. *Sun*	June, July
"Golden yellow" 5	**PYRENEAN CINQUE-FOIL**	**Potentílla Pyrenaìca**	See page 105.		May to Aug.
"Rich yellow" 3 light	**STUART'S PRIMROSE**	****Prímula Stùartii**	See page 105.		Late May, June
"Golden yellow" 5	**BACHE-LOR'S BUTTONS**	**Ranúnculus àcris var. flòre-plèno**	See page 105.		Mid. May to Sept.
"Yellow" 5	**DOUBLE ACONITE-LEAVED BUTTERCUP**	**Ranúnculus aconitifòlius var. lùteus-plènus**	See page 106.		Mid. May to July
"Yellow" 5	**MOUNTAIN BUTTERCUP**	**Ranúnculus montànus**	See page 106.		May to early July
"Yellow" 5	**CREEPING DOUBLE-FLOWERED BUTTERCUP**	**Ranúnculus répens var. flòre-plèno**	See page 106.		May to Aug.
"Yellow" 3 pale	**WEBB'S SCABIOUS OR PIN-CUSHION FLOWER**	**Scabiòsa ochroleùca** *S. Webbiàna*	Smaller than the type. Flowers on long stems. Good for cutting. Neat tufts of grayish green leaves. Border or rock-garden. Prop. by division. Poor soil preferable. Phrygia.	6-10 in. *Sun*	June to early Sept.
"Yellow" 4	**STONE-CROP, WALL PEPPER, LOVE ENTANGLE**	***Sèdum àcre**	See page 106.		Late May, June
"Yellow" 5	**AIZOON STONECROP**	***Sèdum Aizóon**	Starry flowers in dense flat clusters. Color rather poor. Pulpy leaves coarsely toothed. Good for edging or rock-garden. Prop. by seed, spring division or cuttings. Loose soil. Siberia; Japan.	1-2 ft. *Sun*	Mid. June to mid. Aug.
"Yellow" 4 dull & deep	**HYBRID STONECROP**	**Sèdum hỳbridum**	Creeping evergreen. Flowers in clusters, 2-3 in. wide. Small leaves. Good for carpeting. Prop. preferably by division. Sandy soil best. Siberia.	6 in. *Sun*	June

JUNE

Color	English Name	Botanical Name and *Synonyms*	Description	Height and *Situation*	Time of Bloom
"Yellow" 3 dull	MIDDEN-DORF'S STONECROP	*Sèdum Midden-dorfiànum	Creeping tufted evergreen. Starry flowers in flat-topped clusters. Pulpy foliage, deep green turning purple in winter. Good for carpeting. Prop. preferably by division. Sandy soil best. Amurland.	4 in. *Sun*	June, July
"Yellow" 4	SIX-ANGLED STONECROP	Sèdum sexangulàre	Tiny creeper closely resembling S. acre, with starry flowers and delicate pulpy foliage. Prop. by seed or offsets. Ordinary garden soil. Europe.	3-6 in. *Sun*	Early June, July
"Yellow" 8 lighter	CELANDINE POPPY	Stylóphorum diphýllum *Chelidònium d. Papàver Stylóphorum*	See page 106.		May, early June
"Green-ish yellow" 2	GLAUCOUS MEADOW RUE	Thalíctrum glaùcum	Desirable for foliage. Flowers in dense clusters. Fern-like leaves. Wild garden. Prop. in spring by seed and division. Well-drained loam. S. Europe.	2-5 ft. *Half shade*	Late June to late July
"Green-ish yellow" 3	DWARF MEADOW RUE	*Thalíctrum mìnus *T. pur-pùreum, T. saxátile*	Grown for foliage. Insignificant flowers droop in lax feathery panicles. Fine grayish foliage like that of Maiden-hair Fern. Good for cutting. Charming in border or wild garden or as edging. Prop. in spring by seed and division. Well-drained soil. Europe; N. Asia; N. Africa. Var. *adiantifolium* (*T. adiantioides, T. adianthifolium*). A favorite, similar to the type, with foliage even more maidenhair-like. Hort.	1-2 ft. *Half shade*	Late June to mid. July
"Yellow" 4 lighter	GOLDEN ALEXAN-DERS	Tháspium aùreum *T. trifoliàtum var. aùreum*	Wild branching herb. Flowers in umbels. Deeply divided leaves on the stalks and at the base of plant. Naturalize in wild garden. Good moist soil. N. Amer.	1½ ft. *Half shade*	June, July
"Yellow" 5	CAROLINA THER-MOPSIS	*Thermópsis Caroliniàna	Erect lupine-like plant, wild in open roads. Pea-shaped flowers in long dense racemes. Clover-like leaves. Excellent for border. Prop. generally by autumn or spring sown seed. Light well-drained soil. Va.; N. C.	2-4 ft. *Sun or shade*	Early June to mid. July
"Yellow" 4	BEAN-LIKE THER-MOPSIS	*Thermópsis fabàcea *T. montàna*	See page 106.		May, June
"Yellow" 4	ALLEGHANY THER-MOPSIS	*Thermópsis móllis	See page 106.		Mid. May to Aug.
"Yellow"	MOUNTAIN THER-MOPSIS	Thermópsis montàna	See page 106.		May, June
"Yellow" 5	MOUNTAIN GLOBE FLOWER	**Tróllius Europæus *T. globòsus*	See page 109.		Early May to early June

Color	English Name	Botanical Name and *Synonyms*	Description	Height and *Situation*	Time of Bloom
"Orange yellow" 7	JAPANESE GLOBE FLOWER	*Tróllius Japónicus	See page 109.		May, June
"Lemon yellow" 2	LARGE-FLOWERED BELLWORT	*Uvulària grandiflòra	See page 110.		"
"Yellow" 2	NETTLE-LEAVED MULLEIN	*Verbáscum Chàixii *V. orientàle, V. vernàle*	Showy and very hardy. Large flowers in spikes. Plentiful rather coarse green foliage. Good in border or wild garden. Prop. by seed, often self-sown. Any garden soil. Europe.	3 ft. *Sun*	Late June to late July
"Yellow" 3	LONG-LEAVED ITALIAN MULLEIN	*Verbáscum longifòlium *V. pannòsum*	See page 110.		Late May to late June
"Yellow" 2	DARK MULLEIN	*Verbáscum nìgrum	Flowers thickly packed in long racemes at end of stalk. Foliage unattractive in form and color. Border or wild garden. Prop. by seed. Any garden soil. Blooms a second time if cut down early. Europe; W. Asia.	2-3 ft. *Sun*	Early June to late July
"Yellow" 6	YELLOW HORNED VIOLET OR BEDDING PANSY	**Vìola cornùta var. lùtea màjor	See page 35.		Late Apr. until frost
"Yellow" 4	YELLOW OR BARREN STRAW-BERRY	Waldsteìnia fragarioìdes *Dalibarda f.*	See page 113.		May, June
"Orange" 26 more orange	DWARF CHILOE WATER FLOWER	*Gèum Chiloénse var. miniàtum *G. miniàtum*	See page 113.		May, June
"Purplish orange" near 3 & 41	WATER AVENS OR FLOWER	*Gèum rivàle	See page 113.		Late May, June
"Copper" near 7	HYSSOP-LEAVED ROCK ROSE	*Heliánthemum vulgàre var. hyssopifòlium	Shrubby evergreen plant. Flowers like small Wild Roses sometimes saffron-colored, in lax racemes. Glossy low mat of foliage. Valuable for a dry sunny position in rock-garden. Slight protection in winter needed. Prop. by seed, division or cuttings. Sandy loamy soil. Europe.	8-15 in. *Sun*	Early June, July
"Orange" 11 brilliant	DUMORTIER'S DAY LILY	**Hemerocàllis Dumortièrii *H. rùtilans, H. Sieboldii*	First species to bloom. Fragrant lily-shaped flowers in clusters of 2 or 3, smaller than in other species. Foliage grass-like. Replant often. Pretty in clumps and good for cutting. Prop. by division. Any soil. Japan.	1-2 ft. *Sun or half shade*	June, July

Color	English Name	Botanical Name and *Synonyms*	Description	Height and *Situation*	Time of Bloom
"Orange" 12	ORANGE HAWKWEED	**Hieràcium aurantiàcum**	Small clusters of dandelion-like flowers of fine color on leafless stems. Foliage shaggy in tufts near ground. Spreads rapidly and is liable to become a weed. Good ground-cover for sterile waste land or wild garden. Prop. by seed and division. Any soil. W. Europe.	6-24 in. *Sun*	June to Oct.
"Reddish brown" 14	RED-BROWN FLAG	**Ìris cùprea** *I. fúlva*	Several branching stems bearing flowers marked with blue and green. Leaves light green slightly grayish. Swamps, Illinois, S.	2-3 ft. *Sun*	Late June
"Orange" effect 7	OREGON LILY	***Lilíum Columbiànum** *L. parviflòrum, L. Sàyi*	Ineffective unless mixed with other plants. Small turban-shaped flowers dotted with purple; sparse foliage. Border. Bulbous. Prop by offsets, scales, or very slowly by seed. Rich well-drained soil. Avoid direct contact with manure. Ore. to B. C.	1½-3 ft. *Shade*	June, early July
"Orange" 26 orange	ASA GRAY'S LILY	***Lílium Gràyi**	Allied to L. Canadense. Delicate flowers, reddish or brownish, with purple spots inside, 1-9 on a stem. Mass against dense background in border. Bulbous. Prop. by offsets, scales, or very slowly by seed. Light well-drained soil. Avoid direct contact with manure. S. Eastern U. S. A.	2-3 ft. *Half shade*	June, July
"Reddish orange" 26 orange	SPOTTED OR HANSON'S LILY	****Lílium maculàtum** *L. Hánsoni*	Beautiful and stately species. Striking purple-spotted flowers in lax clusters. Bright green leaves in whorls. Effective planted in clumps in Rhododendron beds or massed against background in border. Bulbous. Prop. by offsets or scales. Any light soil. Avoid direct contact with manure. Japan. See Plate, page 216.	3-4 ft. *Sun or half shade*	"
"Deep orange" bet. 12 & 17	ORANGE ICELAND POPPY	***Papàver nudicaùle var. aurantiàcum**	See page 36.		Late Apr. to July, late Aug. to Oct.
"Deep orange" bet. 12 & 17	SMALL ICELAND POPPY	***Papàver nudicaùle var. miniàtum**	See page 36.		Late Apr. to July, mid. Aug. to Oct.
"Pale scarlet orange" 9	HAIRY-STEMMED POPPY	***Papàver pilòsum**	See page 114.		Late May, June
"Red orange" 11	ATLANTIC POPPY	***Papàver rupífragum var. Atlánticum** *P. A.*	See page 114.		Late May to Aug.

Color	English Name	Botanical Name and *Synonyms*	Description	Height and *Situation*	Time of Bloom
"Scarlet" 19 & 4	WILD COLUMBINE	*Aquilègia Canadénsis	See page 36.		Late Apr. to mid. June
"Orange red" 12 & 17	HYBRID CALIFORNIAN COLUMBINE	**Aquilègia formòsa var. hỳbrida *A. Califórnica, var. hỳbrida*	See page 114.		Mid. May to July
"Orange red" 19 & 5	MEXICAN COLUMBINE	*Aquilègia Skínneri	See page 114.		Late May, June
"Orange scarlet" 17	LARGE SCARLET CALIFORNIAN COLUMBINE	*Aquilègia truncàta *A. Califórnica, A. exímea*	See page 36.		Apr. to early June
"Violet crimson" 27 bright	PURPLE POPPY MALLOW	*Callírhoe involucràta	Shaggy trailing plant. Large flowers with white centres. Foliage palmately divided. Plant in border and rock-garden. Blossoms the first year from seed. Prop. also by cuttings. A light rich soil is preferable. Neb. and S. Var. *lineariloba;* (color no. 35 lighter & brilliant). Less shaggy than the type. Large mallow-like flowers varying to pale red in continuous bloom. Charming in the rock-garden, drooping over ledges. Central U. S. A.	9-12 in. *Sun*	June to Sept.
"Maroon" near 33	DARK PURPLE KNAPWEED	Centaurèa atropurpùrea *C. calocéphala*	Strong vigorous plant with silvery foliage and a profusion of globe-shaped flowers which keep well when cut. Makes a choice border plant. Prop. by seed or division. Hungary.	2-3 ft. *Sun*	"
"Crimson" near 26	RED VALERIAN, JUPITER'S BEARD	*Centránthus rùber	Compact bushy plant covered with flowers in thick bold clusters terminating leafy stalks. Gray-green foliage. Good for cutting. Attractive border plant found frequently in old gardens. Prop. by seed and division. Mediterranean Region.	1-3 ft. *Sun*	June, July
"Scarlet" 18	DWARF RED OR NORTHERN SCARLET LARKSPUR	*Delphínium nudicaùle	See page 114.		Late May, June
"Red" 31 or 20 & 27 brilliant	CARTHUSIAN PINK	Diánthus Carthusianòrum *D. atrórubens*	Biennial or perennial. Flowers in varying shades of red borne in tight clusters surmount leafy stalks. Border or rock-garden. Prop. best by division. Rich light soil. Germany; Egypt.	12-18 in. *Sun*	Early June to early July

JUNE

Color	English Name	Botanical Name and *Synonyms*	Description	Height and *Situation*	Time of Bloom
"Deep red" 28 bright to 26	**DARK RED PINK**	****Diánthus cruéntus**	Small flowers in clusters on gray-green stalks. Long narrow pointed leaves. Good border plant. Prop. by seed. Warm not too moist loam. Greece.	1-1½ ft. *Sun*	June, July
"Red" 27 dull	**WOODLAND PINK**	***Diánthus sylvéstris** *D. virgineus*	Small flowers 1-3 on a stalk. Tufts of grass-like leaves. Pretty border plant. Prop. best by division. Rich light soil. S. Western Europe.	1 ft. *Sun*	June
"Red" 26	**COMMON BARREN-WORT, BISHOP'S HAT**	***Epimèdium alpìnum**	See page 114.		Mid. May to early June
"Crim-son" 41 slightly deeper	**BLOOD-RED CRANES-BILL**	****Gerànium sanguíneum**	See page 115.		Late May to mid. July
"Dark red" 28	**DOUBLE DEEP CRIMSON AVENS**	***Gèum atrosanguí-neum var. flòre-plèno**	Very similar to G. Chiloense, but with darker and double flowers. Moist light soil. Hort.	1 ft. *Sun*	June, July
"Orange scarlet" 26 orange	**CHILOE AVENS**	***Gèum Chiloénse** *G. coccíneum* (Hort.)	The best species. Wide-open five-petaled flowers, several on a stem, rise above the leaves, which mostly lie on the ground. Good rock-garden or border plant. Easily cultivated. Prop. by seed and division. Moist soil preferable. Chile. Var. *grandiflorum,* an improved var. A group of this plant makes a brilliant effect in the border or rock-garden. There is a double form which is very popular having a longer flowering season. June to Oct. Hort.	1-2 ft. *Sun*	Late June to early Aug.
"Pur-plish red" 33 or 26 dull	**APACHE PLUME, LONG-PLUMED PURPLE AVENS**	***Gèum triflòrum** *G. ciliàtum*	See page 115.		May, June
"Deep red" 33	**FRENCH HONEY-SUCKLE**	**Hedýsarum coronàrium**	Showy plant. Fragrant pea-like flowers in dense erect spikes. Foliage like that of Wistaria. Border plant. Easily cultivated. Prop. by seed and division. Light well-drained soil. S. Western Europe. Var. *album,* flowers white.	3-4 ft. *Sun*	June, early July
"Red" 26 brilliant	**CORAL OR CRIMSON BELLS**	****Heùchera sanguínea**	Plant of graceful and delicate effect with a very long blooming season. Resembles Mitella. Flowers borne in panicles on delicate stems above the clump of pale green leaves. Good for cutting. Gay and pretty for rock-garden or border. Prop. by seed and division. Any soil. S. Western U. S. A. Var. *splendens;* (color no. 26 deeper	1-1½ ft. *Sun or half shade*	June to late Sept.

Color	English Name	Botanical Name and *Synonyms*	Description	Height and *Situation*	Time of Bloom
			& more brilliant), rich crimson flowers. Hort. Other horticultural vars. range in color from white to dark crimson. *H. Americana,* Alum Root. Slender pyramidal clusters of small greenish or purplish red flowers rise from the tufts of wavy-toothed heart-shaped foliage for which it is chiefly grown. Suitable for the rock-garden or for edging the border. Eastern N. Amer.		
"Orange red" 13 deep & redder	**RED CANADA LILY**	***Lílium Canadénse var. rùbrum**	A picturesque native species. Flowers with dark spots droop in clusters; petals reflexed. Easy to cultivate. Plant in clumps in the border or among shrubs, or naturalize. Bulbous. Prop. by offsets, scales or very slowly by seed. Any well-drained soil. Eastern N. Amer.	1-4 ft. *Sun or half shade*	Early June, July
"Scarlet" 17 deeper	**JAPANESE RED STAR LILY**	***Lílium cóncolor**	Graceful species. Erect star-like flowers in terminal clusters. Good for cutting. Foliage poor. Mass among shrubs or in border. Easy to grow. Bulbous. Prop. by offsets or scales. Slaty soil. Avoid direct contact with manure. China.	1-1½ ft. *Sun or half shade*	June, July
"Orange red" 12 redder & deeper	**THUNBERGIAN LILY**	****Lílium élegans** *L. Dahùricum, L. Thunbergiànum, L. umbellàtum*	A very hardy and useful species. Large flowers borne erect on sturdy leafy stalks. Easy to grow. Brilliant massed in the border. Bulbous. Prop. by offsets or scales. Any light peaty soil. Avoid direct contact with manure. Japan. See Plate, page 215.	1-2 ft, *Sun or half shade*	"
"Scarlet" 16	**TURBAN LILY**	***Lílium pompònum** *L. rùbrum*	Turban-shaped fragrant flowers in regular many-flowered racemes surmount the leafy stem. Mass or scatter in border. Bulbous. Prop. by offsets, scales or very slowly by seed. Light well-drained soil. Avoid direct contact with manure. Alps.	2½-3 ft. *Half shade*	June
"Scarlet" 17 brilliant	**SIBERIAN CORAL LILY**	****Lílium tenuifòlium**	Waxy drooping flowers with recurved petals in clusters of twelve surmount slender stems. Pleasing massed in the border. Prop. by seed and "bud scales." Well-drained light soil. Avoid direct contact with manure. Siberia.	1-2 ft. *Sun*	Late June, July
"Scarlet" bet. 12 & 18	**MALTESE SAGE, JERUSALEM CROSS, SCARLET LIGHTNING**	****Lýchnis Chalcedónica**	Striking though somewhat stalky plant with flowers in brilliant crowded terminal clusters above rather hairy foliage. Excellent near white flowers in the border. Prop. by seed and division. Any rich garden soil. Var. *flore-pleno* a double form. Russia.	2-3 ft. *Sun or shade*	Early June to mid. July

THUNBERGIAN LILY. *Lilium elegans.*

HANSON'S LILY. *Lilium maculatum.*

Color	English Name	Botanical Name and *Synonyms*	Description	Height and *Situation*	Time of Bloom
"Deep crimson" 33 brighter	DEEP CRIMSON MULLEIN PINK OR DUSTY MILLER	*Lýchnis Coronària var. atrosanguínea	Deeply colored flat circular flowers surmount spikes of silvery-gray woolly foliage. Good for cutting. Striking in border. Prop. in spring by seed and division. Any ordinary soil. Europe.	1-2½ ft. *Sun*	June, July
"Red" 18	CHAMPION LAMP FLOWER	*Lýchnis coronàta *L. grandiflòra, L. fúlgens*	Often biennial. Panicles of fringed flowers 1½ in. wide. Protect in winter. Border. Prop. in spring by seed, division or cuttings. Any soil. Var. *speciosa* (*L. speciosa, L. fulgens var. speciosa*). Bushier var. with larger flowers than the type. Border. Japan; China.	1-1½ ft. *Sun or half shade*	June
"Purple red" often 31 brilliant	RED OR MORNING CAMPION	Lýchnis diòica *L. diúrna*	See page 115.		Mid. May to late June
"Red" 11, 17 & 18 brilliant	SHAGGY LYCHNIS	*Lýchnis Haageàna	A good species. Flowers varying between orange-red, scarlet and crimson. A cross between L. fulgens, and L. coronata. Large flowers, in clusters 2 in. across, borne in profusion. For borders. Prop. by seed, division and cuttings. Any light rich soil. There are also pink and white forms. Hort.	8-12 in. *Sun or half shade*	Early June to early Aug.
"Blood red" 27	BRILLIANT GERMAN CATCHFLY	**Lýchnis Viscària var. spléndens	Best var. Flowers in profusion in showy panicles. Foliage grass-like. Pretty for rock-garden or border. Prop. by seed and division. Any light rich soil. Hort.	6-20 in. *Sun*	June
"Scarlet" 18 duller	SCARLET MONKEY FLOWER	*Mímulus cardinàlis	Erect plant with rather snapdragon-like flowers on leafy branching stems. Pretty in border or rock-garden. Prop. by seed, division or cuttings. Moist soil and much watering. Western U. S. A.	2-3 ft. *Sun or half shade*	June to Sept.
"Red" 20 lighter, centre 21 redder	OSWEGO TEA, BEE OR FRAGRANT BALM	**Monárda dídyma	A rather coarse free-growing aromatic herb of unusually fine and brilliant coloring. Effective at a distance, but less attractive at close range. Foliage unimportant. Striking in groups naturalized along shady banks, or planted in roomy border. Apt to prove troublesome by spreading. Prop. by division. Separate often in spring. Ordinary soil. N. Amer. Var. *alba*, a white horticultural form.	1½-2½ ft. *Sun or shade*	Mid. June to early Sept.
"Crimson" 29 turning darker	WHORL-FLOWER	*Morìna longifòlia	Handsome plant with small flowers borne in dense whorls forming showy spikes. Buds white turning pink and finally crimson. Foliage thistle-like. Good in rock-garden or border. Protect in winter. Prop. in autumn by seed and division. Sandy loam. N. India.	2 ft. *Half shade*	June, July

Color	English Name	Botanical Name and *Synonyms*	Description	Height and *Situation*	Time of Bloom
"Scarlet" 24 deeper	SCARLET OURISIA	Ourísia coccínea	Creeping alpine plant with clustered panicles of 12 drooping pentstemon-like flowers with cream-colored anthers rising from a clump of pretty heart-shaped leaves. Protect in winter. Rock-garden. Prop. in spring by division. Sheltered spots in stiff heavy soil. Chile.	6-12 in. *Deep shade*	June, July
"Red"	HERBACEOUS PEONY	**Pæònia vars.	See color "various," page 270.		June
"Crimson" 33 redder etc.	COMMON GARDEN PEONY	**Pæònia officinàlis & vars. *P. fúlgida*	See page 116.		Mid. May to mid. June
"Deep red" 20 richer	FINE-LEAVED PEONY	*Pæònia tenuifòlia	See page 116.		"
"Blood red" 17 redder	BRACTEATE POPPY	*Papàver bracteàtum *P. orientàle var. b.*	See page 116.		Late May to mid. June
"Scarlet" bet. 12 & 19	ORIENTAL POPPY	**Papàver orientàle	Showy and vigorous when once established. Flowers 6-9 in. wide, typically scarlet with black spots; vars. in shades of orange and pink less hardy. Foliage remarkably decorative, but dies down in middle of summer. Excellent in border but must be cut down and concealed by other plants in July. Prop. by seed and division after flowering. Bears transportation badly. Rich garden soil. Asia Minor; Persia. See Plate, pages 219 and 220. **Var. *Parkmanni*, Parkman's Oriental Poppy, a large strong-growing var. with black blotches at the base of the petals. Difficult to obtain. Hort. **Var. *"Royal Scarlet,"* strong-growing var. not very different from type. Hort.	3-3½ ft. *Sun*	Early June to early July
"Light red" 16	RUPIFRAGE POPPY	*Papàver rupífragum	See page 117.		May to Aug.
"Crimson"	CLEVELAND'S PENTSTEMON	*Pentstèmon Clèvelandi	See page 117.		Late May to mid. June
"Purplish red" 41 deeper	HARTWEG'S LARGE-FLOWERED HYBRID PENTSTEMON	Pentstèmon gentianoìdes hỳbrida grandiflòra *P. Hártwegi hỳbrida g.*	Species from which many hybrids have sprung, of erect branching habit. Drooping tubular flowers in long loose clusters. Good for the border. Prop. by seed and division. Any good garden soil. Mexico.	3-4 ft. *Sun*	June to Sept.

ORIENTAL POPPY. *Papaver orientale.*

JUNE LANDSCAPE.

Color	English Name	Botanical Name and *Synonyms*	Description	Height and *Situation*	Time of Bloom
"Red" 35	**DARK BLOOD-RED SILVERY CINQUE-FOIL**	**Potentílla argyrophýlla var. atrosanguínea** *P. atrosan-guínea*	Small strawberry-like flowers sometimes purple borne on branching stems. Leaves silvery beneath. Rock-garden or border. Prop. by seed and division. Dry sandy soil. Himalaya.	2-3 ft. *Sun*	June, July
"Red"	**HYBRID CINQUE-FOIL DOUBLE VARS.**	***Potentílla hýbrida vars.**	Showy double hybrids not strictly hardy. Buttercup-like flowers. *Dr. André;* (color no. 19 & 5), scarlet flowers with yellow margins. *Le Vesuve;* (color no. 12 brighter and darker), light red edged with scarlet or yellow. Pretty strawberry-like foliage. Border. Protect in winter. Prop. by root division in spring. Light sandy soil. Hort.	2 ft. *Sun*	June to Sept.
"Scarlet" bet. 12 & 15	**RUSSELL'S HYBRID CINQUE-FOIL**	***Potentílla Russelliàna**	Rather tender single var., not so showy as the double forms. Butter-cup-like velvety blossoms, rich scarlet, nearly 2 in. across. Foliage strawberry-like. Pretty in the border. Protect slightly in winter. Prop. by division in spring. Light sandy soil. Hort.	1-2 ft. *Sun*	Early June to Aug.
"Pale red" near 22	**ATLANTIC HOUSELEEK**	***Semper-vìvum Atlánticum**	Flowers in clusters on stems clothed with reddish brown leaves rise from a rosette of pale green pulpy foliage. Pretty on walls or dry banks, or as carpeting in rock-garden. Prop. by division or offsets. Sandy soil. Atlas Mts.	10-12 in. *Sun*	June, July
"Red-dish" 38 dull	**COMMON RUNNERED HOUSELEEK**	***Semper-vìvum flagellifórme**	Starry flowers in compact heads, less important than the rosettes of pale green pulpy leaves. Rock-garden, edging and carpeting banks. Prop. by division or offsets. Sandy soil. Siberia.	3-4 in. *Sun*	Early June
"Mauve red" 38 dull	**MOUNTAIN HOUSELEEK**	***Semper-vìvum montànum**	Flowers in compact clusters less interesting than the neat regular rosettes of hairy dark green leaves. Pretty as carpeting in rock-garden. Sandy soil. Alps; Pyrenees.	6 in. *Sun*	Late June
"Red" 24, centre 33	**HOUSE-LEEK, OLD-MAN-AND-WOMAN**	***Sempervì-vum tectòrum**	Very common on roofs in Europe. Attractive starry flowers and rather cactus-like foliage. Pulpy green leaves tipped with reddish brown, in small compact rosettes. Rock-garden, wall or edging. Prop. by offsets. Dry soil. Mts. of Europe and Asia.	1 ft. *Sun*	June, July
"Crim-son" 20 brighter & lighter	**FIRE PINK**	***Silène Virgínica**	Profuse bloomer. Flowers 2 in. wide, in loose clusters, borne near the ground. Rock-garden. Prop. by division or preferably by seed. Sandy loam. U. S. A.	1-2 ft. *Sun*	"

Color	English Name	Botanical Name and *Synonyms*	Description	Height and *Situation*	Time of Bloom
"Deep red" 20, 4 inside	PINK ROOT, WORM GRASS	*Spigèlia Marylándica	Handsome herb wild in rich woods. Tubular flowers, rich red without and greenish yellow within, borne on tufts of slender stems. Border. Loose loam, leaf-mold and peat. Southern U. S. A.	1-2 ft. *Sun or shade*	Late June to early Aug.
"Bright red"	RED SPIDER-WORT	Tradescántia Virginiàna var. coccínea	See page 117.		Late May to late Aug.
"Brownish red" 28 neutral & lighter	ILL-SCENTED WAKE-ROBIN	*Tríllium eréctum *T. pendùlum, T. purpùreum T. fœtidum*	See page 36.		Late Apr. to early June
"Bright red" bet. 20 & 26	COMMON GARDEN OR LATE TULIP	**Tùlipa Gesneriàna	See page 118.		Mid. May to early June
"Deep rose"	PERSIAN CANDYTUFT	Æthionèma Pérsicum	Excellent dwarf plant; shrubby. Similar to Æ. coridifolium and Æ. grandiflorum. Good for edging and for rock-garden. Prop. by seed and cuttings. A sandy soil is necessary. Persia.	9 in. *Sun*	June
"Pink" 32 deeper	MOUNTAIN EVER-LASTING, CAT'S-EAR	Antennària dioìca *Gnaphàlium dioìcum*	Neat alpine plant of spreading habit. Flower heads in crowded clusters. Whitish woolly foliage. Useful for carpeting rockwork or edging. Prop. by seed, more frequently by division in spring. Any soil. *A. tomentosa* of the trade is a form of this with white flowers; much used for edging and carpeting, on account of its silvery foliage. Northern U. S. A.; Britain.	3-6 in. *Sun*	"
"Rose pink" 29 dull	ARETHUSA, INDIAN PINK	Arethùsa bulbòsa	See page 118.		May, June
"Pink" 27 lighter	CUSHION PINK, SEA-SIDE THRIFT, SEA TURF, CLIFF ROSE	**Armèria marítima *A. vulgàris*	See page 118.		Mid. May to mid. June
"Intense pink" 27 lighter	LAUCHE'S THRIFT OR SEA PINK	**Armèria marítima var. Laucheàna *A. L.*	See page 39.		Late Apr. to mid. June
"Pink" 36 deep & green	CHRISTMAS-ROSE-LEAVED MASTER-WORT	Astrántia helleborifòlia	Rather straggling plant with small flowers in tight umbels, having an unpleasant odor Leaves palmately divided. For border or bank. Prop. by division in fall or spring. Any good rather moist soil. Caucasus.	1-2 ft. *Sun*	June

Color	English Name	Botanical Name and *Synonyms*	Description	Height and *Situation*	Time of Bloom
"Pinkish"	BLACK HELLEBORE, GREAT BLACK MASTERWORT	Astrántia màjor	Small striped flowers in tight clusters. Leaves palmately divided. Border or rock-garden. Prop. in fall or spring by division. Any good soil preferably moist. Europe.	1-2 ft. *Sun*	June
"Pink" 29 & white	ENGLISH DAISY	*Béllis perénnis	See page 39.		Mid. Apr. to mid. June
"Purple pink" 40 duller	GRASS PINK ORCHIS	Calopògon pulchéllus *Limodòrum tuberòsum*	Charming native Orchid. Clusters of flowers, 1 in. across, in varying shades of pink and magenta-purple, rise above the grass-like foliage. Choice bog plant and sometimes grown in rock-gardens. Bulbous. Prop. by offsets which do not flower for some years. Moist porous soil. Eastern N. Amer.	12-18 in. *Shade*	June, July
"Pink" 22	CALYPSO	Calýpso boreàlis *C. bulbòsa*	See page 121		May, June
"Pink" 29 lighter	DOUBLE CUCKOO FLOWER OR LADY'S SMOCK	Cardámine praténsis var. flòre-plèno	Belongs to Wallflower order. Varies to white. Flowers in terminal clusters. Deeply cut foliage springing from the root and on the stalks. Border or rock-garden. Prop. by division. Moist soil. Europe.	9-20 in. *Sun or shade*	June
"Purplish pink" near 27	HOARY CEDRONELLA	*Cedronélla càna	Branching sub-shrub. Light whorls of flowers, with blue stamens, form showy spikes. Profuse bloomer. Fragrant silvery foliage. Not very hardy. Winter protection needed. Makes a pleasing effect in elevated position in rock-garden. Mexico.	1-3 ft. *Sun*	June to Sept.
"Deep pink" 31	WHITENED KNAPWEED	Centaurèa dealbàta	Compact bushy plant. Flowers with white or pink centres. Foliage silvery underneath and deep green above. Fairly good border plant. Prop. by seed and division. W. Asia.	8-24 in. *Sun or shade*	Late June to early Aug.
"Rose" 40 duller	ROSY MOUNTAIN BLUET	**Centaurèa montàna var. ròsea	See page 121.		Late May to early July
"Purplish rose" 36 deep	SMALL ROSE OR PINK TICKSEED	*Coreópsis ròsea	Branching species. Daisy-like flowers with yellow centres and rose-colored rays, about ⅜ in. across. Border. Any ordinary soil. S. Eastern U. S. A.	1-2 ft. *Sun*	June to Sept.
"Pink" 39	CROWN-VETCH	*Coronílla vària	Low straggling tufted plant. Flowers pea-shaped, in dense clusters, forming sheets of pretty pink and white bloom. Good for carpeting ledges, etc. Prop. by seed. Any ordinary garden soil, preferably dry. Europe.	1-2 ft. *Sun*	"

Color	English Name	Botanical Name and *Synonyms*	Description	Height and *Situation*	Time of Bloom
"Pink" 29	**CROSS-WORT, FŒTID CRUCIAN-ELLA**	***Crucianélla stylòsa** *Aspèrula ciliàta*	Pretty dwarf plant with a rather unpleasant odor. Flowers small but showy. Shaggy foliage in whorls. Does well on sandy banks; is also used in the border or rock-garden. Prop. by seed and division. Rather light loam. Persia.	6-9 in. *Half shade*	June to Sept.
"Pur-plish pink" 29	**STEMLESS LADY'S SLIPPER**	***Cypripèdium acaùle**	See page 121.		Early May, June
"Deep rose" 29 duller & pale	**ALPINE PINK**	***Diánthus alpìnus**	Alpine plant forming a carpet which is covered with flowers 1 in. across, spotted crimson with dark ring. Dark foliage in shining tufts. Choice rock-garden plant. Guard against drought and wire-worms. Prop. by seed and division. Warm light loam. Russia; S. Eastern Europe.	3-4 in. *Sun*	June, July
"Delicate pink" 27 light	**CHEDDAR PINK**	***Diánthus cæsius**	See page 121.		Late May to early July
"Deep red pink" 31	**MAIDEN PINK**	****Diánthus deltoìdes**	See page 121.		May, June
"Deep rose" 32	**GLACIER PINK**	***Diánthus glaciàlis** *D. neglectus*	Attractive species of low habit; erect and tufted with lovely toothed scentless flowers. Pointed grass-like leaves. Rather difficult to grow. Rock-garden in elevated position. Prop. best by division, also by seed and cuttings. Rich light well-drained soil. Mts. of S. Europe.	3-4 in. *Sun*	June
"Pink" 25, 27, 28, 33, 34 & 35	**BROAD-LEAVED PINK**	****Diánthus latifòlius** (Hort.)	Similar to Sweet William with large double flowers in tight clusters and broader leaves. Plant in the border or rock-garden. Prop. best by division. Rich light soil. Hort.	6-12 in. *Sun*	June to Sept.
"Pur-plish rose" 36	**PALE-FLOWERED PINK**	***Diánthus pallidiflòrus**	Densely tufted with a profusion of flowers. Makes a desirable border plant. Prop. by seed, division or cuttings. Rich light soil. Russia.	6 in. *Sun*	June
"Rose" 36 light	**ROCK PINK**	***Diánthus petræus**	A charming species. Leaves in dense tufts 2 in. high from which rise numerous flower-stalks bearing single flowers. "It escapes the attacks of wire-worms." Suitable for the rock-garden. Prop. by seed and division. Light sandy moist loam. E. Europe.	6 in. *Sun*	Mid. June to Aug.
"Pink" 36, or 37	**SCOTCH, COMMON GRASS, GARDEN OR PHEASANT'S EYE PINK**	****Diánthus plumàrius**	See page 122 and Plate, page 225.		Late May to late June

SCOTCH PINK VARS. *Dianthus plumarius vars.*

 ROBUST EREMURUS. *Eremurus robustus.*

Color	English Name	Botanical Name and *Synonyms*	Description	Height and *Situation*	Time of Bloom
"Pink"	**GARDEN PINK**	****Diánthus plumàrius vars.**	The type of which the following are varieties is described in May. Fringed flowers. *Essex Witch;* fragrant flowers dark pink. A free bloomer; good for cutting. Rock-garden or border. *New Mound;* fragrant double flowers, very pale pink. Useful for cutting. Thrives well in the border. Var. *roseus flore-pleno;* (color no. 29), very fragrant double flowers, clear rose-pink. Useful for cutting and pleasing in the border. Prop. by cuttings or layers. Rich light soil. Hort. See Plate, page 225.	8-12 in. *Sun*	June
"Purplish rose" 31	**SEGUIER'S PINK**	****Diánthus Seguierii**	Scentless flowers borne on leafless stalks. Splendid border plant, producing a striking mass of color. Prop. by seed and cuttings. Rich light soil. Europe; Asia.	1 ft. *Sun*	Late June, July
"Deep rose" 25	**WILD BLEEDING HEART**	****Dicéntra exímia** *Diélytra exímia*	Bears numerous heart-shaped blossoms in long drooping racemes. Foliage fern-like and graceful. Useful in rock-garden and mixed border. Prop. by division. Rich sandy soil. Western N. Y.; Mts. of Va.	1-2 ft. *Half shade*	Early June to Aug.
"Rose" 23	**CALIFORNIA BLEEDING HEART**	***Dicéntra formòsa**	See page 122.		Late May, June
"Rose" 30	**BLEEDING HEART**	****Dicéntra spectàbilis** *Diélytra s.*	See page 40.		Late Apr. to mid. July
"Purplish pink" 32	**RED GAS PLANT OR FRAXINELLA**	****Dictámnus álbus var. rùbra** *D. Fraxinélla var. rùbra*	Not so pretty as D. albus. Bushy plant. Fragrant flowers in racemes surmount the glossy foliage. Do not crowd or disturb often. Flowers last well when cut. Prop. by seed planted when ripe or by division. Moderately rich heavy soil. Hort.	2-3 ft. *Sun or half shade*	June, July
"Rose" 29	**JEFFREY'S SHOOTING STAR, AMERICAN COWSLIP**	***Dodecátheon Jeffreyi**	See page 122.		May, June
"Shell-pink" bet. 24 & 17	**HEDGE-HOG THISTLE, SIMPSON'S CACTUS**	**Echinocáctus Símpsoni**	Beautiful hardy little Cactus which grows in a globe-shaped mass, 3-5 in. in diameter, and is covered with variously colored spines, out of which spring large and beautiful flowers which are sometimes yellowish. Prop. by seed or cuttings. Dry soil in rock-garden. Col.	6-8 in. *Sun*	June to Sept.

Color	English Name	Botanical Name and *Synonyms*	Description	Height and *Situation*	Time of Bloom
"Magenta" 40	GREAT WILLOW HERB, FIRE WEED, FRENCH WILLOW	Epilòbium angustifòlium *Camænèrion angustifòlium*	Native, often seen where woods have been burned off. Showy flowers sometimes pink, in spikes. Leaves resemble those of Willow. Pretty for rough shrubbery and in masses beside water. Prop. by seed and division. Any soil. N. Amer.	3-5 ft. *Half shade*	June to early Aug.
"Peach-colored" 38	ROBUST EREMURUS	*Eremùrus robústus	A handsome species. Gigantic plant with dense spikes, sometimes 4 ft. long, of large starry flowers borne high above the clump of long leaves which disappear after flowering season. Protect in winter and guard against the spring frost. Group in a sheltered situation. Prop. by division. Deep rich sandy soil. Turkestan. See Plate, page 226.	6-10 ft. *Sun*	June, July
"Deep rose" 29	DIFFUSE CENTAURY	Erythræ̀a diffùsa *E. Másoni*	Biennial plant of Gentian family, having a profusion of small bright flowers and shining leaves. Protect in winter. Charming in the rock-garden. Prop. by seed, division or cuttings. Light sandy loam. Europe.	2-3 in. *Sun or half shade*	June
"Pink" 24	ARMENIAN CRANES-BILL	Geránium Armènum *G. Backhousiànum*	Best species. Flowers abundant, large and showy. Leaves springing from root, are deeply divided and have curved contours. Border or rock-garden. Prop. by seed and division. Any good soil. Armenia.	2½-4 ft. *Sun*	June, July
"Rose" 38	ENDRESS'S CRANES-BILL	Geránium Éndressi	See page 122.		Late May to late June
"Light pink" veined with 26	LANCASTER CRANES-BILL	*Geránium sanguíneum var. Lancastriénse	A dwarf var. with purple veined flowers; paler than type. Edge of shrubbery or border. Prop. by seed and division. Any ordinary soil. Europe.	1 ft. *Half shade*	June, July
"Pale pink" 36	CREEPING CHALK-PLANT	*Gypsóphila rèpens *G. prostràta*	Trailing or spreading habit. Myriads of tiny flowers, almost white, in graceful panicles on slender stems. Foliage light green. Excellent for rock-garden. Prop. by seed, division or cuttings. Fairly dry garden soil. Pyrenees; Alps.	6 in. *Sun*	Early June to mid. July
"Pink" 22 light	MANY-PAIRED FRENCH HONEY-SUCKLE	Hedýsarum multijùgum	Pea-like flowers, occasionally purplish, spotted with yellow, in long racemes. Foliage grayish green, composed of many leaflets. Rock-garden. Prop. by seed and division. Light well-drained soil. Armenia.	2-5 ft. *Sun*	June to early Aug.
"Rose"	CHANGE-ABLE ROCK ROSE	Heliánthemum vulgàre var. mutàbile	Shrubby evergreen plant. Pale flowers, yellow at the base, changing quickly to white. Glossy low mat of foliage. Rock-garden. Protect slightly in winter. Prop. by seed, division and cuttings. Sandy loam.	8-12 in. *Sun*	Early June, July

Color	English Name	Botanical Name and *Synonyms*	Description	Height and *Situation*	Time of Bloom
"Purplish rose" bet. 32 & 40	**DELAVAY'S INCARVILLEA**	***Incarvíllea Delaváyi**	Clusters of large trumpet-shaped flowers terminate the flower stalks which rise above a spreading clump of coarse fern-like foliage. A good border plant for sheltered positions. Prop. by seed and division. Rich loam mixed with sand. China.	1-2 ft. *Sun*	Early June to mid. July
"Pink" 29, spots 34	**REDDISH LILY**	***Lílium rubéllum**	See page 122.		Late May, early June
"Pale pink" 36 or 43	**TWIN FLOWER**	**Linnæa boreàlis**	Wild in mossy woods and cold bogs. Trailing evergreen bearing fragrant bell-shaped flowers in pairs on erect stalks rising above a dense matting of tiny leaves. Charming carpeting for rock or bog garden. Prop. by division, rarely by cuttings. Moist peaty soil. Europe; Asia; N. Amer.	4-6 in. *Half shade*	June, July
"Pink" 27 light, 29 pale	**RAGGED ROBIN, CUCKOO FLOWER**	***Lýchnis Flos-cùculi**	Deep or light pink flowers in lax clusters; abundant and continuous in bloom. Lance-shaped leaves. Much grown in old-fashioned gardens. Prop. by seed, division or cuttings. Any ordinary garden soil. Europe; N. Asia. Double vars., red or white.	1-2 ft. *Sun*	"
"Light pink" 29 pale	**DOUBLE CUCKOO FLOWER OR RAGGED ROBIN**	***Lýchnis Flos-cùculi var. pleníssima** *L. p. semper-flòrens*	See page 122.		Late May to late June
"Bright rose" 31	**JUPITER'S FLOWER, UMBELLED LYCHNIS**	***Lýchnis Flós-Jovis** *Agrostémma Flós-Jovis*	Rather small flowers in thick clusters, continuing long in bloom. Foliage soft and downy. Good for cutting and rock-garden. Prop. by seed. Any soil. Europe.	12-18 in. *Sun*	June, July
"Rosy red" 31	**GERMAN CATCHFLY**	***Lýchnis Viscària**	See page 125.		Mid. May to late June
"Deep rose" 29	**ALCEA MALLOW**	***Málva Alcèa**	Large flowers in clusters form masses of effective bloom. Good pale green foliage. Excellent for border, but requires space. Prop. by seed and division. Any soil. Europe.	2-4 ft. *Sun*	June to Sept.
"Pinkish" 45 in effect	**BRADBURY'S MONARDA**	**Monàrda Bradburiàna** *M. amplexi-caùlis*	A rather coarse aromatic herb. Odd flowers with dark spots. Effective in masses on banks or in the border. Prop. by seed and division. Separate often in autumn. Light dry soil. Indiana, South and West.	1½-2½ ft. *Sun*	June
"Purplish pink" 31	**MEXICAN PRIMROSE**	***Œnothèra ròsea**	Sometimes biennial. Shrubby bushy habit. Small flowers. Rock-garden or low border. Prop. by seed and cuttings. Any soil. Mexico.	1-2 ft. *Sun*	June, July

Color	English Name	Botanical Name and *Synonyms*	Description	Height and *Situation*	Time of Bloom
"Light pink"	**SAINFOIN, HOLY CLOVER**	**Onóbrychis satìva** *O. viciæfòlia, Hedýsarum onóbrychis*	Flowers are borne in loose spikes. Leaves small and many. Border plant. Any ordinary soil. Europe.	1-2 ft. *Sun*	June, July
"Pink" 32	**THORNY REST HARROW**	**Onònis spinòsus** *O. arvênsis var. spinòsus*	A dwarf sub-shrub. Showy pea-shaped flowers. Border or margin of shrubbery. Prop. by seed and division. Ordinary garden soil. Europe.	1-2 ft. *Sun*	"
"Purplish pink" bet. 37 & 38	**WILD OR POT MARJORAM**	**Oríganum vulgàre**	Branching plant. Flowers in clusters. Foliage fragrant, used in medicine and for seasoning. Prop. by seed and division in early fall or spring. Any garden soil. Europe.	2 ft. *Sun*	"
"Pink"	**HERBACEOUS PEONY**	****Pæònia vars.**	See color "various," page 268.		June
"Rose" 29	**LARGE ROSY TREE PEONY**	**Pæònia Moután var. ròsea supérba** *P. M. "Reine Elizabeth," "Triomphe de Grand"*	See page 125.		Mid. May to mid. June
"Delicate rose" bet. 26 & 27	**DOUBLE RED TREE PEONY**	**Pæònia Moután var. rùbra plèna**	See page 125.		"
"Pale pink" 22	**ORIENTAL POPPY BLUSH QUEEN**	****Papàver orientàle "Blush Queen"**	Showy and vigorous. Large flowers with deep purple blotch at base of petals. Foliage strikingly decorative, but dies down completely in Aug. Effective massed in border, but supplement with other plants to avoid a bare spot in late summer. Prop. by seed and division after flowering. Difficult to transplant. Rich garden soil. Hort.	2-3½ ft. *Sun*	Early June to early July
"Pink or carmine" 32 varying	**BEARDED PENTSTEMON**	***Pentstèmon barbàtus** *Chelòne barbàta*	Popular species. Graceful tufted plant bearing many erect flower stems from which droop the tubular bearded flowers varying from flesh color to carmine. Leaves smooth and grayish green. Very desirable for rock-garden or border. Prop. by seed and division. Any good garden soil, not too dry or hot. Mexico, N. to Col.	2-3 ft. *Sun*	Early June to mid. July
"Purplish rose" 32	**TUBEROUS JERUSALEM SAGE**	**Phlòmis tuberòsa**	Vigorous plant of coarse habit. Flowers in compact whorls along the branches. Heart-shaped leaves on the stems and in clumps at the base of the plant. Shrubbery or wild garden. Prop. by seed and division. Ordinary soil. Europe; Asia. See Plate, p. 205.	3-6 ft. *Sun*	June, July

JUNE

Color	English Name	Botanical Name and *Synonyms*	Description	Height and *Situation*	Time of Bloom
"Deep rose" 40 lighter	**CAROLINA PHLOX**	***Phlóx Carolìna** *P. ovàta*	Showy plant taller but similar to P. ovata. Terminal clusters of small flowers on erect stalks. Leaves narrow and tapering, springing mostly from the root. Pretty in border. For cultivation see P. ovata. Mts. Pa. to Ala.	1-2½ ft. *Sun*	June, July
"Rose" often 45 lighter	**SHRUBBY SMOOTH-LEAVED PHLOX**	***Phlóx glabérrima var. suffruticòsa** *P. suffruticòsa, P. nítida*	More rigid and with broader leaves than P. glaberrima. Flowers sometimes flesh-colored. Good in border. Prop. by seed, division or cuttings. Moist rich soil. U. S. A.	1-2½ ft. *Sun*	June to early Aug.
"Reddish pink" 31 brilliant	**MOUNTAIN PHLOX**	***Phlóx ovàta** *P. triflòra*	Erect habit, the flowers being borne in small dense terminal clusters above the foliage. Border or rock-garden. Prop. by seed, division and cuttings. Soil rich or moist. Wooded Mts. of S. Eastern U. S. A.	1-1½ ft. *Sun*	June, July
"Rose" 27 lighter & redder	**CRAWLING PHLOX**	***Phlóx réptans** *P. stolonífera*	See page 125.		May, June
"Purplish rose" 40 duller	**FLOWERING WINTER-GREEN, GAY WINGS, FRINGED MILKWORT OR POLY-GALA**	**Polýgala paucifòlia**	See page 125.		"
"Rose" 26	**HYBRID CINQUE-FOIL**	***Potentílla Hapwoodiàna**	Rather tender single var., not so showy as the double forms. Butter-cup-like velvety blossoms. Margins and base of petals deep rose; centre pale rose or pale straw-color. Foliage strawberry-like. Pretty in the border. Protect slightly in winter. Prop. by division in spring. Light sandy soil. Hort.	1-2 ft. *Sun*	Early June to Aug.
"Deep rose" 26	**NEPAL CINQUE-FOIL**	***Potentílla Nepalénsis** *P. formòsa, P. coccínea*	See page 125.		May, June
"Rose" 26	**CORTUSA-LEAVED PRIMROSE**	***Prímula cortusoìdes**	See page 126.		May to early June
"Flesh-color" 22	**MISTASSINI OR DWARF CANADIAN PRIMROSE**	***Prímula Mistassínica** *P. farinòsa var. M., P. pusilla*	See page 126.		May, June
"Rose" 38 more violet	**PRAIRIE SABBATIA**	**Sabbàtia campéstris**	Branching annual or biennial plant belonging to the Gentian family. Solitary flowers borne on leafy stalks. Prop. by spring or fall sown seed. Soil light and dry. N. Amer.	6-15 in. *Sun*	Late June, July

Color	English Name	Botanical Name and *Synonyms*	Description	Height and *Situation*	Time of Bloom
"Rose" about 25	PINK MEADOW SAGE	*Sálvia praténsis var. ròsea	Flowers an in. long, clustered in spikes. Foliage heart-shaped and wrinkled. Border or wild garden. Prop. by seed. Good garden soil. Europe.	2-3 ft. *Sun*	June, early July
"Pink" 38 paler & more violet	ROCK SOAPWORT	*Saponària ocymoìdes	See page 126.		Late May to Aug.
"Purplish pink" often 46	NODDING WOOD HYACINTH	*Scílla festàlis var. cérnua *S. nùtans var. c.*	See page 126.		May, early June
"Pink" 36	PINK WOOD HYACINTH	Scílla festàlis var. ròsea *S. nùtans var. r.*	See page 126.		May to early June
"Flesh-color" 22 dull	FLESH-COLORED SPANISH SQUILL	**Scílla Hispánica var. cárnea *S. campanulàta var. c. S. pátula var. c.*	See page 126.		Late May, June
"Rose" 36	ROSE-COLORED SPANISH SQUILL	**Scílla Hispánica var. ròsea *S. campanulàta, S. pátula*	See page 126.		"
"Rose" 29 centre to 26 edge	FRENCH GALAX	Schizocodon soldanelloìdes	Mountain-loving plant of tufted habit. Prettily fringed flowers, deep rose in the centre, fading almost to white at the edges. Leaves evergreen, slightly tinted with bronze. Rock-garden. Prop. by seed. Sandy peat. Mts. Japan.	5-6 in. *Shade*	June
"Pale pink" 38	AUVERGNE HOUSELEEK	*Sempervìvum Arvernénse	Grown for its foliage. Small starry flowers with red anthers and filaments, in clusters of 8 or 9. Pale green succulent leaves in rosettes. Pretty as carpeting in rock-garden. Prop. by seed or offsets. Sandy soil. France.	6-8 in. *Sun*	June July
"Pale rose" 36 & 11	PURPLE-TIPPED HOUSELEEK	*Sempervìvum calcàreum *S. Califórnicum*	A good species. Thick rosettes of grayish green leaves, tipped with red, more attractive than the clusters of flowers. Pretty as carpeting for rock-garden. Prop. by division or offsets. Sandy soil. Alps of Dauphiny.	10-12 in. *Sun*	June
"Pink" 29 lighter	PINK BEAUTY	Sidálcea malvæflòra var. Lísteri *S. Lísteri*	More or less erect plant. Satiny fringed flowers 2-3 in. across, in terminal clusters. Good for cutting and border. Prop. by seed and division. Any garden soil. Europe.	3 ft. *Sun*	June, July

Color	English Name	Botanical Name and *Synonyms*	Description	Height and *Situation*	Time of Bloom
"Rose pink" 38	**AUTUMN CATCHFLY**	**Silène Scháfta**	Cushion-like plant. Flowers in clusters on stems springing from rosettes of leaves. Rock-garden or edging. Prop. by seed and division. Sandy loam. Caucasus.	4-6 in. *Sun*	June to Sept.
"Deep pink" 24 briliant	**QUEEN OF THE PRAIRIE**	***Spiræa lobàta** *S. palmàta, Filipéndula lobàta, Ulmària rùbra*	Wild in meadows and prairies. Spreading clusters of feathery flowers on erect leafy stems. Beautiful foliage. Effective in masses by water-side. Prop. by seed and division. Moist rich soil. Eastern U. S. A.	2-4 ft. *Half shade*	Early June to early July
"Carmine" 26 deep	**PALMATE-LEAVED MEADOW SWEET**	****Spiræa palmàta** *Ulmària purpùrea Filipéndula purpùrea*	The best species of Spiræa. Broad clusters of brilliant flowers borne on erect stems. Tufted root-leaves palmately divided. Border or water-side. Prop. by seed and division. A fairly rich, moist soil. Japan.	2-4 ft. *Half shade*	Late June, July
"Pink" 22	**ELEGANT PALMATE-LEAVED SPIRÆA**	****Spiræa palmàta var. élegans** *Ulmària rùbra var. élegans*	Decorative plant. Flowers profuse, in feathery clusters. Good for cutting. Fine foliage. Border or water-side. Prop. by seed and division. Moist rich soil. Japan.	1-3 ft. *Half shade*	"
"Rose"	**HILL-LOVING SEA LAVENDER**	**Státice collìna** *S. Besseriàna*	Branching plant. Small flowers in spikes. Foliage grayish. Plant in isolated clumps or in rock-garden. Prop. by seed and division. Sandy well-drained soil. S. Eastern Europe; W. Asia.	1 ft. *Sun*	June, July
"Pink" 29 lighter	**SAXIFRAGE-LIKE TUNICA**	***Tùnica Saxífraga**	Charming little spreading herb. Quantities of tiny flowers varying from pale to dark pink. Minute dark green foliage. Attractive in rock-garden or edge of border. Prop. by seed and division. Any soil. Asia; S. Europe.	6-10 in. *Sun*	"
"Pink" 36	**COMMON VALERIAN, GARDEN HELIOTROPE, ST. GEORGE'S HERB**	**Valeriàna officinàlis**	Very hardy, spreads rapidly. Small aromatic flowers form dense clusters. Foliage fragrant and showy. Border or wild garden. Prop. by seed and division. Any garden soil. Europe; Asia.	2-5 ft. *Sun*	June
"Pink" near 45	**PINK LONG-LEAVED SPEEDWELL**	**Verónica longifòlia var. ròsea**	Hardy in southern New England. Abundant flowers in dense spikes. Good foliage. Border. Prop. by division or cuttings. Garden soil. Europe.	2-3 ft. *Sun*	June, July
"Pink" 29 light	**PINK SPIKE-FLOWERED SPEEDWELL**	***Verónica spicàta var. ròsea**	Conspicuous plant. Flowers in long dense spikes. Pretty in border. Prop. by division. Any garden soil. Hort.	1½-2 ft. *Sun*	"
"Deep purple" 49 darker	**WILD MONKS-HOOD**	****Aconìtum uncinàtum**	Graceful and of almost vine-like habit. Hanging helmet-shaped flowers in loose panicles. Stems slender. Foliage deeply cut. Good border plant. Prop. by seed and division. Any garden soil. Japan; Southern U. S. A.	3-5 ft. *Sun or half shade*	Mid. June to Sept.

Color	English Name	Botanical Name and *Synonyms*	Description	Height and *Situation*	Time of Bloom
"Purple" 47	GLAND BELL-FLOWER	*Adenóphora Lamárckii	Similar to the Campanula. Drooping funnel-form flowers in many-flowered racemes. Border plant. Should not be disturbed when established. Prop. by seed preferably, or by cuttings in spring. Light, rich loam. E. Europe.	1-2 ft. *Sun*	June to Sept.
"Rosy lilac" 43	MT. LEBANON CANDYTUFT	Æthionèma cordifòlium *Ibèris jucúnda*	Dainty flowers in tight clusters on leafy stems. Very pretty for rock-garden. Prop. in spring by seed, in summer by cuttings. A sandy soil is necessary. Lebanon; Taurus.	4-6 in. *Sun*	June
"Lilac" 26 purpler	LARGE-FLOWERED LEBANON CANDYTUFT	Æthionèma grandiflòrum	See page 127.		May to Aug.
"Purple" 47	BROAD-STEMMED ONIONWORT	Állium platycaùle *A. anceps*	See page 127.		May, June
"Purple" 56 or 49	SHARP-SEPALED COLUMBINE	*Aquilègia oxysépala	See page 127.		"
"Purple" 56 or 49	EUROPEAN COLUMBINE	*Aquilègia vulgàris *A. stellàta, A. atràta*	See page 127.		Mid. May to July
"Pinkish laven-der" 37	ALPINE THRIFT	*Armèria alpìna	See page 43.		Late Apr. to mid. June
"Laven-der" near 37	PLANTAIN-LIKE THRIFT	*Armèria plantagínea	See page 128.		Mid. May to late June
"Green-ish purple"	ASARA-BACCA, HAZEL-WORT	Ásarum Europæum	See page 128.		May, June
"Violet" 46	BLUE ALPINE ASTER	**Áster alpìnus	See page 128.		Late May to late June
"Dark purple"	OREGON BOLANDRA	Bolándra Oregàna	A compact neat plant with flowers in flat-topped clusters. Suitable for damp situation in the rock-garden. Recently introduced. Oregon.	15-20 in. *Sun*	June
"Purple" 47 dull	LARGE SELF-HEAL	Brunélla grandiflòra *B. Pyrenaìca*	Flowers in heads rise above the thick foliage. Pretty for rock-garden and for carpeting. Any soil not too dry. Europe.	8-12 in. *Half shade*	June, July
"Purple" 48	ALPINE CALAMINT	Calamíntha alpìna	Branching tufted plant with fragrant flowers, similar to those of Thyme, borne in whorls. Plant in the rock-garden. Prop. by seed, division in spring, or cuttings. Any garden soil. S. Europe.	6 in. *Sun*	June

CARPATHIAN HAIRBELL. *Campanula Carpatica.*

CLUSTERED BELLFLOWER. *Campanula glomerata.*

Color	English Name	Botanical Name and *Synonyms*	Description	Height and *Situation*	Time of Bloom
"Lilac purple" 23 dull	**TOM THUMB CALAMINT**	**Calamíntha glabélla**	Pretty and minute alpine plant of tufted habit, the flowers having an aromatic fragrance. Rock-garden. Prop. by division. Sandy loam. N. Amer.	3 in. *Sun*	June, July
"Purple" 55 & 44	**RUSSIAN BELL-FLOWER**	***Campánula Bononiénsis var. Ruthénica** *C. Ruthênica*	Sixty to a hundred small funnel-shaped flowers in long loose pyramidal spikes. Dark green leaves. A showy border plant. There is a pretty white var. Prop. by seed. Rich well-drained loam. Caucasus; Tauria.	2-3 ft. *Sun*	June
"Violet" 47 shading darker	**CARPA-THIAN HAIRBELL**	****Campánula Carpática**	Very pretty and free flowering species. Neat compact plant bears a continuous wealth of large erect cup-shaped flowers, 1½ in. across, above the pretty clean foliage. Useful for cutting. Invaluable for border or rock-garden. Easily cultivated. Prop. by seed, division or cuttings. Rich well-drained loam. Transylvania. See Plate, page 235. Var. *Hendersoni* (*C. Carpatica turbinata Hendersoni*). Very variable form; in height and shape of flowers resembles var. turbinata, but is more robust. Rich mauve flowers in pyramidal racemes. A handsome hybrid. Hort. Var. *pelviformis*. Widely spreading fragrant flowers, almost 2 in. across, borne freely on branching stems. Pretty foliage like the type. Rock-garden or border. Var. "*Riverslea*"; (color no. 46 deep), stronger var., height 6-8 in.; flowers bluer than the type and 3 in. across. Desirable in rock-garden or border. Hort. Var. *turbinata* (*C. turbinata*), Turban Bell-flower. Height 6-12 in. Good var. lower and more compact than the type Solitary deep purple flowers, 1½-2 in. across, on erect stems. Leaves also larger. There is a fine though rare pale var. Transylvania. Var. "*G. F. Wilson*"; (color no. 44 darker), 6-12 in., a dwarf var., hybrid of var. turbinata and C. pulla. "It has the large flowers of the former and handsome dark foliage of the latter." Hort.	9-18 in. *Sun*	Late June to late Aug.
"Pale blue violet" 44 deep	**GARGANO HAIRBELL**	***Campánula Gargánica**	See page 128.		May to Sept.
"Purple" 48	**DANES' BLOOD, CLUSTERED BELL-FLOWER**	****Campánula glomeràta**	One of the best Campanulas. Erect stems bear funnel-shaped flowers, in dense terminal heads. Charming border plant. There is also a white flowered form. Prop. by seed, division or	1-2 ft. *Sun*	June, July

Color	English Name	Botanical Name and *Synonyms*	Description	Height and *Situation*	Time of Bloom
			cuttings. Rich loam well-drained. Europe; Asia. See Plate, page 236. **Var. *Dahurica,* a common form and excellent plant. Flowers larger than the type in clusters 3 in. thick. Hort.		
"Deep purple" 48	**LARGE-BLOSSOMED BELL-FLOWER**	****Campánula latifòlia var. macrántha** *C. macrántha*	Better known than the type and larger in every way. Flowers nearly as large as Canterbury Bells. Somewhat rank in growth, but a fine border plant. Prop. by seed, division or cuttings. Rich loam well-drained. Central Europe.	3-5 ft. *Sun or shade*	Early June, July
"Violet" 47 shading darker	**PEACH-LEAVED BELL-FLOWER, PEACH BELLS**	****Campánula persicifòlia**	A very charming plant. Many large cup-shaped flowers, 2 in. across, ranging along tall stems, are borne above the tuft of pretty foliage. Flowers a second time. Good for cutting. A graceful plant to group in border or along edge of shrubbery. Prop. by seed and division. Easily cultivated in rich soil. Central and N. Europe. The white var. is especially pleasing. Var. *grandiflora,* larger flowers. Desirable.	1½-3 ft. *Sun*	Early June to mid. July
"Purple" 48	**WALL HAIRBELL**	**Campánula Portenschlagiàna** *C. muràlis*	See page 131.		May, June
"Purple" 39 & 47	**RAINER'S BELL-FLOWER**	***Campánula Ràinerii**	A sturdy dwarf compact plant. Each branch bears a large open erect flower. Rock-garden. Protect from slugs. Prop. by seed, division or cuttings. Rich gritty loam well-drained. Italian Alps.	2-3 in. *Sun*	Mid. June to late July
"Violet" 47 shading to 49	**ENGLISH HAIRBELL, BLUE BELLS OF SCOTLAND**	****Campánula rotundifòlia**	A lovely little plant with drooping bell-flowers on delicate wiry stems. Pretty for borders, though especially suited for crevices in the rock-garden, or naturalization on steep slopes when planted so that the pendent habit is shown. Prop. by seed, division or cuttings. Rich loam, well-drained. N. Temp. Region. Var. *Hostii;* (*C. Hostii*), larger flowers on stouter stems. There is a white form which is not so vigorous, though the flowers are as large. *Var. *soldanæflora;* (*C. soldanella*), the Double Fringed Hairbell. Distinct var., blooms in June. Flowers are half-double and petals much divided. Hort.	6-12 in. *Sun*	June to late Aug.
"Violet" 40	**VAN HOUTTE'S BELL-FLOWER**	****Campánula Van Hoùttei**	A fine species, resembling C. punctata. Solitary bell-shaped flowers 2 in. long nod from the branchlets. Excellent border plant. Prop. by seed, division or cuttings. Rich loam well-drained. Hort.	2 ft. *Sun*	Early June to mid. July

JUNE

Color	English Name	Botanical Name and *Synonyms*	Description	Height and *Situation*	Time of Bloom
"Violet" 47	BLUE SUCCORY OR CUPIDONE	*Catanánche cærùlea	An everlasting. Flower-heads 2 in. wide. Foliage at base of plant. Good for cutting. Prop. by seed and division. Any soil preferably light. Var. *bicolor;* purplish white, centre color no. 41. Var. *alba,* white flowers. S. Europe.	2 ft. *Sun*	June to Sept.
"Purple" 46 shading to 39	MOUNTAIN BLUET OR KNAPWEED	**Centaurèa montàna	Large flowers, resembling the Cornflower, which turn redder as they grow old. Cottony foliage. Excellent for cutting and for the border. Prop. by division. Any good garden soil. Europe.	12-20 in. *Sun*	"
"Purplish" effect 43	RAM'S HEAD LADY'S SLIPPER	*Cyprìpèdium arietìnum	See page 131.		Late May to Aug.
"Deep purple" 49 deep	KASHMIR LARKSPUR	*Delphínium Cashmeriànum	Slender dwarf species. Terminal clusters of flat flowers, 1 in. or more across, which vary somewhat in color. Deeply cut rounded foliage in a tuft at the base. Especially suited for the rock-garden. Prop. best by seed, also by division and cuttings. Well-drained rich soil. Himalayas.	15 in. *Sun*	June, July
"Rosy purple" 39	COMMON OR EASTERN SHOOTING STAR, AMERICAN COWSLIP	*Dodecátheon Mèadia	See page 131.		May, June
"Purplish" 50	BLUE OR BITTER FLEABANE	Erígeron ácris	See page 131.		Mid. May to early June
"Purple" 43	THRIFT-LEAVED FLEABANE	*Erígeron armeriæfòlius	See page 131.		Late May, June
"Bluish purple" 43 deep & bluer	ROBIN'S PLANTAIN, ROSE PETTY	Erígeron bellidifòlius *E. pulchéllus*	See page 131.		Mid. May to July
"Purplish" 32	ROUGH ERIGERON	*Erígeron glabéllus	See page 131.		Late May, June
"Light purple" bet. 43 & 44	VERY ROUGH ERIGERON	*Erígeron glabéllus var. ásper	Tufted plant bearing aster-like flowers. A rough shaggy var. Mass in wild spots or in the border. Prop. by seed and division. Any ordinary garden soil.	6-20 in. *Half shade*	Early June to mid. July
"Violet" 50 dull	BEACH ASTER	*Erígeron glaùcus	See page 131.		Late May, June

JUNE

Color	English Name	Botanical Name and *Synonyms*	Description	Height and *Situation*	Time of Bloom
"Purplish lilac" 47 lighter	**SHOWY FLEABANE**	****Erígeron speciòsus** *Stenáctis speciòsa*	The most desirable species. Vigorous plant of erect habit. Aster-like yellow-centred flowers. Good for cutting. Plant in masses in the rock-garden or border. Prop. by seed and division. Any ordinary soil. N. Pacific Coast. Var. *superbus* has paler flowers and blooms more freely. Hort.	1½-2 ft. *Sun*	June, July
"Violet purple" 39 more violet	**WALL ERINUS**	***Erìnus alpìnus**	See page 132.		May, June
"Light purple"	**GLANDULAR STORK'S OR HERON'S BILL**	**Eròdium macradènium** *E. glandulòsum*	Geranium-like alpine plant. Upper petals larger than the lower, spotted with deep purple. Continuous in bloom. Foliage aromatic. Plant in crevices in rock-garden. Pyrenees.	6-10 in. *Sun*	June to Sept.
"Purplish"	**MANESCAUT'S STORK'S OR HERON'S BILL**	**Eròdium Manescàvi**	Most vigorous and desirable of the family. Showy flowers, like the Geranium, 1½-2 in. across, 7-15 on the stem. Rock-garden or border. Prop. by seed and division. Dry soil not too rich. Pyrenees.	1-1½ ft. *Sun*	June, July
"Lavender" 45 bluer	**OVAL-LEAVED PLANTAIN OR DAY LILY**	***Fúnkia ovàta** *F. cærùlea, F. lanceolàta*	Distinguished for striking foliage. Flowers bell-shaped in limp racemes well above a conspicuous tuft of dark glossy heart-shaped leaves. Requires space. Border. Prop. by newly ripened seeds or more commonly by division. Rich moist soil. Japan; N. China; Siberia. Var. *marginata;* bizarre with white edged leaves, not very attractive. Hort.	1-2 ft. *Half shade*	June, early July
"Pale lilac" 43 lighter	**SIEBOLD'S PLANTAIN OR DAY LILY**	***Fúnkia Sieboldiàna** *F. cucullàta, F. gigantèa, F. glaùca, F. Sieboldii, F. Sinênsis*	The most decorative of the family, especially because of its luxuriant foliage. Resembles F. subcordata, but the flowers drooping and smaller do not rise above the large leaves which are a bluer green. Effective in the border. There is a var. with yellow-edged foliage. Prop. by newly ripened seed, or commonly by division. Cultivation easy, preferably in deep rich soil. Japan.	2-3 ft. *Half shade or shade*	June, July
"Lilac purple" 43	**GOAT'S RUE**	***Galèga officinàlis**	Bushy plant. Flowers pea-shaped in compact racemes, good for cutting. Foliage luxuriant and graceful. Plant groups in the border or wild garden. Prop. by seed and division. Any good soil. Europe; Asia.	2-3 ft. *Sun or half shade*	June to Sept.
"Dark magenta" 41	**BALKAN CRANES-BILL**	***Gerànium Balkànum**	Flowers at ends of long stems. Leaves fragrant. Border or rock-garden. Prop. by seed and division. Any good soil. S. Europe.	1 ft. *Sun*	June

JUNE

Color	English Name	Botanical Name and *Synonyms*	Description	Height and *Situation*	Time of Bloom
"Purple" bet. 39 & 47	**HILL-LOVING CRANES-BILL**	**Geranium collinum** *G. Londessi*	One of the showiest Geraniums. Flowers produced on drooping stems. Foliage insignificant. Second bloom if cut before seeding. Wild garden or border. Prop. by seed and division. Any good soil. E. Europe.	2-3 ft. *Sun*	June, July
"Laven-der" 39	**WILD GERANIUM, WILD OR SPOTTED CRANES-BILL**	**Geranium maculatum**	See page 132.		Early May to July
"Purple" 42	**WALLICH'S CRANES-BILL**	**Geranium Wallichianum**	Trailing plant with large flowers borne freely all summer amidst pretty light green foliage. Rock-garden. Prop. by seed and division. Any good soil. Himalayan Region.	12-15 in. *Sun*	June to Sept.
"Purple" 40 to 47	**ROCKET, SWEET ROCKET, DAME OR DAMASK VIOLET**	***Hésperis matronàlis**	Vigorous. Fragrant flowers in showy spikes. Hairy foliage. Divide often. Pretty for wild garden, shrub-bery or mixed border. Will grow in a cold climate. Prop. by division and cuttings. Rich soil is best. Europe; Siberia.	1-3 ft. *Sun or half shade*	June, July
"Pur-plish" 30 deeper to white	**DOWNY HEUCHERA**	**Heùchera pubéscens** *H. rubifòlia, H. pulveru-lénta*	See page 132.		Late May, June
"Bluish purple" 47	**PYRENEAN DEAD NETTLE**	**Hormìnum pyrenaìcum**	No especial character. Tubular flowers in spikes rise from broad tufts of foliage. Border and excellent for rock-garden. Prop. by seed and divi-sion. Well-drained soil. S. Europe.	6-12 in. *Sun*	June
"Palest purple" 37 pale	**LARGE HOUSTONIA**	**Houstònia purpùrea**	See page 132.		May, early June
"Purple" often 50	**APPEN-DAGED WATER-LEAF**	**Hydrophýl-lum appendiculà-tum**	See page 132.		Mid. May to early June
"Bluish purple" 49	**HYSSOP**	***Hýssopus officinàlis**	Well-known aromatic shrub of culi-nary and medicinal value. Flowers insignificant, sometimes white, in leafy spikes. Narrow leaves. Good as edging for beds or borders. Can be trimmed like Box. Prop. by seed, division or cuttings. Light rather dry loam; give plenty of water. S. Europe; Siberia.	1-2 ft. *Sun*	Mid. June to mid. Aug.
"Delicate lilac" 43 or 36	**GIBRALTAR CANDYTUFT**	***Ibèris Gibraltárica**	See page 135.		May, June
"Pale lilac" 44	**CRESTED DWARF IRIS**	****Ìris cristàta**	See page 135.		Late May, June

Color	English Name	Botanical Name and *Synonyms*	Description	Height and *Situation*	Time of Bloom
"Purple & lavender"	GERMAN IRIS, FLEUR-DE-LIS	**Ìris Germánica vars.	See page 135, and Plate, page 243.		Late May, June
"Violet" 44 deep	GREAT PURPLE OR TURKEY FLAG	**Ìris pállida *I. Junònia, I. Asiática, I. sícula*	See page 135.		"
"Lilac" 50	PLAITED FLAG	**Ìris plicàta *I. aphýlla var. plicàta*	See page 135.		"
"Blue violet" 49	SLENDER BLUE FLAG	*Ìris prismática *I. Virgínica, I. grácilis*	See page 135.		Mid. May to July
"Deep violet" 49	SIBERIAN FLAG	**Ìris Sibírica *I. acùta*	See page 136 and Plate, page 244.		Late May to mid. June
"Bright purple" 56 lighter	LARGER BLUE FLAG	*Ìris versícolor	See page 136.		Late May, June
"Violet purple" 49	ENGLISH IRIS	**Ìris xiphioìdes *I. Ánglica*	Beautiful bulbous Iris flowering at end of Iris season. Flowers resemble I. Kæmpferi, though smaller, the type being purplish violet. Foliage grass-like. Light well-drained soil enriched. It is well to put sand about the bulb. Pyrenees. There are white, lavender, deep blue, reddish purple, and striped vars. *Mont Blanc* is a pretty white form.	1-2 ft. *Sun*	Late June, July
"Purplish" 45	VARIEGATED NETTLE	Làmium maculàtum *L. purpùreum* (Hort.)	See page 136.		Mid. May to late July
"Purple" near 37 & 40	BLACK PEA OR BITTER VETCH	Láthyrus nìger *Órobus nìger*	Of the Pea order. Not a climber. Small flowers in clusters of 6 or 8. Compound leaves, pale green, or when dry, black. Good ground cover for border. Prop. by seed and division. Any soil. Central Europe.	1-2 ft. *Shade*	June, July
"Lilac" 47 light	KENILWORTH IVY, MOTHER-OF-THOUSANDS	Linària Cymbalària	Tender perennial ivy-like creeper, killed by frost, which perpetuates itself by seed. Flowers small, foliage pretty. Useful to trail over walls and in odd corners. Protect slightly in winter. Prop. by seed and division. Moist soil. There is a pretty white var. Europe.	4 in. *Half shade*	June to Sept.
"Purple" bet. 47 & 48	TALL BLUE-FLOWERED PERENNIAL LUPINE	**Lupìnus polyphýllus *L. grandiflòrus*	Pea-shaped flowers in long spikes rise above the handsome clump of satiny palmate leaves. Good old-fashioned border plant. Prop. by seed and division. Any garden soil. Cal.	2-5 ft. *Sun*	June, July

FLEUR-DE-LIS. *Iris Germanica.*

 SIBERIAN FLAG. *Iris Sibirica.*

Color	English Name	Botanical Name and *Synonyms*	Description	Height and *Situation*	Time of Bloom
"Purplish" near 29 duller	SLENDER-BRANCHED PURPLE LOOSESTRIFE	Lýthrum vírgatum	Flowers borne in threes, forming loose racemes on long leafy stems. Shrubbery or border. Prop. by division. Any good garden soil. Europe; Asia.	2-3 ft. *Sun*	June, July
"Violet" 44 pinker	SHARP-LEAVED BEARD-TONGUE	*Pentstèmon acuminàtus	Erect glaucous plant. Flowers lilac, changing to violet, in slender interrupted panicles on leafy stems. Border. Prop. by seed and division. Any good garden soil. S. Western U. S. A.	1-2 ft. *Sun*	June, early July
"Light purple" 47	DIFFUSE PENTSTEMON OR BEARD-TONGUE	**Pentstèmon diffùsus	Half shrubby plant. Tube-shaped flowers in showy open spikes. Winter protection of leaves advisable. Excellent border plant. Prop. by seed and division. Any soil. N. Western Amer	1-2 ft. *Sun*	Early June to early July
"Lilac purple" 45	SLENDER BEARD-TONGUE	*Pentstèmon grácilis	See page 136.		Late May to early July
"Rosy purple" 47	OVAL-LEAVED PENTSTEMON	Pentstèmon ovàtus *P. glaucus*	See page 136.		Late May to late June
"Pale violet" 47, edge white	DOWNY PENTSTEMON	*Pentstèmon pubéscens	See page 136.		Late May to mid. July
"Lavender" 44	ONE-SIDED PENTSTEMON OR BEARD-TONGUE	**Pentstèmon secundiflòrus	Numerous tubular flowers, opening broadly, in a long florescence. Long and narrow grayish foliage. One of the showiest Pentstemons. Effective in border or rock-garden. Protect in winter. Prop. by seed and division. Any soil. Col.	12-18 in. *Sun*	June, July
"Deep violet"	VIOLET PRAIRIE CLOVER	Petalostèmon violàceus *Kuhnístera purpùrea*	Broad bushy plant bearing many showy flower-spikes. Foliage finely divided. Plant in rock-garden or border. Prop. by seed and cuttings. Any light soil. Ind. to Tex.	1½-3 ft. *Sun*	June until frost
"Bluish purple" bet. 44 & 50	AMERICAN JACOB'S LADDER, CHARITY	**Polemònium cærùleum	See page 139.		Mid. May to July
"Purplish" 43 light	EAR-LEAVED PRIMROSE	*Prímula auriculàta *P. longifòlia*	See page 139.		May, June
"Deep blue-purple"	ROUND-HEADED HIMALAYAN PRIMROSE	*Prímula capitàta	See page 139.		May, early June
"Pale purple" 43 to white	TOOTH-LEAVED PRIMROSE	*Prímula denticulàta	See page 139.		"

Color	English Name	Botanical Name and *Synonyms*	Description	Height and *Situation*	Time of Bloom
"Dark purple" 44 deeper	KASHMIR TOOTH-LEAVED PRIMROSE	*Prímula denticulàta var. Cachemiriàna *P. Cachmeriàna*	See page 139.		May, early June
"Lilac purple" 43 pale	BIRD'S-EYE PRIMROSE	*Prímula farinòsa	See page 139.		"
"Purplish" near 30 deeper to white	JAPANESE PRIMROSE	**Prímula Japónica	See page 139 and Plate, page 247.		Late May to Aug.
"Violet purple" 52 more violet	ROSETTE MULLEIN	*Ramónda Pyrenaìca *Ramóndia Pyrenaìca*	See page 139.		Mid. May to July
"Purple" 44 bluer	ROSEMARY, OLD MAN	*Rosmarìnus officinàlis	Tender aromatic sub-shrub. Small flowers in short racemes. Leaves of culinary and medicinal value. Often seen in old-fashioned borders. Protect in winter. Prop. by seed, cuttings or layers. Dry light soil. Mediterranean Region.	2-4 ft. *Sun*	June, July
"Bluish violet" 44	TWO-COLORED SAGE	Sálvia bícolor	See page 140.		Late May, June
"Blue purple" 44 deep	VERVAIN SAGE	Sálvia Verbenàcea *S. spèlmina, S. Spièmanni*	Flowers occasionally whitish. Border. Prop. by seed, Rich soil. Great Britain; Asia.	2-3 ft. *Sun*	June, July
"Bluish purple" near 50	SMALL OR LILAC-FLOWERED SCABIOUS	Scabiòsa Columbària	Branching plant with flowers in globular heads, $1\frac{1}{2}$ in. wide. Good for cutting. Plant in picking garden. Prop. by seed and division. Any garden soil. Europe.	1-2 ft. *Sun*	June to Oct.
"Lilac" 43	LILAC WOOD HYACINTH	*Scílla festàlis var. lilacìna	See page 140.		May, early June
"Purple" 43 deep	WIDOW'S CROSS, BEAUTIFUL STONE-CROP	Sèdum pulchéllum	Wild on rocks. Pretty trailing species. Flowers small and closely packed on many branches. Tiny succulent leaves turning reddish or purplish. Carpeting for border or rock-garden. Prop. by seed or offsets. Any ordinary soil. U. S. A.	3-6 in. *Sun*	June, July
"Pinkish purple" 43	CUSHION PINK, MOSS CAMPION	*Silène acaùlis	Mossy dwarf herb with small flowers just peeping out from a dense mass of pale green foliage. Rock-garden. Prop. by seed and division. Sandy loam. Europe.	2 in. *Sun*	"
"Rich purple"	LARGE-FLOWERED BLUE-EYED GRASS	Sisyrínchium grandiflòrum *S. Doùglasii*	See page 140.		May, June

JAPANESE PRIMROSE. *Primula Japonica.*

 WOOD BETONY. *Stachys Betonica.*

Color	English Name	Botanical Name and *Synonyms*	Description	Heght and *Situation*	Time of Bloom
"Purple red" 41	WOOD BETONY	Stàchys Betónica *Betónica officinàlis*	Flowers in showy spikes. Prettily shaped dark green leaves. Border or wild garden. Prop. by seed and division. Moist soil preferable. Europe; Asia Minor. See Plate, page 248.	1-2½ ft. *Sun*	Late June, July
"Violet" often 45	LARGE-FLOWERED WOUND-WORT	*Stàchys grandiflòra *Betónica ròsea*	Large flowers in whorls, occasionally pinkish. Broad heart-shaped leaves. Good in border. Prop. by seed and division. Ordinary garden soil. Asia Minor.	1-1½ ft. *Sun*	Mid. June to late July
"Purple" near 37	WOOLLY WOUND-WORT	Stàchys lanàta	Effective plant. Flowers in thick showy interrupted spikes. Foliage silvery and woolly, in low tufts. Border or as edging. Prop. by seed and division. Ordinary soil. Tauria to Persia.	1-1½ ft. *Sun*	"
"Lilac" 43	MOTHER OF THYME, CREEPING THYME	Thỳmus Serpýllum	Creeping evergreen. Tiny flowers in whorls. Aromatic leaves, slightly downy, in dense masses. Good in rock-garden for covering arid ground, or as an edging. Prop. by division. Flourishes in poor soil. Europe; Asia; N. Africa. Among the most desirable forms are: var. *argenteus*, bright green foliage slightly streaked with silver. Var. *aureus*, golden leaves, especially beautiful in spring. Var. *citriodorous*, Lemon Thyme, 2-3 in. high, very small leaves, lemon-like fragrance. Var. *lanuginosus*, downy gray foliage. Effective for compact carpeting.	3-4 in. *Sun*	Mid. June to mid. Aug.
"Pale lilac" 37	COMMON GARDEN THYME	Thỳmus vulgàris	Aromatic herb of dense growth valued in cookery. Small flowers in terminal spikes. Bright pale green foliage. Attractive as carpeting for banks, rock-garden or border. Prop. by seed and division Any soil. Europe.	1-2 in. *Sun*	June, July
"Purplish" 44, 48 or 49	COMMON SPIDER-WORT	*Tradescántia Virginiàna *T. Virgínica*	See page 140.		Late May to late Aug.
"Purple" 46	LONG-LEAF-STALKED TRILLIUM	Tríllium petiolàtum	See page 140.		May, early June
"Purplish" 46	HERBACEOUS PERIWINKLE OR MYRTLE	Vínca herbàcea	See page 141.		Late May, June
"Violet" 47 or 49	HORNED VIOLET, BEDDING PANSY	**Vìola cornùta	See page 48.		Late Apr. until frost

Color	English Name	Botanical Name and *Synonyms*	Description	Height and *Situation*	Time of Bloom
"Lilac" usually 50	BIRD'S-FOOT VIOLET	*Vìola pedàta	See page 141.		May, June
"Ultra-marine blue" 54 pale & dull	CRISPED METALLIC BUGLE	*Ajùga metállica var. críspa	See page 141.		"
"Blue" 46	BUGLE	**Ajùga réptans	See page 141.		Early May to mid. June
"Sky blue" 51	AZURE ONION-WORT	Állium azùreum	Flowers, striped with a darker shade, in dense clusters. Large leaves springing from the root. Easily cultivated. Plant in masses in border or wild garden. Bulbous. Prop. by seed or offsets. Well-drained soil. Siberia.	1-2 ft. *Sun*	June, July
"Light blue" 58 lighter	NARROW-LEAVED AMSONIA	*Amsònia angustifòlia *A. ciliàta*	See page 142.		May, June
"Light blue" 58 pale	AMSONIA	**Amsònia Tabernæ-montàna *A. latifòlia, A. salicifòlia, Tabernæ-montàna Amsònia*	See page 142.		Late May, early June
"Dark blue" 62	BARRE-LIER'S ALKANET	*Anchùsa Barrelièri	See page 142.		May, June
"Blue" 54	CAPE ALKANET	*Anchùsa Capénsis	See page 142.		Late May to early July
"Blue" 54, buds 41	ITALIAN ALKANET	*Anchùsa Itálica	See page 142.		Late May to mid. July
"Blue" 46 & white	ALPINE COLUM-BINE	*Aquilègia alpìna	See page 142.		May, June
"Blue" near 47	LONG-SPURRED COLUM-BINE	**Aquilègia cærùlea *A. leptocèras, A. macrántha*	See page 145 and Plate, page 251.		Mid. May to July
"Deep blue" 63 & white	ALTAIAN COLUM-BINE	**Aquilègia glandulòsa	See page 145.		May, June
"Lilac blue" 44 blue & white	STUART'S COLUM-BINE	*Aquilègia Stùarti	See page 145.		Mid. May to July

LONG-SPURRED COLUMBINE. *Aquilegia cærulea.*

ORIENTAL LARKSPUR. *Delphinium formosum.*

JUNE

Color	English Name	Botanical Name and *Synonyms*	Description	Height and *Situation*	Time of Bloom
"Blue" 63	BLUE WILD OR FALSE INDIGO	**Baptísia austràlis *B. cærùlea, B. exaltàta*	See page 145.		Late May to mid. June
"Bluish" white & 57	LOOSE-FLOWERED BORAGE	*Boràgo Laxiflòra	Somewhat coarse plants, this being a pretty species with drooping flowers, sometimes purplish, in racemes. Easily cultivated and naturalized in rough places. Prop. by seed, division or cuttings in spring. Any garden soil. Corsica.	1½-2 ft. *Sun*	June
"Pale blue" 62 lighter	TUFTED HAIRBELL	*Campánula cæspitòsa	See page 145.		May to Aug.
"Purplish blue" 46	GREAT BELL-FLOWER	*Campánula latifòlia	Six to fifteen large erect hairy flowers in loose spikes. Border or margin of shrubbery. Prop. by seed and division. Rich loam well-drained. Var. *eriocarpa* is also a good plant. Central and N. Europe.	3-4 ft. *Sun or shade*	June to early Aug.
"Blue" 60	UNDIVIDED-LEAVED VIRGIN'S BOWER	*Clématis integrifòlia	Herbaceous species of erect bushy habit. Large solitary drooping flowers terminate leafy stalks. Border and rock-garden. Prop. by seed and cuttings. Rich deep soil. Asia; Europe.	2 ft. *Sun*	Mid. June to Aug.
"Blue" 61	MUSK LARKSPUR	*Delphínium Brunoniànum	Showy panicles of pretty flowers. Leaves three- or five-parted. Suitable for rock-garden or border. Prop. by seed, division or cuttings. Deep friable soil. Thibet.	6-18 in. *Sun or half shade*	June, July
"Blue" 60	CAUCASIAN LARKSPUR	*Delphínium Caucásicum	A dwarf species with 3-4 flowers on the stem. For cultivation, etc., see D. Cashmerianum. Caucasus.	3-4 in. *Sun*	"
"Blue" 61	BEE LARKSPUR	**Delphínium elàtum *D. alpìnum, D. pyramidàle*	Numerous flowers marked with violet, in long wands. Finely divided foliage. If plants are cut back they will flower again. Good for border. Prop. by seed, division or cuttings. Deep friable or sandy soil, well enriched. Europe.	2-6 ft. *Sun or half shade*	June to Sept.
"Blue" usually 54	ORIENTAL LARKSPUR	**Delphínium formòsum	One of the best species. Flowers with violet spurs, varying from deep to light blue, and closely set on tall spikes, rise from the handsome dark divided foliage. New flowers will come if the old are removed or whole plant cut down and top-dressing applied. Divide every 3 years. Charming in groups in border or in beds of shrubs, or naturalized. Prop. by seed, division, and cuttings, which will root without bottom heat. Blooms first year from seed sown the previous autumn or in hot-beds in March.	2-3 ft. *Sun or half shade*	June, July

Color	English Name	Botanical Name and *Synonyms*	Description	Height and *Situation*	Time of Bloom
			Deep friable or sandy loam well enriched. Armenia. See Plate, page 252. Var. *cœlestinum;* (bet. color no. 51 & 57), a hybrid of a lovely light blue; very desirable. Flowers in large open spikes.		
"Blue" 62 or 54	HYBRID LARKSPUR	**Delphìnium hỳbridum	Flowers in dense racemes. There are many vars. double and half double. Beautiful in border or front of shrubbery. See D. formosum. Hort. See Plate, page 255. Var. *Barlowi;* (color 62 more violet), a large-flowered semi-double var. with deep blue flowers having brownish centres. Hort.	3-4 ft. *Sun or half shade*	June, July
"Deep blue" effect 46	LARGE-FLOWERED DRAGON'S-HEAD	*Dracocé-phalum grandiflòrum *D. Altaiênse*	Quickly passing tubular flowers, which blossom well only in damp seasons, in dense spikes barely rising above the bright green foliage. Border plant. Prop. by seed and division. Moist sandy loam. N. Asia.	1 ft. *Half shade*	"
"Blue" 62 more violet	NODDING DRAGON'S-HEAD	*Dracocé-phalum nùtans	See page 146.		May, June
"Amethyst blue" 63 lighter	AMETHYST SEA HOLLY	**Erýngium amethýstinum	Odd plant of thistle-like appearance. Round heads of flowers on blue stalks. Deeply cut spiny foliage. Useful for sub-tropical effects and for border. Prop. by seed or division. Light soil is best. Dalmatia; Croatia.	1-3 ft. *Sun*	June to early Sept.
"Blue" 62	GENTIANELLA, STEMLESS GENTIAN	*Gentiàna acaùlis	See page 146.		May, June
"Dark blue" 60	ALPINE GENTIAN	Gentiàna alpìna	See page 146.		"
"Light blue" 60 deeper & brilliant	VERNAL GENTIAN	*Gentiàna vérna	See page 51.		Apr. to early June
"Violet blue" 46 light	IBERIAN CRANES-BILL	Geranium Ibèricum	Showy plant. Flowers in loose clusters. Leaves deeply divided and rather decorative. Shrubbery, border or rock-garden. Prop. by seed and division. Any good garden soil. Caucasus. Var. *platypetalum* (*G. platypetalum*). Flowers deeper and richer in color and larger than type, (bet. color nos. 47 & 48).	1-1½ ft. *Shade*	June to Sept.
"Dark blue"	DUSKY-FLOWERED CRANES-BILL, MOURNING WIDOW	Geranium phæum	A distinct species with very dark almost black flowers sparsely spotted with white. Border. Prop. by seed and division. Any good soil. Europe	2 ft. *Shade*	Early June to mid. July

HYBRID LARKSPUR. *Delphinium hybridum.*

SPANISH IRIS. *Iris Xiphium.*

Color	English Name	Botanical Name and *Synonyms*	Description	Height and *Situation*	Time of Bloom
"Blue" 46 bluer	MEADOW CRANES-BILL	*Gerànium praténse	Flowers large, in pairs. Foliage decorative. Border. Prop. by seed and division. Any good soil. Europe; Siberia. *Var. *flore-pleno;* (color no. 46 bluer), shorter than the type, bearing darker colored very double flowers in profusion in early June and July. Often blossoms again in autumn. Border. Europe.	1½-2½ ft. *Shade*	June to Sept.
"Light blue" 52 deep & dull	HAIR-FLOWERED GLOBE DAISY	*Globulària trichosántha	See page 146.		Late May to Aug.
"Blue" 53	GLOBE DAISY	*Globulària vulgàris	Dense globular flower-heads surrounded by a tuft of leaflets. Rock-garden or margin of border. Prop. by seed and division. Moist peaty well-drained soil. S. Europe.	6-12 in. *Half shade*	June to Sept.
"Pale blue" 57	BLUETS, INNOCENCE, QUAKER LADY	*Houstònia cærùlea	See page 149.		May, early June
"Light blue" 61	AMETHYST HYACINTH	*Hyacínthus amethýstinus	See page 149.		"
"Bright blue or lilac" 60 or 49	SWORD-LEAVED FLAG	Ìris ensàta *I. biglùmis, I. oxypétala, I. fràgrans*	Small flowers, throat yellow, inner petals upright and slender. Glaucous foliage. Prop. by division after flowering. Rich well-drained soil. Russia; Caucasus; Japan.	1-3 ft. *Sun*	June
"Pale blue" effect 44	MISSOURI OR WESTERN BLUE FLAG	*Ìris Missouriénsis *I. Tolmieàna*	See page 149.		Late May, June
"Blue" bet. 54 & 56	NEGLECTED FLAG	**Ìris neglécta	See page 149.		Late May to early June
"Blue"	EASTERN SIBERIAN IRIS	**Ìris Sibírica var. orientàlis *I. S. var. hæmatophýlla, I. S. var. sanguìnea, I. hæmatophýlla, I. sanguìnea*	A fine var. Flowers larger and of deeper color than the type; not lasting so long in bloom. New leaves bronze-tinted. Excellent compact border plant. Prop. by division. Rich dry or wet soil. Siberia.	1-2½ ft. *Sun*	June, early July
"Deep blue" bet. 54 & 56	SPURIOUS IRIS	*Ìris spùria	See page 149.		Late May to late June

Color	English Name	Botanical Name and *Synonyms*	Description	Height and *Situation*	Time of Bloom
"Violet blue" 46 & 2	SPANISH IRIS	**Ìris Xíphium *I. Hispánica, I. spectábilis*	A lovely Iris. Large delicate flowers poised on slender stems, in varying shades of violet, blue and purple, with long narrow petals yellow in the middle. They run into pure yellow, white and many pretty combinations of color in the vars. Leaves slender and inconspicuous, disappearing when bulbs are ripe. Plant in sheltered places and where other foliage gives support. Excellent for cutting and easily cultivated. Bulbous. Prop. by offsets. Divide often. Loose friable loam well enriched. Spain; N. Africa. See Plate, page 256.	1-2 ft. *Sun*	Mid. June to July
"Blue" 61	AUSTRIAN FLAX	*Lìnum Austrìacum *L. perénne var. Austrìacum*	Small flowers in constant bloom. Leaves narrow. Border. Prop. by seed and division. Light rich soil. Austria.	1-2 ft. *Sun*	June to Sept.
"Blue" 54 lighter	PERENNIAL FLAX	**Lìnum perénne	See page 149.		Mid. May to Aug.
"Blue" 62	GENTIAN-BLUE CROMWELL	Lithospér-mum prostràtum	See page 150.		May, June
"Blue" 62 greener	NOOTKA LUPINE	*Lupìnus Nootka-ténsis	See page 150.		Late May to early July
"Light blue" bet. 47 & 52	COMMON WILD LUPINE	*Lupìnus perénnis	Native plant. Pea-shaped flowers, varying to white, in long spikes. Foliage palmate. Good for wild garden. Prop. by seed and division. Any soil, preferably sandy. Canada; Atlantic States.	1-2 ft. *Sun*	June, July
"Blue" 51	SIBERIAN LUNGWORT	*Merténsia Sibírica	See page 150.		May, early June
"Deep sky blue" 58 brighter	EARLY FORGET-ME-NOT	*Myosòtis dissitiflòra	See page 52.		Late Apr. to July
"Bright blue" 58 brighter	TRUE FORGET-ME-NOT	*Myosòtis palústris	See page 150.		May, June
"Blue" 57	EVER-FLOWERING FORGET-ME-NOT	*Myosòtis palústris var. sempérflorens	See page 153.		May to Sept.
"Blue" 58 darker	WOOD FORGET-ME-NOT	*Myosòtis sylvática	See page 153 and Plate, page 259.		May, early June

WOOD FORGET-ME-NOT. *Myosotis sylvatica.*

 FAN-SHAPED COLUMBINE. *Aquilegia flabellata.*

JUNE

Color	English Name	Botanical Name and *Synonyms*	Description	Height and *Situation*	Time of Bloom
"Blue" 58 brilliant	ALPINE WOOD FORGET-ME-NOT	*Myosòtis sylvática var. alpéstris *M. alpêstris*	See page 153.		May, early June
"Blue" 53	LARGE-FLOWERED CATMINT	Népeta Macrántha	Best of the Catmints. Erect branching habit. Border. Prop. by spring sown seed or division. Light well-drained soil. Siberia.	3-4 ft. *Sun*	Late June to early Sept.
"Pur-plish blue" 46	LARGE SMOOTH BEARD-TONGUE	*Pentstèmon glàber *P. Górdoni, P. speciòsus*	Many large tubular flowers, opening widely at the mouth, on unbranched grayish stalks. Foliage broad and long running up the flower stems. Handsome for the border. Prop. by seed and division. Moist rather rich soil. West of the Missouri.	1-2 ft. *Sun*	June, early July
"Bright blue" 46	BLUE SMOOTH BEARD-TONGUE	*Pentstèmon glàber var. cyanánthus *P. cyánthus*	See page 153.		Late May, June
"Lilac blue" 46	LARGE-FLOWERED BEARD-TONGUE	Pentstèmon grandiflòrus	Spikes of flowers resembling the Foxglove, 2 in. long, on gray-green stalks. Foliage broad. Winter protection of leaves advisable. Border. Prop. by seed and division. Any soil. Wisconsin, S. and W.	2-3 ft. *Sun*	Early June to early July
"Pur-plish blue" 46 & 41	SHOWY PENTSTE-MON OR BEARD-TONGUE	**Pentstèmon spectábilis	Handsome showy species. Many flowers, 1 in. long, in long inflorescence. Foliage somewhat grayish. Very effective border plant. Prop. by seed and division. Any good soil. New Mexico.	2-2½ ft. *Sun*	Early June to mid. July
"Pur-plish blue" 46 dull	MICHELI'S HORNED RAMPION	*Phyteùma Michélii	Neat habit. Curious flowers in spherical clusters. Long narrow leaves. Suitable for the rock-garden. Prop. by seed or division in spring. Any good garden soil. Sardinia.	6-8 in. *Sun*	Late June, July
"Pale blue" 52	DWARF JACOB'S LADDER	**Polemò-nium hùmile *P. Richard-sonii*	Very good species. Low alpine plant. Clusters of bell-shaped flowers with golden anthers. Good in border or in groups in rock-garden. Prop. by seed and division. Rich loam. Arctic Regions.	6 in. *Half shade*	June, July
"Light blue" bet. 46 & 52	GREEK VALERIAN	*Polemònium réptans	See page 55.		Late Apr. to early June
"Azure blue"	HIMA-LAYAN VALERIAN	*Polemònium réptans var. Hima-layànum *P. grandi-flòrum, P. cærùleum var. g.*	See page 153.		Late May to early July

JUNE

Color	English Name	Botanical Name and *Synonyms*	Description	Height and *Situation*	Time of Bloom
"Deep violet blue" near 47	MEADOW SAGE	**Sálvia praténsis	The most popular hardy blue-flowered Salvia. Flowers clustered in large spikes. Foliage in clumps. Excellent in border. Prop. by seed. Any soil. Europe.	2-3 ft. *Sun*	June, early July
"Blue" 53 light	PIN-CUSHION FLOWER	**Scabiòsa Caucásica	The best free-blooming vigorous species. Rayed flowers on long stems in flat heads with pale centres. Insignificant grayish foliage. Excellent for cutting. Border or picking garden. Protect in winter. Prop. by seed and division. Fairly good garden soil. Caucasus Mts.	1½-2 ft. *Sun*	June, July
"Pale blue" often bet. 43 & 44	GRASS-LEAVED SCABIOUS	*Scabiòsa graminifòlia	Graceful habit. Flowers occasionally lavender. Silvery white foliage. Rock-garden. Prop. by seed and division. Fairly good garden soil. S. Europe.	1-1½ ft. *Sun*	June to Oct.
"Bluish" near 39 paler	WOODLAND SCABIOUS	*Scabiòsa sylvática	Thick bushy plant. Solitary flowers on long stems. Good for cutting. Leaves large. Border or picking garden. Prop. by seed or division. Ordinary garden soil. Europe.	1-2 ft. *Sun*	Early June to late Sept.
"Blue" 32, turns 61	PRICKLY COMFREY	Sýmphytum aspérrimum	See page 154.		Late May to mid. July
"Blue" bet. 60 & 61	ANGEL'S OR BIRD'S EYES, GERMANDER SPEEDWELL	**Verónica Chamædrys	See page 154.		Late May, June
"Blue" 46	HOARY SPEEDWELL	**Verónica incàna *V. cándida, V. neglécta*	Vigorous plant. Small blue flowers in numerous slender spikes. Downy grayish foliage. Excellent for border and rock-garden. Prop. by division. Garden soil. S. Western Europe; N. Asia.	1-2 ft. *Sun*	Mid. June to late July
"Light blue" 50	COMMON SPEEDWELL	Verónica officinàlis	See page 154.		May to Aug.
"Deep blue" near 47	SCALLOPED-LEAVED SPEEDWELL	*Verónica pectinàta	See page 154.		May, June
"Purplish blue" 54	ROCK SPEEDWELL	**Verónica rupéstris *V. fruticulòsa*	See page 154.		Mid. May to late June
"Bright blue" 61 duller	SPIKE-FLOWERED SPEEDWELL	**Verónica spicàta	One of the best species. Dense spikes of flowers generally clear blue, sometimes pale pink. Downy foliage. Excellent and invaluable for border. Prop. by seed and division. Any good garden soil. Europe; N. Asia.	2-2½ ft. *Sun*	Early June, July

JUNE

Color	English Name	Botanical Name and *Synonyms*	Description	Height and *Situation*	Time of Bloom
"Pale blue" 53 pale	BASTARD SPEEDWELL	*Verónica spùria *V. paniculàta, V. amethýstina*	See page 154.		Mid. May to June
"Intense blue" bet. 60 & 61	HUNGARIAN SPEED-WELL, SAW-LEAVED SPEEDWELL	**Verónica Teùcrium	See page 155.		Late May to early June
"Blue" near 47	BROAD-LEAVED HUNGARIAN SPEEDWELL	Verónica Teùcrium var. latifòlia	Tall var. with leaves larger than the type. A good border plant. Prop. by seed or division. Any garden soil. Europe.	2-3 ft. *Sun*	June, July
"Violet blue"	PEA-LIKE VETCH	Vícia oroboìdes *Òrobus lathyroìdes*	Striking pea-shaped flowers in dense clusters. Leaves very pointed. Border or wild garden. Prop. by seed. Any garden soil. Siberia.	1-3 ft. *Sun*	June
"Purplish blue" 53	ARROW-LEAVED VIOLET	Vìola sagittàta *V. dentàta*	See page 155.		May, early June
"Parti-colored" 6 shading from 19 to 14	GREAT-FLOWERED GAILLARDIA	**Gaillárdia aristàta *G. grandiflòra*	Daisy-like flowers in profusion, petals yellow tipped shading to reddish brown in centre; last well and are desirable for cutting. Foliage deficient. Group in the border. Prop. by seed, division or cuttings. Blossoms the first season and continues in bloom after frost. Light well-drained porous soil. Western U. S. A. Var. **grandiflora*, best of species; flowers larger than type.	1½-3 ft. *Sun*	June to Nov.
Often 46 deep with 1 & 35	BROWN-FLOWERED IRIS	*Ìris squàlens	See page 155.		Late May, June
Parti-colored	HERBACEOUS PEONY	**Pæònia vars.	See color "various," page 270.		June
Parti-colored	DOUBLE HYBRID CINQUE-FOIL	*Potentílla hỳbrida vars.	Somewhat trailing and rather tender double hybrids which bloom more or less all summer. Velvety buttercup-like flowers produced in profusion. Strawberry-like foliage. Effective in border and rock-garden. Protect in winter. Prop. by root division in spring. Light sandy soil. *Eldorado;* purple suffused and edged with yellow. *Victor Lemoine;* (mixed color nos. 17 brilliant & 6), red and yellow. *William Rollinson;* (color nos. 12 & 15), height 12-18 in., reddish brown and orange; foliage beautiful. Hort.	2 ft. *Sun*	June, July

Color	English Name	Botanical Name and *Synonyms*	Description	Height and *Situation*	Time of Bloom
Various often 21	LONG-LEAVED BEAR'S BREECH	*Acánthus longifòlius	One of the hardiest species. Flowers white to purplish brown, in loose spikes. Ornamental chiefly for its large beautifully shaped bright green leaves which are longer than in A. mollis. Looks well in the border or rock-garden. Protect in winter. Prop. by seed or division in spring or fall. Light well-drained loam. S. Europe.	3-4 ft. *Sun*	June, July
Various	FAN-SHAPED COLUMBINE	*Aquilègia flabellàta	Compact growing species with showy purple, bright lilac or white flowers. Spurs short and incurved. Handsome foliage. Prop. by seed. It prefers a light well-drained but moist loam. Japan. See Plate, page 260.	1-1½ ft. *Sun*	June
White to 48 duller	DOUBLE-FLOWERED EUROPEAN COLUMBINE	*Aquilègia vulgàris var. flòre-plèno	See page 156.		Mid. May to July
Usually 21 or 2	WILD GINGER, CANADA SNAKEROOT	Ásarum Canadénse	Creeping plant with curious flowers, chocolate brown, greenish purple, or yellow, borne close to the ground and hidden by cyclamen-like leaves. Good under trees and easily cultivated in rich moist soil. Woods, Eastern U. S. A.	3-9 in. *Sun or shade*	Late June to early Aug.
36, 39, 47, 48, 43, 45 etc.	CANTERBURY BELLS	**Campánula Mèdium	An old favorite. Biennial plant of erect bushy habit bears spreading racemes of large bell-shaped flowers, contracted at the mouth, in shades of blue, purple or white and single or double. Border. Protect in winter. Prop. by seed which can be sown indoors. The plants set out in May will then flower the first year. Rich well-drained loam. S. Europe. See Plate, page 265. Var. *calycanthema*, commonly called Cup and Saucer or Hose and Hose. This form is the most frequently cultivated. The calyx is colored like the flower giving a cup-and-saucer effect. Blooms in July. Hort.	1½-4 ft. *Sun*	Late June, July
48, or white spotted with 45	NOBLE BELL-FLOWER	**Campánula nóbilis	Nodding spotted flowers 3 in. long, reddish violet, cream-colored or white, are crowded near the ends of the branchlets. Broad foliage. Very handsome in border with good background. Prop. by seed and division. Rich well-drained loam. China.	2 ft. *Sun*	Mid. June to Aug.
36, 31, 27, 32	RED CHRYSANTHEMUM	**Chrysánthemum coccíneum *Pyrèthrum hỳbridum, P. ròseum*	Charming flowers like large Daisies, which vary from carmine through pink to white. Feathery foliage. A profuse bloomer. The old flowers should be cut off. Good for picking. Very attractive in clumps or masses in the border. Prop. by division.	1-2 ft. *Sun*	June, July

CANTERBURY BELLS. *Campanula Medium.*

SWEET WILLIAM. *Dianthus barbatus.*

Color	English Name	Botanical Name and *Synonyms*	Description	Height and *Situation*	Time of Bloom
			Any good well enriched garden soil. There are many double and single vars. Caucasus.		
62, 57	**CHINESE LARKSPUR**	****Delphínium grandiflòrum var. Chinénse**	A lovely slender-stemmed bushy species. Numerous long-spurred flowers deep or light blue or white. Lower petals and spur sometimes violet and upper petals yellow. Pretty, deeply divided leaves. Charming in masses. If the plant is cut back after flowering it will bloom again. Prop. by seed, division and cuttings. Deep friable or sandy soil, well enriched. China.	1-2 ft. *Sun*	June, July
33, 34, 35, etc.	**SWEET WILLIAM**	****Diánthus barbàtus**	Vigorous quickly spreading plant which blooms profusely. Flowers in flat-topped clusters, pink, white, red and parti-colored. Single and double vars. Good for cutting. Single dark vars. are prettiest. An old garden favorite. Prop. by seed. Any soil. Central and S. Europe. See Plate, page 266.	10-18 in. *Sun*	"
Usually bet. 32 & 39	**COMMON FOXGLOVE**	****Digitàlis purpùrea** *D. tomentòsa*	Usually a biennial plant but often self-perpetuating. Tubular flowers varying from deep purplish pink to white, often prettily spotted, droop from tall flower stalks. Foliage in clumps at the base. The white is by far the prettiest kind. Very effective in clumps among shrubs in the wild garden or border, or naturalized in masses. If cut down it will keep on flowering somewhat through the summer. Prop. by seed. Rich light soil preferable. Europe. See Plates, pages 166 and 176. **Var. *gloxiniæflora,* (*D. gloxinoides, D. gloxiniæflora*). A robust var. larger in every way than the type. Hort. The trade offers var. *alba* which is not spotted.	2-3 ft. *Sun or half shade*	June, early July
Sometimes 43 dull	**VIRGINIA WATER-LEAF**	**Hydrophýl-lum Virgínicum**	See page 156.		Mid. May to July
Various	**GERMAN FLAG, FLEUR-DE-LIS**	****Ìris Germánica vars.**	See page 156.		Late May, June
Various	**JAPANESE IRIS**	****Ìris lævigàta** *Ì. Kæmpferi*	One of our most beautiful and effective plants which forms vigorous clumps. Flowers large and flat, sometimes 10 in. across, ranging in color from white to deep blue and plum color, sometimes mottled or deeply veined. Narrow erect leaves. Beautiful in masses beside water or in the border. Water freely during flower-	2-3 ft. *Sun*	June, July

Color	English Name	Botanical Name and *Synonyms*	Description	Height and *Situation*	Time of Bloom
			ing season. Easily cultivated. Prop. by seed or division. Any good soil, enriched with well-rotted manure. Some of the best vars. are: *Kagarabi;* (white shaded with color no. 41), *Kyodai-san, Kigan-no-misao, Kaku yako-ro Shippo, Samidare, Tora-odori, Oyo-do, Kumo-no-isho, Geisho-in, Mana-dsuru,* white with golden centre. East Siberia; Japan. See Plate, page 269.		
Bet. 38 & 44 or white	**BITTER-ROOT, SPATULUM**	**Lewísia rediviva**	Peculiar plant allied to Portulaca. Flowers, varying from deep rose to white, 1-2 in. across, rise from a tuft of succulent leaves. Plant shrivels after flowering but is not dead. Prop. by seed and spring division. Moist well-drained soil. Mts. Western U. S. A.	2-4 in. *Sun*	June, July
1 spotted brown, 11 to 14	**WILD YELLOW OR CANADA LILY**	****Lílium Canadénse**	Well known native species found in moist meadows and bogs. Spotted flowers with reflexed petals, varying from yellow to red, droop in a circle and surmount the graceful stems around which the leaves grow in whorls. Easy to cultivate. Charming scattered among shrubs, massed in border or to naturalize. Bulbous. Prop. by offsets or scales. Light well-drained soil. Avoid direct contact with manure. Eastern N. Amer.	1-4 ft. *Sun or half shade*	"
Often 32 deeper	**TURK'S CAP OR MARTAGON LILY**	***Lílium Martagon** *L. Dalmáticum*	Vigorous and picturesque species with turban-shaped fragrant flowers in lax racemes, varying in color from white to purplish red. Foliage decorative. Mass against high background in border. Bulbous. Prop. by offsets, scales or very slowly by seed. Any light soil. Avoid direct contact with manure. Europe; Asia.	2½-5 ft. *Sun*	June
Often 33 brilliant	**MULLEIN PINK, DUSTY MILLER, ROSE CAMPION**	***Lýchnis Coronària** *Agrostêmma Coronària, Coronària tomentòsa*	Woolly plant with striking mullein-like foliage. Flowers 1½ in. wide, varying from white to rich crimson, form a striking contrast to the pale silvery leaves below. Effective in rock-garden or border. Prop. by seed. Any soil. Europe; Asia.	1-2½ ft. *Sun*	June, July
Various	**HERBACEOUS PEONY**	****Pæònia vars.**	Peonies are perhaps the most beautiful of perennials. The following is a carefully chosen list of some of the best hort. vars. in the market, double or semi-double excepting in the Japanese vars. The flowers are large and handsome and the foliage effective. Beautiful plants for the border and especially striking for distant effects,	3-4 ft. *Sun or half shade*	June

JAPANESE IRIS. *Iris lavagata.*

HERBACEOUS PEONY VARS. *Pæonia vars.*

Color	English Name	Botanical Name and *Synonyms*	Description	Height and *Situation*	Time of Bloom
			edging walks, beds of shrubs, etc. A half shady position is desirable. Usually prop. by division in early autumn. Gross feeders, they like a deep moist loam enriched with cow manure. See Plate, page 270. **White vars.**—***Duke of Wellington;* a strong grower with large fragrant sulphur white blossoms. *Duchesse de Nemours;* one of the best white Peonies, cup-shaped flowers, sulphur white with greenish reflections fading to pure white. *Festiva alba;* large cup-shaped flowers, glossy and cream white with a few carmine spots. A dwarf var. ***Festiva maxima;* an early and very free bloomer. Flowers enormous, snow-white, with a few purplish carmine spots in centre. One of the best and most vigorous of Peonies. ***Marie Lemoine;* large convex bloom, ivory white petals lightly edged with pink. One of the best late bloomers. *Queen Victoria;* large full blossom, cream white, centre petals tipped with red blotches. Excellent for cutting. **Yellow vars.**—***Solfaterre;* sulphur yellow; a good kind. **Red vars.**—***Edouard André;* vigorous plant with large round flowers, deep crimson shaded black with a metallic lustre. *General Jacqueminot;* fragrant, color of the Jacqueminot Rose. ***Richardson's Rubra superba;* brilliant deep crimson; the latest of all Peonies to bloom but fragrant and lasting long. **Pink vars.**—*Edward Crousse;* large rather bright pink flowers, slightly lilac tinted. *Alexander Dumas;* brilliant pink interspersed with white, salmon and creamy buff. *Béranger;* mauve pink, broad cup-shaped flowers. *Duchesse de Nemours;* very large and fragrant, bright clear violet pink, with lilac tinted centre. One of the earliest to flower, and especially good for cutting. *Humea alba;* scarce var., each blossom having several shades of pink. *Jennie Lind;* clear rose pink, fragrant and long lasting. *Richardson's Dorchester;* large delicate flesh pink flowers which are late blooming. *Souvenir d' Exposition Universelle;* clear cherry pink; large, fragrant and enduring. **Parti-colored vars.**—***Golden Harvest;* large fragrant tricolor flowers,		

Color	English Name	Botanical Name and *Synonyms*	Description	Height and *Situation*	Time of Bloom
			guard petals and centre blush pink, collar pinkish white and a few petals tipped and striped with light crimson; general effect creamy pink. ***Marie Lemoine;* early and fragrant, delicate China pink shading to ivory white at the tips. *Marguerite Gerard;* immense flowers, flesh color fading to creamy white with petals spotted with carmine purple. *Marie Stuart;* anemone-shaped flower, with a ring of bright clear pink petals and centre of sulphur white. **Single Japanese vars.**—These are very charming. ***Ophir;* dark carmine. ***Vesta;* purplish red. ***Crystal.* ***Diana;* blush pink with creamy white centre. ***Neptune;* shell pink. **Topaz;* deep rose, shading lighter at the margin.		
Various	**WHITE FLOWERED PEONY**	****Pæònia albiflòra vars.**	See page 159.		Late May to mid. June
Various	**TREE PEONY**	**Pæònia Moután** *P. arbòrea*	See page 159		Mid. May to mid. June
Often 33 redder	**COMMON GARDEN PEONY**	****Pæònia officinalis vars.**	See page 159.		"
Various	**LARGE JAPANESE PETASITES**	**Petasìtes Japónica var. gigantèa**	Large composite flowers, purple varying to white, in clusters. Round felt-like leaves with wrinkled margins, 3½-4 ft. broad. Stalk and buds edible. Good for subtropical effects. Sachaline Islands.	6 ft. *Sun*	June, July
Various	**EARLY OR SUMMER-FLOWER-ING PHLOX**	***Phlóx glabérrima var. suffruticòsa**	Some good early vars. are: *Arnold Turner;* white with red centre. *Beauty of Mindon;* white rose tinted. *Burns;* rosy purple. *Circle;* pinkish white with deeper centre. *Modesty;* lilac pink, large flowers. *Miss Lingard;* (centre color no. 43 deep), white with a faint pinkish eye. A good var.	1-2 ft. *Sun*	"
Various	**WILD SWEET WILLIAM**	***Phlox maculàta**	Resembles P. paniculata, but is not so much cultivated. Fragrant flowers, comparatively small, varying from purplish to white, in compact oval clusters. Leaves thick and smooth. Pretty in border. Prop. by seed, division and cuttings. Moist rich soil. Penn. S. and W.	1-2½ ft. *Sun*	"
Various	**DOWNY PHLOX**	***Phlóx pilòsa** *P. aristàta*	See page 160.		May, June

JUNE

Color	English Name	Botanical Name and *Synonyms*	Description	Height and *Situation*	Time of Bloom
Often bet. 46 & 47	SAGE	Sàlvia officinàlis	Popular sub-shrub. Flowers in whorls vary in size and in color from purple and blue to white. Leaves whitish and rather downy. S. Europe. Var. *tenuior*, flowers blue; used as kitchen herb. Mediterranean Region.	1-2 ft. *Sun*	June, early July
Often 39 or 38	EGYPTIAN OR GYPSIES' ROSE	*Scabiòsa arvénsis *S. vària*	Straggling plant. Flowers in convex heads, pink lilac or blue, about 1½ in. wide. Good for cutting. Foliage insignificant. Reserve garden. Prop. by seed and division. Fairly good garden soil. Europe.	1-2½ ft. *Half shade*	Early June to mid. Aug.
Various	COMMON BLUEBELL of England, WOOD HYACINTH	*Scílla festàlis *S. nùtans, S. nonscrípta, S. cérnua*	See page 160.		May, early June
Various	SPANISH SQUILL OR JACINTH, BELL-FLOWERED SQUILL	**Scílla Hispánica *S. campanulàta, S. pátula*	See page 160.		Late May, June
Often mixed or 39, 25, or 3 & green, etc.	DARWIN TULIP	**Tùlipa "Darwin"	See page 163.		Late May to early June
Various	COMMON GARDEN OR LATE TULIP	**Tùlipa Gesneriàna vars.	See page 164.		Mid. May to early June
25, 42 lighter, white	PURPLE MULLEIN	*Verbáscum phœníceum *V. ferrugineum*	Distinct and valuable species. Flowers, varying from purplish to pinkish, open best in wet weather. Leaves green, scarcely downy. Border. Prop. by seed. Any garden soil. S. Europe; Asia.	2-3 ft *Shade*	June, July
Various	PANSY, HEARTSEASE	**Vìola trícolor	See page 68.		Mid. Apr. to mid. Sept.

JULY

WHITE TO GREENISH

Color	English Name	Botanical Name and *Synonyms*	Description	Height and *Situation*	Time of Bloom
"White"	SNEEZE-WORT	Achillèa Ptármica	See page 167.		June to mid. Sept.
"White"	DOUBLE SNEEZE-WORT	**Achillèa Ptármica var. "The Pearl"	See page 167 and Plate, page 276.		June to Oct.
"White"	SIBERIAN MILFOIL OR YARROW	*Achillèa Sibírica *A. Mongòlica, A. ptarmicoìdes*	Large flowers in dense clusters on stiffly erect stems. Useful for cutting. Good border plant. Prop. generally by division, also by seed, and cuttings in spring. Dry soil. Siberia.	1½-2 ft. *Sun*	July to Oct.
"Nearly white"	WHITE MONKS-HOOD OR OFFICINAL ACONITE	**Aconìtum Napéllus var. álbum *A. Taúricum var. álbum, A. pyramidàle var. álbum*	Effective plant. Large and showy helmet-shaped flowers in racemes on erect stems. Handsome deeply divided leaves. Roots and flowers poisonous. Good for border and for rocky banks. Prop. by division. Rich soil preferable. Europe; Asia; N. Amer.	3-4 ft. *Sun or shade*	Mid. July to early Sept.
"White"	WHITE POTANNINI'S GLAND BELL-FLOWER	*Adenóphora Potannìni var. álba	Very similar to the Campanula. Shrubby plant with spikes of drooping bell-shaped flowers. Border plant. Should not be disturbed when established. Prop. by seed or cuttings in spring. Rich loam, well-drained. Turkestan.	2-3 ft. *Sun*	July, Aug.
"White" 39 under petals	NARCISSUS-FLOWERED ANEMONE	Anemòne narcissiflòra *A. umbellàta*	See page 71.		May to Aug.
"White"	CANADA ANEMONE	*Anemòne Pennsylvánica *A. Canadênsis, A. dichótoma*	See page 71.		Mid. May to early July
"Cream white" tinged with 39	SNOWDROP WIND-FLOWER	**Anemòne sylvéstris	See page 11.		Late Apr. to mid. July
"Greenish white"	TALL OR VIRGINIAN ANEMONE	Anemòne Virginiàna	See page 167.		June, July
"White"	ST. BERNARD'S LILY	*Anthéricum Liliàgo	Small lily-like flowers in loose racemes. Grass-like foliage springs from the root. Needs winter protection. Charming in the border. Bulbous. Prop. by seed, division of the root, and by stolons. S. Europe.	2-3 ft. *Sun*	Mid. July to early Aug.

A MIDSUMMER GARDEN

276 DOUBLE SNEEZE-WORT. *Achillea Ptarmica var. "The Pearl."*

Color	English Name	Botanical Name and *Synonyms*	Description	Height and *Situation*	Time of Bloom
"White"	ST. BRUNO'S LILY	*Anthéricum Liliástrum *Paradìsea L.*	See page 72.		Late May to early July
"White"	LARGE-FLOWERED SANDWORT	Arenària graminifòlia *A. procera*	Alpine plant. Flowers in loose downy panicles. Long grass-like leaves, rough-edged. Very desirable for rock-garden. Prop. by seed, division or cuttings. Any garden soil. Caucasus.	6-10 in. *Sun*	July
"White"	GALIUM-LIKE WOODRUFF	Aspérula Galioìdes	See page 168.		June, July
"White"	BRANCHING ASPHODEL	*Asphódelus álbus	See page 168.		"
"White"	BOUQUET STAR-FLOWER, YARROW-LEAVED STARWORT	*Áster ptarmicoìdes *Chrysópsis álba*	Dwarf early-blooming species, flowering in profusion. Good for cutting and for the border. Prop. by seed and division. U. S. A.	18 in. *Sun*	July, Aug.
"White"	SMALL WHITE ASTER	*Áster vimíneus	See page 168.		June, July
"Creamy white"	FALSE GOAT'S BEARD	**Astílbe decándra *A. biternàta*	See page 168.		Early June to early July
"White"	JAPANESE FALSE GOAT'S BEARD	**Astílbe Japónica *Hoteìa J., H. barbàta, Spiræa J.*	See page 168.		Mid. June to mid. July
"White"	LARGE WHITE WILD INDIGO	*Baptísia leucántha	See page 171.		Early June to mid. July
"Pinkish white"	PLUME POPPY	**Boccònia cordàta *B. Japónica*	Striking and picturesque plant. Small flowers in plume-like panicles, borne well above the handsome large-leaved dull-green foliage. Wild garden or shrubbery; for subtropical effects. Apt to prove troublesome in border. Prop. by seed or suckers. Rich soil is best. China; Japan.	3-8 ft. *Sun*	Early July to early Aug.
"White"	ALLIARA-LEAVED BELL-FLOWER	*Campánula alliariæfòlia *C. lamiifòlia, C. macrophýlla*	Vigorous plant. Bell-shaped flowers droop generally on one side of the stem. Large heart-shaped hairy leaves. Pretty in the border. Prop. by seed, division or cuttings. Rich loam well-drained. Caucasus.	1½-2 ft. *Sun*	July
"White"	WHITE CARPATHIAN HAIRBELL	**Campánula Carpática var. álba	See page 171.		Late June to late Aug.

JULY

Color	English Name	Botanical Name and *Synonyms*	Description	Height and *Situation*	Time of Bloom
"White"	WHITE TURBAN BELL-FLOWER	**Campánula Carpática var. turbinàta álba *C. turbinàta var. álba*	More dwarf and compact than the type with larger flowers, 1½-2 in. across, solitary on erect stems. Grayish leaves. Charming for rock-garden or border. Prop. by seed, division or cuttings. Rich loam, well-drained. Hort.	6-12 in. *Sun*	July, Aug.
"White"	MILK-WHITE BELL-FLOWER	*Campánula lactiflòra	Erect branching plant of stiff habit. Panicles of milky-white flowers tinged with blue. Good in border. Prop. by seed, division or cuttings. Rich loam, well-drained. Caucasus.	2½-6 ft. *Sun or shade*	"
"White"	WHITE PEACH-LEAVED BELL-FLOWER	**Campánula persicifòlia var. álba	See page 171.		Early June to early July
"White"	BACK-HOUSE'S PEACH-LEAVED BELL-FLOWER	**Campánula persicifòlia var. Báckhousei	See page 171.		June, July
"White"	MOER-HEIM'S PEACH-LEAVED BELL-FLOWER	**Campánula persicifòlia var. Moerheimi	See page 171.		"
"White"	SPOTTED BELL-FLOWER	*Campánula punctàta	See page 171 and Plate, page 279.		June to late July
"White"	WHITE CHIMNEY-PLANT OR STEEPLE BELLS	*Campánula pyramidàlis var. álba	Striking plant. Numerous open bell-shaped flowers close to the stem, in pyramidal racemes. Best treated as biennial. Effective in the border or in isolated groups. Protect in winter. Prop. by seed, division or cuttings. Rich loam, well-drained. Europe.	4-6 ft. *Sun*	July, Aug.
"White"	WHITE BLUE BELLS OF SCOTLAND	*Campánula rotundifòlia var. álba	See page 171.		June to Sept.
"White"	WHITE MOUNTAIN BLUET OR KNAPWEED	**Centaurèa montàna var. álba	See page 75.		Late May to early July
"White"	WHITE JUPITER'S BEARD	Centránthus rùber var. álbus	See page 172.		June, July
"Cream white"	TARTARIAN CEPHALA-RIA	*Cephalària Tatárica	Rather like C. alpina in habit. Scabious-like flowers. Good for the back of border. Prop. by seed and division. Any soil. Europe; Asia; Africa.	6 ft. *Sun*	Early July to Sept.

SPOTTED BELLFLOWER. *Campanula punctata.*

BLACK SNAKEROOT. *Cimicifuga racemosa.*

Color	English Name	Botanical Name and *Synonyms*	Description	Height and *Situation*	Time of Bloom
"White"	BIEBERSTEIN'S MOUSE-EAR CHICKWEED	Cerástium Bièbersteinii	See page 172.		June, July
"Cream white"	WHITE SMOOTH TURTLE HEAD	*Chelòne glàbra *C. oblìqua var. álba*	Pentstemon-like flowers, often rose-tinted, in spikes terminate leafy stalks. Good in border. Prop. by seed, division in the spring, or by cuttings. Rich garden soil. Hort.	1-2½ ft. *Half shade*	July, Aug.
"White"	DR. JAMES'S SNOW-FLOWER	Chionophila Jamesii	See page 172.		June, July
"White"	LARGE-FLOWERED WHITE-WEED	**Chrysánthemun máximum	See page 172.		"
"White"	SHASTA DAISY	*Chrysánthemum "Shasta Daisy"	See page 173.		June to Sept.
"White"	TURFING DAISY	*Chrysánthemum Tchihátchewii	See page 173.		June, July
"Creamy white"	BLACK SNAKEROOT	*Cimicífuga racemòsa *C. serpentària*	Long feathery spike-like racemes of flowers disagreeable in odor, are borne high above the handsome dark foliage. Effective against a dark background, in a moist shady corner. Prop. by seed and division. Any soil. See Plate, page 280. Var. *dissecta,* foliage more deeply divided. N. Amer. **Cimicifuga fœtida var. simplex,* (*C. simplex*) is also grown and is said to be "tall and handsome," having a "fine dense raceme." Kamtschatka.	3-8 ft. *Sun or shade*	July, early Aug.
"White"	WHITE HERBACEOUS VIRGIN'S BOWER	**Clématis récta *C. eréct*a	See page 173.		Early June to mid. July
"White"	RAMONDIA-LIKE CONANDRON	Conándron ramondioìdes	See page 173.		June, July
"White"	HEART-LEAVED COLEWORT	*Crámbe cordifòlia	See page 173.		"
"White"	MOSQUITO OR CRUEL PLANT	Cynánchum acuminatifòlium *Vincetóxicum acuminàtum, V. Japónicum*	Rather ungraceful plant, noted for its clusters of small flowers in which insects are trapped. Foliage rather large, grayish underneath. Wild garden. Prop. by seed. Any soil. Japan.	1-2 ft. *Sun*	July

Color	English Name	Botanical Name and *Synonyms*	Description	Height and *Situation*	Time of Bloom
"White"	**DAHLIA**	****Dáhlia vars.**	See color "various," page 359.		July to late Oct.
"White"	**WHITE CAROLINA LARKSPUR**	***Delphínium Caroliniànum var. álbum** *D. Caroliniànum var. álbidum*	Racemes of flowers, good for cutting. Second bloom possible if first flowers are cut off. Deeply divided foliage. Attractive in border. Prop. in spring or autumn by seed, division or cuttings. Transplant every 3 or 4 years. Deep rich soil, sandy and loamy. U. S. A.	2-3 ft. *Sun*	July
"White"	**WHITE LARGE-FLOWERED LARKSPUR**	****Delphínium grandiflòrum var. álbum**	Very pretty var. with dense racemes of flowers good for cutting. Pretty foliage. Attractive in border. Prop. in spring or autumn by seed, division or cuttings. Deep rich soil, sandy and loamy. Hort.	1-2 ft. *Sun*	July, Aug.
"White"	**SAND PINK**	***Diánthus arenàrius**	See page 174.		Early June to early July
"White"	**SPREADING PINK**	***Diánthus squarròsus**	See page 174.		June, July
"Cream white"	**DIAPENSIA**	**Diapénsia lappónica**	Shrubby evergreen. Solitary cup-like flowers on stalks densely sheathed below with narrow leaves. Rock-garden. Prop. by division. Moist peaty sandy soil. N. Europe; N. Asia; N. Amer.	1-2 in. *Sun*	July
"White"	**GAS PLANT, BURNING BUSH, FRAXINELLA, DITTANY**	****Dictámnus álbus** *D. Fraxinélla*	See page 174.		June, July
"White"	**WHITE FOXGLOVE**	****Digitàlis purpùrea var. álba** *D. tomentòsa var. a.*	See page 174.		June, early July
"Purplish white"	**FULLER'S TEASEL**	**Dípsacus Fullònum**	See page 177.		June, July
"White"	**CYCLAMEN POPPY**	**Eomècon chionántha**	Poppy-like flowers, 2-3 in. across, with bright yellow anthers, several flowers on the stem. Heart-shaped cyclamen-like leaves, bright yellowish green. Increases rapidly by running rootstocks. Border or rock-garden. Prop. by division. Moist soil. China.	1-2 ft. *Sun*	July, Aug.
"White"	**WHITE GREAT WILLOW HERB OR FIRE WEED**	***Epilòbium angustifòlium var. álbum** *E. spicàtum var. a.*	See page 177.		June to early Aug.

JULY

Color	English Name	Botanical Name and *Synonyms*	Description	Height and *Situation*	Time of Bloom
"Greenish white"	MOTTLED SWAMP-ORCHIS, FALSE LADY'S SLIPPER	Epipáctis Royleàna *E. gigantèa*	See page 177.		June, July
"White"	HIMALAYAN EREMURUS	*Eremúrus Himalaìcus	See page 177.		"
"White"	HORSEWEED, BUTTERWEED	Erígeron Canadénsis	See page 177.		June to Sept.
"White"	RATTLESNAKE-MASTER, BUTTON SNAKEROOT	Erýngium aquáticum *E. yuccæfòlium*	See page 177.		June to Oct.
"White"	WHITE SEA HOLLY	*Erýngium plànum var. álbum	Thistle-like plant. Large spiny leaves. Striking for subtropical effect in the border on account of its ornamental foliage. Prop. by seed or division. Soil, light and sandy. Europe.	1-2 ft. *Sun*	July to mid. Aug.
"White"	TALL THOROUGH-WORT OR BONESET	Eupatòrium altíssimum	Woolly branching plant. Flowers in dense flat heads, bloom profusely. Good for cutting. Effective in border. Prop. by seed and division. Any soil. S. Eastern U. S. A.	3-5 ft. *Sun*	July, Aug.
"White"	THOROUGH-WORT, BONESET, INDIAN SAGE	Eupatòrium perfoliàtum	Vigorous downy plant of disagreeable odor. Flowers in dense flat heads. Leaves wrinkled, of medicinal value. Border. Prop. by seed and division. Any soil preferably moist. N. Amer.	3-5 ft. *Sun*	"
"White"	FLOWERING SPURGE	Euphórbia corollàta	What appear to be small flowers, though really tiny leaves, are borne in profusion on branching stems. Much used for cutting. Border or reserve garden. Prop. by division or cuttings. Any light soil. Eastern U. S. A.	1½-3 ft. *Sun*	"
"Pure white"	SUBCORDATE DAY LILY	*Fúnkia subcordàta *F. álba, F. cordàta, F. Japónica, F. liliiflòra, F. macrántha*	Best Funkia. Fragrant lily-like flowers in one-sided racemes, rise above the big clump of large handsome foliage. Effective in border or along edge of shrubbery. Prop. by division. Rich soil. Japan.	1½ ft. *Shade*	"
"White"	GALAX, COLTSFOOT, WAND PLANT	Gàlax aphýlla	An evergreen grown chiefly for its foliage. Flowers in slender wands rise on leafless stems above a neat spreading clump of bright green leaves which turn to reddish bronze in winter. Foliage excellent for cutting. Desirable as carpeting in rock-garden, for bog-garden or for border of Rhododendrons. Prop. by division. Cool moist peaty loam. S. Eastern U. S. A.	6-9 in. *Shade*	July

Color	English Name	Botanical Name and *Synonyms*	Description	Height and *Situation*	Time of Bloom
"White"	**WHITE GOAT'S RUE**	***Galèga officinàlis var. álba** *G. Pérsica*	See page 178.		June, July
"White"	**NORTHERN BEDSTRAW**	**Gàlium boreàle** *G. septentrionàle*	See page 178.		Early June to mid. July
"White"	**WHITE OR GREAT HEDGE BEDSTRAW, WILD MADDER**	**Gàlium Mollùgo**	See page 178.		Early June to late Aug.
"White"	**SYLVAN BEDSTRAW**	***Gàlium Sylváticum**	Myriads of tiny flowers in panicles on prostrate stems. Good for cutting, and with the mist-like effect of Gypsophila paniculata mingles well with other flowers. Good in border. Prop. by seed and division. Ordinary soil. S. Europe.	1½-3 ft. *Sun*	July, Aug
"White"	**CAPE OR GIANT SUMMER HYACINTH**	****Galtònia cándicans** *Hyacínthus cándicans*	Striking spikes of fragrant bell-shaped flowers, like large Snowdrops, rise above hyacinth-like foliage. Excellent and effective in border with good background. Protect in very cold climates. Bulbous. Prop. by offsets or slowly by seed. Light rich soil. S. Africa.	3-5 ft. *Sun or half shade*	"
"White"	**WINTERGREEN, CHECKERBERRY, BOXBERRY**	**Gaulthèria procúmbens**	Familiar creeping evergreen. Waxy flowers droop beneath glossy dark green foliage. Scarlet berries edible and fragrant. Beautiful for dense carpeting in rock-garden. Prop. by seed, division, cuttings or layers. Soil rather moist, peaty or sandy. Eastern U. S. A.	4-6 in. *Half shade*	"
"White"	**WHITE BLOOD CRANES-BILL**	***Geránium sanguíneum var. álbum**	See page 77.		Late May to mid. July
"White"	**AMERICAN IPECAC**	***Gillènia stipulàcea** *Porteránthus stipulàcus*	See page 178.		June, early July
"White"	**BOWMAN'S ROOT, INDIAN PHYSIC**	***Gillènia trifoliàta** *Porteránthus trifoliàtus*	See page 178.		"
"White"	**SWORD LILY**	****Gladìolus vars.**	See color "various," page 365.		July to Oct.

JULY

Color	English Name	Botanical Name and *Synonyms*	Description	Height and *Situation*	Time of Bloom
"Greenish white"	RATTLE-SNAKE PLANTAIN	**Goodyèra pubéscens**	Dwarf Orchid. Flowers in dense spikes rise from a tuft of foliage. Beautiful leaves, velvety green, delicately veined with silver. Rock-garden or for naturalization. Prop. by cutting, including a piece of root. Moist soil, peat and leaf-mold. N. Amer.	3 in. *Shade*	July, Aug.
"Pinkish white"	POINTED-LEAVED CHALK-PLANT	***Gypsóphila acutifòlia**	Resembles G. paniculata. Tiny flowers in spreading feathery panicles. Good for cutting. Scanty foliage. Border or rock-garden. Prop. by seed, division or cuttings. Fairly dry limy soil. Caucasus.	3-4 ft. *Sun*	Mid. July to late Aug.
"Pinkish white"	CERAS-TIUM-LIKE CHALK-PLANT	***Gypsóphila cerastioìdes**	Downy creeping plant. Large flowers, red-veined, in clusters. Leaves softly hairy on both surfaces. Desirable in rock-garden. Prop. by seed, division or cuttings. Fairly dry limestone soil. Himalayas.	3-6 in. *Sun*	July
"White"	ELEGANT CHALK-PLANT	***Gypsóphila élegans**	See page 178.		June, July
"Pinkish white"	BABY'S BREATH	****Gypsóphila paniculàta**	Dense spreading bush. Numerous tiny flowers in light feathery panicles; delicate foliage. Good for cutting. Excellent in border or large rock-garden. Resists drought. Prop. by seed, division or cuttings. Fairly dry garden soil. Europe. Var. *flore-pleno,* flowers double.	2-3 ft. *Sun*	July, Aug.
"White"	STEVEN'S CHALK-PLANT	***Gypsóphila Stèveni** *G. glaùca*	See page 178.		June, early July
"White"	UMBEL-FLOWERED SUN ROSE	**Heliánthemum umbellàtum**	See page 178.		June, July
"Greenish"	HERNIARY, RUPTURE-WORT	**Herniària glàbra**	Rapidly spreading plant of dense growth, forming a turfy mat with clusters of small flowers. Foliage evergreen, reddish in winter. One of the best plants for carpeting in poor soil. Prop. by seed and division. Any soil. Europe; Asia.	2 in. *Sun*	July, Aug.
"White"	DOUBLE WHITE SWEET ROCKET	***Hésperis matronàlis var. álba plèna**	See page 179.		June, July
"White"	WHITE CORAL BELLS	**Heùchera sanguínea var. álba** *H. álba*	See page 179.		June to late Sept.
"Whitish" faintly	HAIRY ALUM ROOT	**Heúchera villòsa** *H. caulèscens*	See page 179.		Late June to Sept.

32

Color	English Name	Botanical Name and *Synonyms*	Description	Height and *Situation*	Time of Bloom
"Cream white"	**BROAD-LEAVED WATER-LEAF**	**Hydrophýllum Canadénse**	See page 179.		June, July
"White"	**WHITE VARIEGATED NETTLE**	**Làmium maculàtum var. álbum** *L. álbum*	See page 81.		Mid. May to late July
"White"	**GOLD-BANDED OR JAPAN LILY**	****Lílium auràtum**	Extremely showy. Flowers nearly a foot wide, thickly dotted with purple and marked with central yellow bands. Effective massed in border or scattered in Rhododendron bed. Plant deep in well-spaded peaty well-drained soil, and avoid direct contact with manure. Short-lived bulb. Prop. by offsets, scales, or very slowly by seed. Japan. See Plate, page 287.	2-4 ft. *Sun or half shade*	Mid. July to mid. Aug.
"White"	**MADONNA LILY**	****Lílium cándidum**	See page 180.		June, July
"White"	**GIANT LILY**	***Lílium gigantèum**	A large imposing plant with fragrant funnel-shaped flowers on gigantic stems well-clothed with large heart-shaped leaves. When conditions are right, not hard to grow. Prop. by scales or offsets. Light well-drained peaty soil. Avoid direct contact with manure. Himalayan Region.	4-10 ft. *Sun or half shade*	July, early Aug.
"White"	**COMMON TRUMPET LILY**	***Lílium longiflòrum**	A profusion of fragrant waxy trumpet-shaped flowers, slightly drooping. Pretty in border or along edge of shrubbery. Bulbous. Prop. by offsets, scales, or very slowly by seed. Any light soil. Avoid direct contact with manure. Japan; China. Var. *eximium* (*L. Harrisii, L. eximium*), Bermuda or Easter Lily, has greater profusion of larger flowers and more leaves.	1-3 ft. *Sun or half shade*	"
"White"	**WALLICH'S LILY**	***Lílium Wallichiànum**	Rather hard to cultivate. Very large fragrant funnel-form flowers shading into green, grow horizontally on leafy stalks. Good in margin of shrubbery. Bulbous. Prop. by offsets, scales, or very slowly by seed. Light rich soil. Avoid direct contact with manure. Himalayas.	4-6 ft. *Half shade*	July
"White"	**WHITE PERENNIAL FLAX**	***Lìnum perénne var. álbum**	See page 81.		Mid. May to Aug.
"White"	**WHITE MANY-LEAVED LUPINE**	****Lupìnus polyphýllus var. albiflòrus** *L. p. var. álbus, L. grandiflòrus var. a.*	See page 180.		June, July

GOLD-BANDED LILY. *Lilium auratum.*

WHITE BALLOON FLOWER. *Platycodon grandiflorum var. album.*

SINGLE WHITE MALTESE CROSS. *Lychnis Chalcedonica var. alba.*

Color	English Name	Botanical Name and *Synonyms*	Description	Height and *Situation*	Time of Bloom
"White"	DOUBLE WHITE OR EVENING CAMPION	*Lýchnis álba var. flòre-plèno *L. vespertìna var. flòre-plèno*	Fragrant double flowers, larger than Pinks, and opening in the evening. Grow in loose clusters. Common in old-fashioned gardens. Prop. by seed, division or cuttings. Any soil. Europe.	1-2 ft. *Sun*	Mid. July to mid. Sept.
"White"	SINGLE & DOUBLE WHITE MALTESE CROSS	**Lýchnis Chalcedónica vars. álba & álba plèna	See page 180 and Plate, page 288.		June to early Aug.
"White"	WHITE MULLEIN PINK OR DUSTY MILLER	*Lýchnis coronària var. álba	See page 180.		June, July
"White"	SIEBOLD'S LYCHNIS	*Lýchnis coronàta var. Sièboldii *L. S., L. fúlgens var. S.*	See page 180.		"
"White"	JAPANESE LOOSE-STRIFE	**Lysimàchia clethroìdes	See page 180.		Mid. June to late July
"White"	WHITE MUSK MALLOW	*Málva moschàta var. álba	Showy plant. Large single fragrant flowers 2 in. across in clusters. Foliage sweet-smelling. Good in the border. Prop. by seed and division. Any soil.	1-2 ft. *Sun or shade*	July to early Sept.
"White"	DOUBLE SCENTLESS CAMOMILE	*Matricària inodòra var. pleníssima *M. i. var. ligulòsa, var. múltiplex, M. grandiflòra, Chrysánthemum i. var. flòre-plèno*	See page 183.		June to Sept.
"Yellowish white"	BALM	Melíssa officinàlis	See page 183.		June to early Aug.
"Whitish"	VARIEGATED ROUND-LEAVED MINT	Méntha rotundifòlia var. variegàta	See page 183.		June, July
"White" tinged with 25	MICHAUX'S BELL-FLOWER	Michaùxia campanuloìdes	Stately plant, similar to Campanula, with drooping purple-tinged flowers scattered along the stalks. Large rough leaves. Good biennial for background of border. Protect slightly in winter. Prop. by seed just ripe. Light rich soil. Levant.	4-5 ft. *Sun*	July

JULY

Color	English Name	Botanical Name and *Synonyms*	Description	Height and *Situation*	Time of Bloom
"Dull white"	WHITE WILD BERGAMOT	*Monárda fistulòsa var. álba	Rather coarse open flowers and inconspicuous foliage. Striking in masses along banks or in wild places. Prop. by division; separate often in spring. Ordinary soil. N. Amer.	2-2½ ft. *Sun*	July
"White" turns to 36	STEMLESS EVENING PRIMROSE	*Œnothèra acaùlis Œ. Taraxacifòlia	See page 82.		Late May to Aug.
"White"	WHITE REST-HARROW	Onònis arvénsis var. álba *O. spinòsa var. álba*	Spreading plant of branching habit covered with a mass of pea-shaped flowers. Useful for banks and wild parts of rock-garden. Prop. by seed and division. Any soil. Europe.	1-1½ ft. *Sun*	Mid. July to early Aug.
"Silvery white"	ORIENTAL POPPY SILVER QUEEN	**Papáver orientàle "Silver Queen"	See page 184.		Early June to early July
"White"	AMERICAN FEVERFEW, PRAIRIE DOCK	Parthénium integrifòlium	See page 184.		June to Sept.
"White"	HARMALA RUE	Peganum Harmala	See page 184.		June, July
"White"	FOXGLOVE BEARD-TONGUE	**Pentstèmon lævigàtus var. Digitàlis *P. Digitàlis*	See page 187.		Early June to mid. July
"White"	PERENNIAL PHLOX	**Phlóx paniculàta *P. decussàta*	See color "various," page 370.		July to Oct.
"White"	WHITE FALSE DRAGON-HEAD	*Physostègia Virginiàna var. álba *P. Virgínica var. álba*	Pretty but rather stiff in effect; grows in clumps. Tubular flowers in dense racemes terminate leafy erect stalks. Frequent division is necessary. Border or naturalization. Prop. by division in the spring. Any ordinary. soil, preferably moist and rich. Hort.	1-3 ft. *Sun*	Early July, Aug.
"White"	WHITE BALLOON FLOWER	**Platycòdon grandiflòrum var. álbum *Campánula g. var. a. Wahlenbérgia g. var. a.*	See page 187 and Plate, page 288.		June to Oct.
"White"	WHITE JAPANESE PRIMROSE	*Prímula Japónica var. álba	See page 187.		June, July
"White"	ROUND-LEAVED WINTER-GREEN, INDIAN LETTUCE	Pýrola rotundifòlia	See page 187.		"
"Green"	PYRENEAN MIGNONETTE	Resèda glaùca	See page 188.		"

JULY

Color	English Name	Botanical Name and *Synonyms*	Description	Height and *Situation*	Time of Bloom
"White"	RODGER'S BRONZE-LEAF	Rodgérsia podophýlla	See page 188.		June, early July
"White"	PEARL-WORT	Sagìna subulàta *S. pilìfera, Spérgula pilìfera, S. subulàta*	Tufted alpine evergreen, starred with numerous small flowers. Tiny leaves. Excellent for forming a velvety carpeting, especially in the shade on level soil. Prop. by division. Corsica.	4 in. *Shade*	July, Aug.
"White"	SILVERY CLARY, SILVER-LEAVED SAGE	*Sálvia argéntea	See page 188.		June, early July
"White"	WHITE MEADOW SAGE	*Sálvia praténsis var. álba	See page 188.		"
"White"	WHITE SYLVAN SAGE	Sálvia sylvéstris var. álba	Flowers in showy spikes. Oblong leaves. Border. Prop. by seed. Any soil. Europe.	3-3½ ft. *Sun*	July
"Whit-ish"	CANADIAN OR WILD BURNET	Sanguisórba Canadénsis	See page 188.		Mid. June to late July
"Green-ish"	BURNET	Sanguisórba mìnor *Potèrium Sanguisórba*	See page 188.		Late June to mid. July
"White"	PYRAMIDAL COTYLEDON SAXIFRAGE	*Saxífraga Cotylèdon var. pyramidàlis	See page 188.		June, July
"White"	WHITE CAUCASIAN SCABIOUS	*Scabiòsa Caucásica var. álba	See page 189.		"
"White"	WHITE WOODLAND SCABIOUS	*Scabiòsa sylvática var. albiflòra	See page 189.		Early June to late Sept.
"Cream white"	WHITE STONECROP	*Sèdum álbum	Pretty creeper. Starry flowers in clusters. Dense pulpy pale green foliage. Good as carpeting for rock-garden or margin of border. Prop. preferably by division. Sandy soil. Europe; N. Asia.	4-6 in. *Sun or half shade*	Mid. July to late Aug.
"Pinkish white"	SPANISH STONECROP	Sèdum Hispánicum	Creeping plant. Starry flowers in clusters. Succulent leaves, very narrow, grayish turning reddish. Good for carpeting. Prop. preferably by division. Sandy soil best. Europe.	3-4 in. *Sun*	July
"White"	MONRE-GALENSIS STONECROP	Sèdum Monregalénse *S. cruciàtum*	See page 189.		June, July

Color	English Name	Botanical Name and *Synonyms*	Description	Height and *Situation*	Time of Bloom
"White"	NEVIUS' STONECROP	*Sèdum Nevii	Prostrate plant with erect flower stems. Rosettes of succulent evergreen foliage. Excellent for edging, and for rock-garden. Prop. by division. Sandy soil. Eastern U. S. A.	3-5 in. *Sun*	July
"Greenish"	ROSEROOT, ROSEWORT	Sèdum ròseum *S. Rhodìola*	See page 189.		June, July
"White"	DOUBLE SEASIDE CATCHFLY	Silène marítima var. flòre-plèno	See page 189.		"
"White"	GOAT'S BEARD	**Spiræa Arúncus *A. sylvèster*	See page 190.		June, early July
"Creamy white"	ASTILBE-LIKE MEADOW SWEET	**Spiræa astilboìdes *S. Arúncus var. a. Arúncus a., Astìlbe a., A. Japónica*	See page 190.		"
"Cream white"	KAMTSCHATKAN MEADOW SWEET	*Spiræa Camtschática *S. gigantèa, Filipéndula Camtschática, Ulmària C.*	Irregular flower-clusters crown erect stems and rise from a tuft of palmate root leaves. Good in the border or beside water. Prop. by seed and division. Fairly rich moist soil. Kamschatka.	5-10 ft. *Half shade*	July
"Creamy white"	FINGERED SPIRÆA	*Spiræa digitàta	See page 190.		Early June to early July
"Yellowish white"	DROPWORT	*Spiræa Filipéndula *F. hexapétala, Ulmària F.*	See page 190.		June, early July
"Muddy white"	WHITE PALMATE-LEAVED MEADOW SWEET	*Spiræa palmàta var. álba	See page 190		Late June, July
"Cream white"	ENGLISH MEADOW SWEET, MEADOW QUEEN, HONEY SWEET	*Spiræa Ulmària *Filipéndula Ulmària, Ulmària pentapétala, U. palùstris*	See page 190.		June, July
"White"	WHITE COMMON SEA LAVENDER OR MARSH ROSEMARY	Státice Limònium var. álba	Clusters of small flowers in spikelets crown numerous slender branches. Leaves spring from the root. Rock-garden and border. Prop. by seed and division. Deep rich soil. Europe; N. Asia.	$1\frac{1}{4}$ ft. *Sun*	July, Aug.

JULY

Color	English Name	Botanical Name and *Synonyms*	Description	Height and *Situation*	Time of Bloom
"White"	HOFMANN'S SYMPHYANDRA	Symphyándra Hofmanni	Branching plant somewhat resembling the Campanula. Large leafy clusters of drooping bell-shaped flowers varying to blue. Pretty in rock-garden or border. Prop. by seed, division or cuttings. Dry soil. Bosnia.	1-2 ft. *Half shade*	July
"Creamy white" often 37	COMMON COMFREY	Sýmphytum officinàle	See page 93.		Late May to mid. July
"White"	FEATHERED COLUMBINE	*Thalíctrum aquilegifòlium	See page 93.		"
"White"	MOUNTAIN WILD THYME	Thýmus Serpýllum var. montànus *T. m., T. Chamædrys*	See page 193.		Early June to mid. Aug.
"White"	WHITE SPIDERWORT	Tradescántia Virginiàna var. álba	See page 94.		Late May to late Aug.
"Yellowish white" 2	HUNGARIAN CLOVER	Trifòlium Pannónicum	See page 193.		Late June to mid. July
"White"	WHITE SPIKE-FLOWERED SPEEDWELL	**Verónica spicàta var. álba	See page 193.		Early June, July
"White"	WHITE HORNED VIOLET OR BEDDING PANSY	**Vìola cornùta var. álba	See page 23		Late Apr. until frost
"Creamy white" 2 greener	ADAM'S NEEDLE, BEAR OR SILK GRASS, THREADY YUCCA	**Yúcca filamentòsa	See page 194.		June, July
"Yellow" 5	WOOLLY-LEAVED MILFOIL	*Achillèa tomentòsa	See page 96.		Late May to mid. Sept.
"Pale yellow" 4	EGYPTIAN MILFOIL OR YARROW	Achillèa Tournefórtii *A. Ægyptìaca*	Alpine plant. Flowers in small flat clusters. Foliage silvery and fern-like. Excellent for covering dry bare places. Prop. by seed, division and cuttings. Any garden soil. Greece.	12-18 in. *Sun*	July to Oct.
"Yellow" 4 pale	PYRENEAN MONKSHOOD	Aconìtum Anthòra *A. Pyrenàicum*	See page 194.		June, July

Color	English Name	Botanical Name and *Synonyms*	Description	Height and *Situation*	Time of Bloom
"Yellow" 2	PALE YELLOW WOLFSBANE	*Aconìtum Lycóctonum *A. barbàtum, A. ochroleùcum, A. squarròsum*	Narrow helmet-shaped flowers, sometimes whitish, in racemes. Foliage deeply cut. Roots poisonous. Good border plant. Prop. by seed and division. Prefers rich soil. Europe; Siberia.	3-4 ft. *Sun or half shade*	July to early Sept.
"Yellow" 4	PIGMY SUN-FLOWER	Actinélla grandiflòra	See page 194.		June, early July
"Yellow" 6 very deep	DOTTED PICRA-DENIA	Actinélla scapòsa *Picradènia scapòsa*	Hairy alpine plant. Numerous aster-like flowers 1 in. across. Leaves spring from root. Good for rock-garden. Prop. by seed and division. Light soil. Col.	6-15 in. *Sun*	July, Aug.
"Yellow" 6 lighter	CROWN-BEARD	*Actinómeris squarròsa	Coarse branching plant bearing numerous daisy-like flowers with irregular rays; several on the stem. Plant in wild garden or with shrubs. Prop. by division. U. S. A.	4-8 ft. *Sun*	Mid. July to late Aug.
"Yellow" 3	PYRENEAN ADONIS	*Adònis Pyrenàica	Solitary showy and brilliant buttercup-like flowers. Foliage very finely divided. Adapted for rock-garden or border. Prop. by seed and division. Any ordinary garden soil, rather moist. Pyrenees.	1-1½ ft. *Sun or half shade*	July
"Yellow" 6	SILVERY MADWORT	Alýssum argéntium *A. alpêstre*	See page 194.		June to early Aug.
"Yellow" 4	BEAKED MADWORT	Alýssum rostràtum *A. Wièrzbickii*	See page 194.		Early June to early Aug.
"Yellow" 5	GOLDEN MARGUE-RITE	*Ánthemis Kélwayi *A. tinctòria var. Kélwayi.*	See page 194.		Mid. June to Oct.
"Yellow" 2, centre 6	GOLDEN MARGUE-RITE, ROCK CAMOMILE	*Ánthemis tinctòria	See page 96.		Mid. May to Oct.
"Yellow" 2	YELLOW CANADIAN COLUMBINE	*Aquilègia Canadénsis var. flaviflòra, *A. C. var. flavêscens, A. cærùlea var. f.*	See page 24.		Late Apr. to early July
"Yellow" 3 & 2	GOLDEN-SPURRED COLUMBINE	**Aquilègia chrysántha *A. leptocèras var. c.*	See page 96.		Late May to late Aug.
"Orange yellow" 4	MOUNTAIN TOBACCO OR SNUFF	*Árnica montàna	Large daisy-shaped flowers. Leaves broad and oval springing from the root or growing in pairs on the stalk. Rock-garden or border. Prop. by seed, or generally by division. Europe.	1 ft. *Sun*	July, Aug.

Color	English Name	Botanical Name and *Synonyms*	Description	Height and *Situation*	Time of Bloom
"Yellow" 4	TRUE ASPHODEL, KING'S SPEAR	*Asphodelìne lùtea *Asphódelus lùteus*	See page 195.		June, July
"Yellow" 4	YELLOW MILK VETCH	Astrágalus alopecuroìdes	See page 195.		"
"Yellow"	CHINESE MILK VETCH	Astrágalus Chinénsis	See page 195.		"
"Yellow"	GALEGA-LIKE MILK VETCH	Astrágalus galegifòrmis	See page 195.		Early June to early July
"Yellow" 6	WILLOW-LEAVED OX-EYE	*Bupthál-mum salicifòlium *B. grandi-flòrum*	See page 195.		June, July
"Yellow" 6	SHOWIEST OX-EYE	*Bupthál-mum speci-osíssimum	See page 195.		"
"Yellow" 5	SHOWY OX-EYE	*Bupthál-mum speciòsum *B. cordifòlium*	See page 195.		"
"Yellow"	ITALIAN CANNA	**Cánna vars.	See color "various," page 357.		July to late Sept.
"Yellow" 6	WILD SENNA	*Cássia Marylándica	Handsome shrub-like plant. Numerous small clusters of flowers with chocolate colored anthers. Ornamental light green compound foliage. Prop. by seed and division. Easily cultivated even in poor soil. New Eng. and West.	3-5 ft. *Sun*	July, Aug.
"Yellow" 5	BABYLO-NIAN CENTAURY	Centaurèa Babylónica	Picturesque strong-growing plant. Numerous globe-shaped flowers in spikes. Grown chiefly for the silvery foliage. Useful for back of border and margin of shrubbery. Prop. by seed. Sandy soil is best. Levant.	6-12 ft. *Sun*	July
"Yellow" 5	BLUE-LEAVED CENTAURY OR STAR THISTLE	Centaurèa glastifòlia	Grows vigorously and is free-flowering. Silvery thistle-like flowers. Border. Prop. by seed and division. Easily cultivated in ordinary soil. Caucasus.	3-4 ft. *Sun*	July, Aug.
"Yellow" 5	SHOWY CENTAURY OR KNAPWEED	**Centaurèa macrocéphala	The best and most showy Centaurea. Very large round flowers. Excellent foliage. Good for cutting. Effective in border or shrubbery. Prop. by seed. Armenia.	$2\frac{1}{2}$-3 ft. *Sun*	Mid. July to Sept.
"Yellow" 1 & 21 light	LEMON-COLOR MOUNTAIN BLUET	Centaurèa montàna var. citrìna *C. m. var. sulphùrea*	See page 99.		Late May to early July

Color	English Name	Botanical Name and *Synonyms*	Description	Height and *Situation*	Time of Bloom
"Pale yellow"	RUSSIAN KNAPWEED	Centaurèa Ruthénica	Large thistle-like flowers. Dark green ornamental leaves deeply divided. Good in border. Prop. by seed and division. Any garden soil. Orient.	3-4 ft. *Sun*	July
"Pale yellow" 4 very pale	ALPINE CEPHALARIA	*Cephalària alpìna	See page 196.		Late June to late July
"Yellow" 3	GOLD JOINT	Chrysógonum Virginiànum	See page 196.		June, July
"Yellow" 3	ERECT SILKY CLEMATIS	*Clématis ochroleùca *C. sericea*	A plant of erect somewhat bushy habit. Solitary flowers. Silky foliage. Pretty in border or rock-garden. Winter mulching desirable. Prop. by seed and cuttings. Rich deep soil. Eastern U. S. A.	1-2 ft. *Sun*	July, Aug.
"Yellow" 6 paler	LARKSPUR TICKSEED	*Coreópsis delphinifòlia	Compact branching plant. Flowers have dark brown centres and 6-10 yellow rays. Smooth foliage. Good border plant. Prop. by seed. Easily cultivated in any soil. Dry woods, Va. to Ala.	1-3 ft. *Sun*	Mid. July to mid. Sept.
"Yellow" 5	LARGE-FLOWERED TICKSEED	**Coreópsis grandiflòra	See page 196 and Plate, page 297.		June to Sept.
"Yellow" 6	LANCE-LEAVED TICKSEED	**Coreópsis lanceolàta	See page 196 and Plate, page 298.		"
"Yellow" 5	STIFF TICKSEED	*Coreópsis palmàta *C. præcox*	See page 196.		June, July
"Yellow" 6 deeper	STAR TICKSEED	*Coreópsis pubéscens *C. auriculàta*	Vigorous branching species, more leafy than C. delphinifolia. The flowers have yellow rays with a band of purplish brown encircling the brown or yellow disk. Thick rather glossy foliage. Good for wild garden or border. Prop. by seed. Easily cultivated in any soil. Rich woods, Va. to Ill. and La.	1-4 ft. *Sun*	Mid. July to late Sept.
"Yellow" 6 paler	WHORLED TICKSEED	*Coreópsis verticillàta *C. tenuifòlia*	A pretty species, showy and slightly branching. Small and pretty flowers 1-2 in. across, with yellow rays jagged at the tip, and dull yellow disks are borne erect on wiry stems. Foliage fine, feathery and dark. Good for the border. Prop by seed. Easily cultivated in any ordinary soil. Dry soil, Ontario to N. C. and Ark.	1-3 ft. *Sun*	July, Aug.
"Yellow"	DAHLIA	**Dáhlia vars.	See color "various," page 359.		July to late Oct.

LARGE-FLOWERED TICKSEED. *Coreopsis grandiflora.*

LANCE-LEAVED TICKSEED. *Coreopsis lanceolata.*

JULY

Color	English Name	Botanical Name and *Synonyms*	Description	Height and *Situation*	Time of Bloom
"Yellow" 2	ZALIL'S LARKSPUR	*Delphínium Zàlil *D. sulphù-reum, D. hÿbridum var. s.*	See page 196.		June, July
"Yellow" 2	YELLOW FOXGLOVE	*Digitàlis ambígua *D. grandi-flòra, D. ochroleùca*	See page 197.		Early June to mid. July
"Yellow" 2 & 1	RUSTY FOXGLOVE	*Digitàlis ferrugínea *D. aùrea*	Biennial plant. Long dense racemes of rusty tubular flowers are borne on very leafy stalks. Border. Prop. by seed and division. Any rich light soil. S. Europe.	4-6 ft. *Sun or shade*	July
"Cream color" 2	WOOLLY FOXGLOVE	*Digitàlis lanàta	See page 197.		Early June to late July
"Bright yellow" 4, anthers 13	BUNGE'S EREMURUS	*Eremùrus Búngei	See page 197.		June, July
"Yellow" 6 lighter	WOOLLY BAHIA	Eriophÿllum cæspitòsum *Actinélla lanàta, Bàhia l.*	See page 198.		"
"Yellow" 4	COMMON GIANT FENNEL	*Férula commùnis	Tiny flowers in flat-topped clusters on branching stalks are borne well above the mound of glossy finely-cut foliage for which the plant is chiefly grown. Plant in bold groups on lawns or edge of shrubberies. Prop. by fresh seed. Rich deep open soil. Mediterranean Region.	6-10 ft. *Sun*	July
"Yellow" 4	LADY'S BEDSTRAW	Gàlium vèrum	See page 198.		June to Sept.
"Yellow" bet. 5 & 6	DYER'S GREEN-WEED, BASE BROOM	*Génista tinctòria	See page 198.		Late June, July
"Yellow" 5	YELLOW GENTIAN	Gentiàna lùtea	Flowers in flat-topped clusters. Border or rock-garden. Prop. by seed. Leave undisturbed. Light rich soil. Europe, W. Asia.	2-3 ft. *Half shade*	July, Aug.
"Yellow"	SWORD LILY	**Gladìolus vars.	See color "various," page 365.		July to Oct.
"Yellow" 4	BROAD-LEAVED GUM-PLANT	Grindèlia squarròsa	See page 198.		Late June to Sept.

Color	English Name	Botanical Name and *Synonyms*	Description	Height and *Situation*	Time of Bloom
"Yellow" 5	BIGELOW'S SNEEZE-WEED	*Helènium Bígelovii	Daisy-like flowers, about 2 in. across with brown centres, either solitary or few on slender stems. Good for cutting. Useful, and easily cultivated in border. Prop. by seed, division or cuttings. Any soil, preferably rich and moist. N. Amer.	2-3 ft. *Sun*	July, Aug.
"Yellow" 6	BOLANDER'S SNEEZE-WEED OR SNEEZE-WORT	*Helènium Bolánderi	See page 198.		June to Sept.
"Yellow" 5	PURPLE-HEADED SNEEZE-WEED	*Helènium nudiflòrum *Leptopoda brachypoda*	Fragrant daisy-like flowers, with drooping petals, sometimes striped with brownish purple; centre, brown or purplish. Effective in the border. Prop. by seed, division or cuttings. Moist rich soil preferable. S. Eastern U. S. A.	1-3½ ft. *Sun*	July to Oct.
"Yellow" 2 to 5	ROCK OR SUN ROSE	*Heliánthemum vulgàre	See page 199.		Early June, July
"Yellow" 5	HAIRY SUNFLOWER	**Heliánthus móllis	The most desirable of the perennial sunflowers. A vigorous plant with large solitary flowers about 3 in. across, and rough grayish leaves. Effective in the back of border. Prop. by seed and division. Any garden soil. U. S. A. See Plate, page 301.	2-5 ft. *Sun*	July, Aug.
"Yellow" 5	STIFF SUNFLOWER	*Heliánthus rígidus *H. Missouriênsis*	Distinct species. Showy flowers with reflexed petal tips borne in panicles. Valuable in shrubbery or border or for naturalization in the wild garden. Prop. by division; divide every 2 years. Dry soil, not too heavy. Western U. S. A.	1-4 ft. *Sun*	"
"Orange yellow" 6	PALE-LEAVED WOOD SUNFLOWER	Heliánthus strumòsus	Rayed flowers about 3 in. across are borne on branching stems. Rather rough grayish foliage. Shrubbery. Var. *mollis* (*H. macrophyllus*), leaves downy underneath. Good for wild garden. For cultivation, etc., see H. rigidus. N. Amer.	3-7 ft. *Sun*	Mid. July to late Sept.
"Yellow" 5	THROATWORT SUNFLOWER	Heliánthus trachelifòlius	Branching plant with rough stalks. Pure green foliage. Not so showy as some other kinds. Plant among shrubs. For cultivation., etc., see H. rigidus. Central U. S. A.	3-5 ft. *Sun*	Mid. July to early Sept.
"Yellow" 6	OX-EYE, FALSE SUNFLOWER	*Heliópsis læ̀vis *H. helianthoìdes*	Branching plant bearing a profusion of sunflower-like blossoms about 2 in. across. Excellent for cutting. Effective in shrubbery or border. Prop. generally by division. Light soil, not too moist. Eastern N. Amer.	3-5 ft. *Sun*	Mid. July to late Sept.

HAIRY SUNFLOWER. *Helianthus mollis.*

ST. JOHN'S WORT. *Hypericum Moserianum.*

JULY

Color	English Name	Botanical Name and *Synonyms*	Description	Height and *Situation*	Time of Bloom
"Yellow" 6, centre 7	PITCHER'S OX-EYE OR FALSE SUN-FLOWER	*Heliópsis lævis var. Pitcheriàna *H. Pitcheriàna*	Dwarf bushy var., spreading 3 feet, distinguished by its deeper coloring and free blooming. Cup-shaped flowers in loose panicles. Good for the border, for dry situations, and for cutting. For cultivation, etc., see H. lævis. Hort.	2-3 ft. *Sun*	Mid. July to late Sept.
"Yellow" 6 deep	LEMON LILY	**Hemerocállis flàva	See page 199.		June, early July
"Orange yellow" 8	MIDDENDORF'S YELLOW DAY LILY	**Hemerocállis Middendorfii	See page 199.		Late June, July
"Yellow" 5	LESSER YELLOW DAY LILY	*Hemerocállis mìnor *H. gramìnea, H. graminifòlia*	Fragrant lily-like flowers in clusters. Good for cutting. Dark green grass-like foliage. Attractive on banks of streams or in the border. Prop. by division. Any moist rich soil. Sìberia.	2 ft. *Half shade*	July, Aug.
"Yellow" 6 lighter	THUNBERG'S YELLOW DAY LILY	**Hemerocállis Thúnbergii	See page 199.		Late June, July
"Golden yellow" 5	SHAGGY HAWKWEED	Hieràcium villòsum	See page 199.		June to mid. Aug.
"Yellow" 5	AARON'S BEARD, ROSE OF SHARON	Hypéricum calycìnum	Rapidly spreading sub-shrub, almost evergreen. Large flowers, 3 in. across, with conspicuous stamens and red anthers. Dark glossy leaves. Used in England for carpeting under trees. Border or shrubbery. Protect in winter. Prop. by cuttings of root or ripe wood. Prefers sandy loam. S. Eastern Europe.	1 ft. *Sun or half shade*	July, Aug.
"Yellow" bet. 5 & 6	GOLD FLOWER, ST. JOHN'S WORT	**Hypéricum Moseriànum	A very attractive sub-shrub bearing handsome single flowers, 2 in. across with a circle of conspicuous orange stamens. Foliage dark and fine. Protect in winter. Pleasing in the border, though never producing a mass of color as some flowers drop while others open. Prop. by seed and cuttings. Light warm soil preferable. Hort. See Plate, page 302.	2 ft. *Sun or half shade*	"
"Yellow" 5	SWORD-LEAVED ELECAMPANE	*Ínula ensifòlia	Vigorous plant and continuous bloomer with large aster-like flowers and rather coarse foliage. Useful for rock-garden or border. Prop. by seed and division. Any good soil. Europe; Asia.	6-8 in. *Sun*	Mid. July to Sept.

Color	English Name	Botanical Name and *Synonyms*	Description	Height and *Situation*	Time of Bloom
"Deep yellow" 7	GLANDULAR FLEA BANE OR INULA	*Ínula glandulòsa	A plant of coarse habit. Large aster-like flowers with fringed half-drooping petals. Leaves scattered along the stalks and in clumps at their base. Good border plant. Prop. by seed and division. Any soil. Caucasus.	2-3 ft. *Sun*	July, early Aug.
"Yellow" 6	ELECAMPANE	*Ínula Helènium	Vigorous somewhat coarse plant. Flowers with fringe-like petals. Large rough leaves downy beneath. Naturalize in wild places. Prop. by seed and division. Easily cultivated in ordinary soil. Europe; Siberia.	3-4 ft. *Sun*	Mid. July to late Aug.
"Yellow" effect 11	TUCK'S FLAME FLOWER	*Kniphòfia Túckii	See page 200.		June to Sept.
"Yellow"	EDELWEISS	*Leontopòdium alpìnum *Gnaphàlium L.*	See page 200		June, July
"Yellow" 5	GRAY-HEADED CONE-FLOWER	*Lépachys pinnàta *Ratìbida p., Rudbéckia p.*	See page 200.		June to mid. Sept.
"Yellow" 7 dull	YELLOW CANADA LILY	*Lílium Canadénse var. flàvum	See page 200.		June, July
"Yellow" 5	THUNBERGIAN LILY ALICE WILSON	**Lílium élegans var. "Alice Wilson"	Excellent var. with flowers almost lemon-colored terminating leafy stems. Easy to grow. Mass in border or along margin of shrubbery. Bulbous. Prop. by offsets or scales. Rather peaty soil. Avoid direct contact with manure. Japan.	1-2 ft. *Sun or shade*	July
"Yellow" 6	YELLOW THUNBERGIAN LILY	**Lílium élegans var. alutàceum *L. e. var. Armeniàcum, var. citrìnum*	Dwarf var. with spotted flowers terminating erect leafy stalks. Charming massed in margin of border. Bulbous. Prop. by offsets or scales. Peaty soil well enriched. Avoid direct contact with manure. Japan.	8-10 in. *Sun or half shade*	"
"Yellow" 3	CAUCASIAN LILY	*Lílium monadélphum *L. Cólchicum, L. Szovitziànum*	See page 200.		June, July
"Yellow" 5	PARRY'S LILY	**Lílium Párryi	See page 200.		"
"Golden yellow" 6 more orange	HUMBOLDT'S LILY	*Lílium pubérulum *L. Bloomeriànum, L. Califórnicum, L. Humboldtii*	See page 200.		"

JULY

Color	English Name	Botanical Name and *Synonyms*	Description	Height and *Situation*	Time of Bloom
"Buff" 1	NANKEEN LILY	**Lílium testàceum *L. excêlsum, L. Isabellìnum*	See page 203.		Mid. June to mid. July
"Yellow" 3	DALMATIAN TOAD-FLAX	*Linària Dalmática	See page 203.		Early June to early July, late July to late Aug.
"Yellow" 3 & 7	MACEDONIAN TOAD-FLAX	Linària Macedónica	See page 203.		June, July
"Yellow" 5	YELLOW FLAX	*Lìnum flàvum	See page 203.		"
"Yellow" 3	BIRD'S-FOOT TREFOIL, BABIES' SLIPPERS	Lòtus corniculàtus	See page 203.		June to Oct.
"Yellow" 5	MONEY-WORT, CREEPING JENNY	*Lysimàchia nummulària	See page 203.		Early June to late July
"Yellow" 5	SPOTTED LOOSE-STRIFE	*Lysimàchia punctàta *L. verticulàta*	See page 203.		June, July
"Yellow" bet. 5 & 6	BULB-BEARING LOOSE-STRIFE	Lysimàchia stricta *L. terréstris*	See page 203.		"
"Yellow" 4	COMMON YELLOW LOOSE-STRIFE	*Lysimàchia vulgàris	See page 204.		"
"Yellow" 5	WELSH POPPY	Meconópsis Cámbrica	See page 204.		"
"Yellow" 3	LARGE-FLOWERED BIENNIAL EVENING PRIMROSE	*Œnothèra biénnis var. grandiflòra *Œ. Lamarckiàna*	See page 204.		June to Sept.
"Yellow" 5	SUNDROPS	**Œnothèra fructicòsa	See page 204.		June, July
"Yellow" 5	FRASER'S EVENING PRIMROSE	**Œnothèra glaùca var. Fràseri *Œ. Fràseri*	See page 204.		June to Sept.
"Lemon yellow" 3	LINEAR-LEAVED EVENING PRIMROSE	Œnothèra lineàris *Œ. fructicòsa var. l., Œ. ripària*	See page 204.		June to early Aug.

JULY

Color	English Name	Botanical Name and *Synonyms*	Description	Height and *Situation*	Time of Bloom
"Yellow" 3	MISSOURI PRIMROSE	**Œnothèra Missouriénsis *Œ. macrocàrpa, Megaptèrium M.*	See page 204.		June to early Aug.
"Yellow" 5	TAURIAN GOLDEN DROP	Onósma stellulàtum var. Taùricum *O. Taùricum*	Hairy evergreen growing in thick tufts. Drooping roundish tubular flowers in clusters, almond scented. Rock-garden. Prop. by cuttings. Thrives in light well-drained soil of good depth. Greece.	6-12 in. *Sun*	July, Aug.
"Pale yellow" 6	MANY-SPINED OPUNTIA	Opúntia Missouriénsis *O. fèrox, O. splèndens*	Perfectly hardy and very effective. Low spreading Cactus with grayish and "reddish brown spines" and obovate joints. Flowers 2-3 in. across. Good for rock-garden, ledges, etc. Prop. by cuttings more easily than by seed. Any soil, well-drained. Prairies, W. Central U. S. A.	1 ft. *Sun*	"
"Yellow" 6	WESTERN PRICKLY PEAR	Opúntia Rafinésquii *O. mesacántha*	The handsomest species. Spreading Cactus of prostrate habit with small spines in tufts. Showy red-centred flowers. Fragile but useful for sheltered situation in the rock-garden. "Protect from slugs by dressing of soot." Prop. by joints rooted in sand under glass. Will grow in rocky places where soil is too thin for other plants. Dry well-drained soil. Central U. S. A.; Tex.	1 ft. *Sun*	"
"Dull yellow" 3 duller	BARBERRY FIG, COMMON PRICKLY PEAR	Opúntia vulgàris *O. Opúntia*	See page 207.		June to Sept.
'Pale yellow" 2	CROWDED PENTSTEMON, OR BEARD-TONGUE	*Pentstèmon confértus	See page 207.		June, early July
"Yellow" 6 to 13	JERUSALEM SAGE	Phlòmis fruticòsa	See page 207.		June, July
"Yellow" 3	SILVERY CINQUEFOIL	Potentílla argéntea	See page 105.		Early May to early July
"Yellow" 1	SILVERY-LEAVED CINQUEFOIL OR FIVE-FINGER	Potentílla argyrophýlla *P. insignis*	See page 207.		June, July
"Lemon yellow" 3	CALABRIAN CINQUEFOIL	Potentílla Calabra	See page 105.		Late May to early July

CREEPING DOUBLE-FLOWERED BUTTERCUP. *Ranunculus repens. var. flore-pleno.*

 SHOWY CONE-FLOWER. *Rudbeckia speciosa.*

Color	English Name	Botanical Name and *Synonyms*	Description	Height and *Situation*	Time of Bloom
"Yellow" 3	SHRUBBY CINQUE-FOIL OR FIVE-FINGER	Potentílla fruticòsa	See page 207.		June to Sept.
"Golden yellow" effect 7	HYBRID CINQUE-FOIL OR FIVE-FINGER	*Potentílla "Gloire de Nancy"	See page 208.		"
"Yellow" 3	LARGE-FLOWERED CINQUE-FOIL OR FIVE-FINGER	Potentílla grandiflòra	See page 208.		June, July
"Golden yellow" 5	PYRENEAN CINQUE-FOIL	Potentílla Pyrenaìca	See page 105.		May to Aug.
"Golden yellow" 5	BACHE-LOR'S BUTTONS	Ranúnculus àcris var. flòre-plèno	See page 105.		Mid. May to Sept.
"Yellow" 5	MOUNTAIN BUTTER-CUP	Ranúnculus montànus	See page 106.		May to early July
"Yellow" 5	CREEPING DOUBLE-FLOWERED BUTTERCUP	Ranúnculus répens var. flòre-plèno	See page 106 and Plate, page 307.		May to Aug.
"Yellow" 5	TALL OR GREEN-HEADED CONE-FLOWER	*Rudbéckia laciniàta	Large daisy-like flowers with few much recurved petals, and greenish centres, on branching stems. Good in wild garden or border. Prop. by seed and division. Any garden soil; requires plenty of moisture. Canada S. to Fla., W. to Rocky Mts.	2-7 ft. *Sun or half shade*	Mid. July to late Aug.
"Yellow" bet. 5 & 6	GOLDEN GLOW	**Rudbéckia laciniàta var. flòre-plèno	Very showy free-flowering plant with large double flowers. May be cut after flowering for a second bloom. Popular for massing as it makes a very brilliant effect. Prop. by division. Any garden soil; requires plenty of moisture. Hort.	2-10 ft. *Sun*	Late July to late Sept.
"Yellow" 6, brown centre	SHOWY CONE-FLOWER	*Rudbéckia speciòsa *R. áspera*	Compact in growth. Masses of yellow-rayed flowers, 3-4 in. across, with velvety maroon centres. Excellent border plant. Prop. by division. Thrives in ordinary garden soil. U. S. A. See Plate, page 308.	1-3 ft. *Sun or half shade*	July, early Aug.
"Deep yellow" 6	THIN-LEAVED CONE-FLOWER	**Rudbéckia tríloba	One of the best Rudbeckias. Dense bushy plant. Numerous rayed flowers form a brilliant mass of color. Clean foliage. Associate with Larkspurs in shrubbery or border. Flowers the first year from seed. Biennial, perpetuated by self-sown seed. Any soil. Eastern U. S. A.	2-5 ft. *Sun or shade*	July, Aug.

JULY

Color	English Name	Botanical Name and *Synonyms*	Description	Height and *Situation*	Time of Bloom
"Yellow" 4 green	RUE, HERB OF GRACE	Rùta gravèolens	Common Rue. Not a very pretty plant with panicles of small fragrant flowers, and much divided leaves. Plant in sheltered position and protect in winter. Prop. by seed and division. Asia; S. Europe.	1½-2 ft. *Sun*	July
"Pale yellow" 2	JUPITER'S DISTAFF	*Sálvia glutinòsa	Flowers in whorls on erect stems. Foliage poor in color. Lower leaves very large, upper ones smaller. Border. Prop. by seed in early summer or spring. Rich soil. Europe; Asia.	3 ft. *Sun*	Mid. July to early Aug.
"Yellow" 3 pale	WEBB'S SCABIOUS OR PIN-CUSHION FLOWER	Scabiòsa ochroleùca *S. Webbiàna*	See page 208.		June to early Sept.
"Yellow" 5	AIZOON STONECROP	*Sèdum Aizóon	See page 208.		Mid. June to mid. Aug.
"Golden yellow" 4	ORANGE STONECROP	*Sèdum Kamtscháticum	Starry flowers in flat clusters. Bright pulpy evergreen foliage. Good for compact carpeting. Prop. preferably by division. Sandy soil best. E. Asia.	4-9 in. *Sun or half shade*	July, Aug.
"Yellow" 3 dull	MIDDENDORF'S STONECROP	*Sèdum Middendorfiànum	See page 209.		June, July
"Yellow" 2 greenish	STONE-HORE, STONE ORPINE, TRIP-MADAM	*Sèdum refléxum	Trailing evergreen. Flowers marked with pale chocolate grow on stems becoming erect at time of bloom. Pulpy closely tufted leaves. Excellent for carpeting. Prop. preferably by division. Dry soil. Europe.	8-10 in. *Sun*	Early July to early Aug.
"Yellow"	CRESTED STONE-HORE, STONE ORPINE OR TRIP-MADAM	Sèdum refléxum var. cristàtum *S. monstròsum, S. robûstum*	Dwarf var. of trailing habit, with stems grouped in fan-shaped or crested manner. For cultivation, etc., see S. reflexum.	3 in. *Sun*	"
"Yellow" 4	SIX-ANGLED STONECROP	Sèdum sexangulàre	See page 209.		Early June, July
"Yellow" 2	HENS-AND-CHICKENS, HOUSELEEK	*Sempervìvum globíferum *S. sobolíferum*	Small starry flowers in meagre clusters. Pulpy foliage, bright green tinged with red, in small cactus-like rosettes. Plant in clumps in rock-garden or on stone-wall. Prop. by division or offsets. Any sandy soil. European Alps.	6-9 in. *Sun*	July, Aug.
"Pale yellow" 2 bright	HAIRY HOUSELEEK	*Sempervìvum hírtum	Small star-like flowers in panicles. Rosettes of fringed pulpy leaves. Pretty foliage plant for rock-garden. Prop. by seed or offsets. Sandy soil. Middle Europe.	6-9 in. *Sun*	"

GOLDENROD. *Solidago sempervirens.*

FEN RUE. *Thalictrum flavum.*

Color	English Name	Botanical Name and *Synonyms*	Description	Height and *Situation*	Time of Bloom
"Yellow" 5	DORIAN GROUND-SEL OR RAGWEED	Senècio Dòria	Showy erect plant. Rayed flowers in loose clusters. Root-leaves grayish green. Border or wild garden. Prop. by seed, division or cuttings. Any good loam. S. Europe.	3-4 ft. *Sun*	July, Aug.
"Yellow" 5	ROUGH ROSINWEED	Sílphium aspérrimum	Large sunflower-like blossoms on rough stalks. Coarse foliage thickly produced at the top of the plant. Good for naturalization. Prop. by seed and division. Any soil. N. Amer.	2-5 ft. *Sun*	July to early Sept.
"Yellow" 5	COMPASS PLANT, PILOT WEED	**Sílphium laciniàtum	Foliage extremely decorative on a large scale. Vigorous growth. Sunflower-like blossoms, facing the east droop above a luxuriant mass of prickly leaves, cool grayish green and oak-like. Plant in clumps of two or three in roomy wild garden or shrubbery. Prop. by seed and division. Any garden soil. Prairies, U. S. A.	6 ft. *Sun*	Mid. July to mid. Sept.
"Yellow" 6 lighter	CUP PLANT, INDIAN CUP	*Sílphium perfoliàtum	Sunflower-like flowers. Coarse foliage. Good in wild garden and amidst shrubbery. Prop. by seed and division. Easily cultivated in any garden soil. Prairies, U. S. A.	4-8 ft. *Sun*	"
"Yellow" 6 lighter	WHORLED ROSINWEED	Sílphium trifoliàtum	Sunflower-like blossoms in round clusters surmount very leafy stems. Large leaves, generally slightly shaggy. Mass in wild garden. Prop. by seed and division. Any rich soil. Eastern U. S. A.	4-7 ft. *Sun*	Mid. July to mid. Sept.
"Yellow" 5 to 6	GOLDEN-ROD	*Solidàgo	Strikingly effective. Erect leafy stems bear minute flowers in plumy panicles. Effective planted with Asters in shrubbery. Will bloom after frost. May spread too much. Prop. by division. Grows in any soil, but improves under good treatment. Among the best vars. offered for sale are:—*Canadensis,* charming foliage. *Canadensis var. glabrata;* (color no. 5). ***Drummondii;* (color no. 5). *Latifolia;* (bet. 5 & 6). *Nemoralis. Patula;* (color no. 4 deeper). **Petiolaris;* (color no. 5). ***Rigida;* (color no. bet. 5 & 6). Coarse foliage. *Sempervirens. Shortii;* 5 ft. high. *Speciosa;* (color no. bet. 5 & 6). *Ulmifolius.* See Plate, page 311.	2-5 ft. *Sun*	Late July to early Oct.
"Yellow" 2	FALSE RHUBARB, FEN RUE	Thalíctrum flàvum	Pyramidal clusters of pale greenish yellow flowers. Pretty fern-like foliage. Rock-garden. Prop. in spring by seed and division. Well-drained loam. Europe; Asia. See Plate, page 312.	2-4 ft. *Half shade*	July, Aug.

Color	English Name	Botanical Name and *Synonyms*	Description	Height and *Situation*	Time of Bloom
"Greenish yellow" 2	GLAUCOUS MEADOW RUE	Thalíctrum glaùcum	See page 209.		Late June to late July
"Greenish yellow" 3	DWARF MEADOW RUE	*Thalíctrum mìnus *T. purpùreum, T. saxátile*	See page 209.		Late June to mid. July
"Yellow" 4 lighter	GOLDEN ALEXANDERS	Tháspium aùreum *T. trifoliàtum var. a.*	See page 209.		June, July
"Yellow" 5	CAROLINA THERMOPSIS	*Thermópsis Caroliniàna	See page 209.		Early June to mid. July
"Yellow" 4	ALLEGHANY THERMOPSIS	*Thermópsis móllis	See page 106.		Mid. May to Aug.
"Yellow"	CROCUS-FLOWERED BLAZING STAR	**Tritònia crocosmæflòra vars. *Montbrètia crocosmæflòra*	Lovely slender branching plants with long spike-like racemes of flowers 2 in. across. Tall stiff narrow foliage springs from near the bulb. Protection of mulch or indoor wintering is necessary. Very gay and attractive for the border. Bulbous. Plant in Apr. or May and prop. by scales and offsets. Rich well-drained soil. Vars.: **Rayon d'or;* (color no. bet. 6 & 7 brilliant), large ochre yellow flowers with brown blotches at the base of the tube. **Gerbe d'or;* (color no. 7 more orange), brilliant golden yellow. **Soleil couchant;* (color no. 7 tinged with 18), a profusion of small rich golden yellow flowers. Hort. There are orange and red vars.	2-3 ft. *Sun*	July to Oct.
"Yellow" 2	NETTLE-LEAVED MULLEIN	*Verbáscum Chaixii *V. orientàle, V. vernàle*	See page 210.		Late June to late July
"Yellow" 2	DARK MULLEIN	*Verbáscum nìgrum	See page 210.		Early June to late July
"Deep yellow" 3	OLYMPIAN MULLEIN	**Verbáscum Olýmpicum	Most conspicuous of the Mulleins. Large flowers in dense evenly branching spikes. The foliage bold in character is covered with white down. Requires about 3 years to bring into bloom. Prop. by seed. Any garden soil. Greece. See Plate, page 315.	6-10 ft. *Sun*	July

OLYMPIAN MULLEIN. *Verbascum Olympicum.*

BUTTERFLY WEED. *Asclepias tuberosa.*

JULY

Color	English Name	Botanical Name and *Synonyms*	Description	Height and *Situation*	Time of Bloom
"Yellow" 6	YELLOW HORNED VIOLET OR BEDDING PANSY	**Vìola cornùta var. lùtea màjor	See page 35.		Late Apr. until frost
"Orange" 9	GOLDEN PERUVIAN LILY	*Alstrœmèria aurantiàca *A. aùrea*	Beautiful large lily-like flowers, marked with brown and green, 10 to 15 in an umbel. Foliage gray-green. Somewhat difficult to grow. Do not disturb when established and protect in winter. Tuberous plant, yet may be propagated by seed. Well-drained soil mixed with decayed vegetable matter. Chili.	2-4 ft *Half shade*	July, Aug.
"Red orange" shading 10 to 12	BUTTERFLY WEED, PLEURISY ROOT	**Asclèpias tuberòsa	Most striking and flamboyant in color and one of the hardiest of plants. Flowers in large flat clusters. When naturalized it makes a brilliant effect. Excellent for the border and among shrubs. Prop. by division, rarely by seed. Any soil. N. Amer. See Plate, page 316.	2-3 ft *Sun*	Early July to early Aug.
"Orange" 13	BLACKBERRY OR LEOPARD LILY	*Belemcánda Chinénsis *B. punctàta, Pardánthus Sinênsis, P. Chinênsis, Íxia punctàta*	Similar to the Iris. Crimson-spotted flowers an inch long, passing quickly. Lance-shaped leaves in tufts. Seeds resembling Blackberries. Prop. by seed and division. Rich and sandy soil. China; Japan.	2-4 ft. *Sun*	July, Aug.
"Orange" near 7	DOUBLE ORANGE DAISY	*Erígeron aurantìacus	A showy species with solitary daisy-like flowers. Pretty, massed in wild garden or border. Prop. by seed and division. Any garden soil. Turkestan.	9 in. *Half shade*	"
"Flame color" 12	FISCHER'S HORNED POPPY	*Glaùcum Físcheri	Poppy-like flowers bloom briefly in rapid succession. Snow-white woolly foliage. Excellent as an edging. Prop. by division. Best to treat as a biennial. Any garden soil. Hort.	9-12 in. *Sun*	July
"Copper" near 7	HYSSOP-LEAVED ROCK ROSE	*Heliánthemum vulgàre var. hyssopifòlium	See page 210.		Early June, July
"Orange" 10 deeper	ORANGE DAY LILY	*Hemerocállis aurantìaca	Showy profusion of trumpet-shaped flowers are borne in clusters on long stems. Grass-like foliage. Striking in border or rock-garden. Prop. by seed and division. Any moist rich soil. Japan. **Var. *major;* larger than the type in every way and very fine with wide-open flowers of a deeper color and glaucous foliage. Japan.	2½-3 ft. *Half shade*	July, early Aug.
"Orange" 11 brilliant	DUMORTIER'S DAY LILY	**Hemerocállis Dumortièrii *H. rùtilans, H. Sièboldii*	See page 210.		June, July

Color	English Name	Botanical Name and *Synonyms*	Description	Height and *Situation*	Time of Bloom
"Tawny orange" 14 brighter	BROWN DAY LILY, MAHOGANY LILY	*Hemero-cállis fúlva *H. dísticha*	A larger plant than H. flava. Flowers lily-shaped, in clusters of 6 to 12; good for cutting. Foliage grass-like. Attractive in masses especially in semi-wild spots and on the edge of water. Easily cultivated. Prop. by division. Any moist rich soil. Europe; Asia. See Plate, page 319. *Var. *Kwamso,* Double Orange Lily; (color no. 9), double flowered and sometimes with variegated foliage. Handsome "for edgings and rock-garden." *Var. *flore-pleno;* a fine var. with a red spot on each petal. Var. *variegata;* leaves striped with white. Hort. vars.	2-3 ft. *Sun or half shade*	July, early Aug.
"Or-ange" 12	ORANGE HAWK-WEED	Hieràcium aurantìacum	See page 211.		June to Oct.
"Or-ange" effect 7	OREGON LILY	*Lílium Columbiànum *L. parviflò-rum, L. Sàyi*	See page 211.		June, early July
"Apri-cot" 10 pinker	SHINING THUNBER-GIAN LILY	**Lílium élegans var. fúlgens *L. Bâteman-niæ, L. fúlgens, L. sanguí-neum*	One of the best and strongest vars. Very easy to grow. Flowers smaller than the type, unspotted and of a particularly charming color, generally 3 on the stalk. Remarkably effective massed in border or on margin of shrubbery. Rich light peaty soil. Avoid direct contact with manure. Japan.	1-3½ ft. *Sun or shade*	Mid. July to early Aug.
"Warm apricot" 11	WALLACE'S THUNBER-GIAN LILY	**Lílium élegans var. Wállacei *L. Wállacei*	Rather pale flowers, spotted with black, generally two on the stalk. Excellent in masses on the edge of shrubbery, or in the border or shady part of rock-garden. Light rich soil. Avoid direct contact with manure. Japan.	1-2½ ft. *Sun or half shade*	July
"Or-ange" 26 orange	ASA GRAY'S LILY	*Lílium Gràyi	See page 211.		June, July
"Reddish orange" 26 orange	SPOTTED OR HANSON'S LILY	**Lílium maculàtum *L. Hánsoni*	See page 211.		"
"Reddish orange" 12 redder	WILD ORANGE-RED OR PHILADEL-PHIA LILY	*Lílium Philadél-phicum	The most characteristic and widely distributed of our native Lilies. Charming delicate flowers of a beautiful color spotted with black, on graceful leafy stalks. Good for margin of Rhododendron bed or in border. Bulbous. Prop. by offsets, scales, or very slowly by seed. Light rich soil. Avoid direct contact with manure. N. Amer.	1-3 ft. *Sun or half shade*	July, Aug.

BROWN DAY LILY AND PERENNIAL PHLOX. *Hemerocallis fulva and Phlox paniculata.*

TIGER LILY. *Lilium tigrinum.*

Color	English Name	Botanical Name and *Synonyms*	Description	Height and *Situation*	Time of Bloom
"Reddish orange" bet. 19 & 20	AMERICAN TURK'S CAP LILY	**Lílium supérbum	Native in meadows and marshes. Delicate drooping flowers, having pointed reflexed petals, spotted within, in a pyramidal panicle of about 20. Charming, scattered in margin of Rhododendron beds or massed in border. Bulbous. Prop. by offsets, scales, or very slowly by seed. Any well-drained soil. Avoid direct contact with manure. Eastern N. Amer.	3-6 ft. *Sun or half shade*	Early July to early Aug.
"Orange" 12	TIGER LILY	**Lílium tigrìnum	A showy and vigorous plant. Racemes of drooping flowers with reflexed purple-spotted petals. Dark glossy foliage. Old-fashioned and useful giving a stately effect in the border. Naturalize near shrubbery or along a stone wall. Bulbous. Prop. by offsets, scales, or bulbs. Any well-drained soil. Avoid direct contact with manure. China. See Plate, page 320. Var. **splendens;* (color no. 16 deeper), one of the best Lilies bearing elongated racemes of about 25 flowers. Very effective clumped in border. Var. *flore-pleno;* double flowered and more thickly spotted. China; Japan.	2-5 ft. *Sun or half shade*	Mid. July to Sept.
"Red orange" 11	ATLANTIC POPPY	*Papàver rupífragum var. Atlánticum *P. A.*	See page 114.		Late May to Aug.
"Orange" 11	RUSSELL'S CINQUE-FOIL OR FIVE-FINGER	Potentílla Russelliàna	Large single flowers. Pretty compound leaves. Protect in winter. Not strictly hardy. Border. Prop. by division in spring. "Heavy soil." Hort.	2 ft. *Sun*	July, Aug.
"Orange" 6 to 18	POTTS' BLAZING STAR	**Tritònia Póttsii *Montbrètia Póttsii*	Vigorous and showy species. Star-like yellow flowers, flushed with brick red, in waving racemes. Stiff linear leaves. Protect with mulch or house in winter. Excellent for the border. Bulbous. Plant in Apr. or May and prop. by scales or offsets. Rich well-drained soil. S. Africa.	2-3 ft. *Sun*	July to Oct.
"Red" 27	RED YARROW OR MILFOIL	*Achillèa Millefòlium var. rùbrum	Numerous tiny flowers in broad flat clusters. Foliage finely divided. Effective on edge of shrubbery or in border. Prop. by division and cuttings. Any soil. S. Europe; Asia; N. Amer.	1-3 ft. *Sun*	Mid. July to mid. Sept.
"Violet crimson" 27 bright	PURPLE POPPY MALLOW	*Callírhoe involucràta	See page 212.		June to Sept.

Color	English Name	Botanical Name and *Synonyms*	Description	Height and *Situation*	Time of Bloom
"Red"	FRENCH OR CROZY CANNA	**Cánna vars.	See color "various," page 357.		July to late Sept.
"Red"	ITALIAN CANNA	**Cánna vars.	See color "various," page 358.		"
"Maroon" near 33	DARK PURPLE KNAPWEED	Centaurèa atropurpùrea *C. calocéphala*	See page 212.		June to Sept.
"Crimson" near 26	RED VALERIAN, JUPITER'S BEARD	*Centránthus rùber	See page 212.		June, July
"Red"	DAHLIA	**Dahlia vars.	See color "various," page 359.		Late July to late Oct.
"Orange scarlet" 18 more orange	SOUTHERN SCARLET LARKSPUR	*Delphínium cardinàle	Flowers in showy spike-like racemes. Resembles P. nudicaule, but the flowers are finer and clusters larger. Leaves deeply divided. Handsome among shrubs or in the border. Protect in winter. Prop. by seed, division or cuttings. Deep friable sandy soil, well enriched. Cal.	2-3 ft. *Sun*	July, Aug.
"Red" 31 or 20 & 27 brilliant	CARTHUSIAN PINK	Diánthus Carthusianòrum *D. atrórubens*	See page 212.		Early June to early July
"Deep red" 28 bright to 26	DARK RED PINK	**Diánthus cruéntus	See page 213.		June, July
"Crimson" 33	SEROTIN'S PURPLE CONE-FLOWER	*Echinàcea purpùrea var. serótina *E. intermèdia*	A fine form though of somewhat coarse habit. Large daisy-like flowers, 3-6 in. across, in varying shades of rose and purple with prominent centres. Border. Prop. by division. Any soil, preferably rich and sandy. N. Amer.	2-3½ ft. *Sun*	July, Aug.
"Red" 41 slightly deeper	BLOOD-RED CRANES-BILL	**Geránium sanguíneum	See page 115.		Late May to mid. July
"Dark red" 28	DOUBLE DEEP CRIMSON AVENS	*Gèum atrosanguíneum var. flòre-plèno	See page 213.		June, July
"Orange scarlet" 26 orange	CHILOE AVENS	*Gèum Chiloénse *G. coccíneum,* (Hort.)	See page 213.		Late June to early Aug.
"Red"	SWORD LILY	**Gladìolus vars.	See color "various," page 365.		July to Oct.
"Deep red" 33	FRENCH HONEY-SUCKLE	Hedýsarum coronàrium	See page 213.		June, early July

JULY

Color	English Name	Botanical Name and *Synonyms*	Description	Height and *Situation*	Time of Bloom
"Red" 26 brilliant	CORAL OR CRIMSON BELLS	**Heùchera sanguínea	See page 213.		June to late Sept.
"Bright red" 18	BULB-BEARING LILY	*Lílium bulbíferum	A desirable plant. Clusters of brown-spotted flowers sometimes shading to orange. The scattered leaves bear small bulbs. Pretty in border. Bulbous. Prop. by offsets or scales, very slowly by seed. Light rich soil. Avoid direct contact with manure. Europe.	2-4 ft. *Half shade*	July, Aug.
"Orange red" 13 deep & redder	RED CANADA LILY	*Lílium Canadénse var. rùbrum	See page 214.		Early June to Aug.
"Orange red" 16	SOUTHERN RED LILY	*Lílium Cátesbæi	Not easily cultivated in the North. Bright flowers spotted with dark purple. Bulbous. Prop. by offsets, scales, or very slowly by seed. Light well-drained soil, sandy and gravelly. Avoid contact with manure. S. Eastern U. S. A.	1-2 ft. *Sun or half shade*	July
"Scarlet" 18 deeper	SCARLET MARTAGON OR RED LILY	*Lílium Chalcedóni-cum	Gay nodding turban-shaped flowers, about 3 in. across, in small clusters, surmount leafy stalks. Good massed in border. Bulbous. Prop. by offsets, scales or very slowly by seed. Light well-drained soil. Avoid contact with manure. Greece.	3-4 ft. *Sun or half shade*	"
"Scarlet" 17 deeper	JAPANESE RED STAR LILY	*Lílium cóncolor	See page 214.		June, July
"Orange red" 12 redder & deeper	THUNBER-GIAN LILY	**Lílium élegans *L. Dahùri-cum, L. Thun-bergiànum, L. umbellàtum*	See page 214.		"
"Bright red" 17, centre 5	PANTHER LILY	**Lílium pardalìnum *L. Califòrni-cum*	A fine species. Drooping orange-centred flowers with purple-spotted reflexed petals. Foliage in whorls. Good for margin of shrubbery, etc. Bulbous. Prop. by offsets, scales, or very slowly by seed. Light well-drained soil. Avoid contact with manure. Cal.	2-3 ft. *Half shade*	July
"Scarlet" 17 brilliant	SIBERIAN CORAL LILY	**Lílium tenuifòlium	See page 214.		Late June, July
"Scarlet" 33 warmer	SHINING CARDINAL FLOWER	*Lobèlia fúlgens *L. cardinàlis* (Hort.) *L. formòsa*	Strikingly brilliant flowers, large, deeper color and more showy than the Cardinal Flower. Long narrow bronze-tinted leaves on the flower stalks. Effective in clumps against green. Protect in winter; not strictly hardy. Prop. by seed or cuttings. Mexico.	2-3 ft. *Sun*	July, Aug.

Color	English Name	Botanical Name and *Synonyms*	Description	Height and *Situation*	Time of Bloom
"Scarlet" bet. 12 & 18	MALTESE CROSS, JERUSALEM SAGE, SCARLET LIGHTNING	**Lýchnis Chalcedónica	See page 214.		Early June to mid. July
"Deep crimson" 33 brighter	DEEP CRIMSON MULLEIN PINK OR DUSTY MILLER	*Lýchnis Coronària var. atrosanguínea	See page 217.		June, July
"Red" 11, 17 & 18 brilliant	SHAGGY LYCHNIS	*Lýchnis Haageàna	See page 217.		Early June to early Aug.
"Red" 24 redder	BLOOD-RED AMARYLLIS	Lycòris sanguínea	Half-hardy bulb. Large flowers, 3-4 in. wide, in umbels, bloom after the long narrow foliage has disappeared. Japan.	1-3 ft. *Sun*	July, Aug.
"Rosy red" 18	HALL'S AMARYLLIS	*Lycòris squamígera *Amarýllis Hállii*	Hardy bulb. Immense fragrant lily-like flowers, striped with yellow, bloom after the long narrow foliage has disappeared. Showy in border. Prop. by offsets. Var. *purpurea* is particularly hardy. Japan.	1-3 ft. *Sun*	"
"Scarlet" 18 duller	SCARLET MONKEY FLOWER	*Mímulus cardinàlis	See page 217.		June to Sept.
"Red" 20 lighter, centre 21 redder	OSWEGO TEA, BEE OR FRAGRANT BALM	**Monárda dídyma	See page 217.		Mid. June to early Sept.
"Crimson" 29 turning darker	WHORL-FLOWER	*Morìna longifòlia	See page 217.		June, July
"Bright red" 27	ROUND-LEAVED REST HARROW	Onònis rotundifòlia	Half shrubby plant with bright pea-shaped blossoms striped with deep rose. Compound leaves scattered along the branching stems. Rough part of rock-garden or border. Prop. by seed and division. Any soil. Europe.	$1\frac{1}{2}$ ft *Sun*	July
"Scarlet" 24 deeper	SCARLET OURISIA	Ourísia coccínea	See page 218.		June, July
"Scarlet" bet. 12 & 19	ORIENTAL POPPY	**Papàver orientàle	See page 218.		Early June to early July
"Light red" 16	RUPIFRAGE POPPY	*Papàver rupífragum	See page 117.		May to Aug.

Color	English Name	Botanical Name and *Synonyms*	Description	Height and *Situation*	Time of Bloom
"Scarlet" 18	TORREY'S BEARDED PENTSTEMON	**Pentstèmon barbàtus var. Tórreyi *P. Tórreyi*	A very attractive kind. Graceful plant which forms a dense clump bearing many erect stems from which droop brilliant tubular flowers. Gray-green leaves in clumps at the base. A charming border plant. Prop. by seed and division. Any soil not too dry and hot. Col.; N. Mexico.	4-5 ft. *Sun*	Early July to early Aug.
"Purplish red" 41 deeper	HARTWEG'S LARGE-FLOWERED HYBRID PENTSTEMON	Penstèmon gentianoìdes hỳbrida grandiflòra *P. Hártwegi hỳbrida g.*	See page 218.		June to Sept.
"Red"	PERENNIAL PHLOX	**Phlóx paniculàta vars. *P. decussàta*	See color "various," page 370.		July to Oct.
"Vermilion" 18 redder	CAPE FUCHSIA	*Phygèlius Capénsis	Effective sub-shrubby plant. Long pendent tube-shaped flowers in branching racemes terminate the many stalks which rise above the thick foliage. Not very hardy and needs protection. Prop. by seed or cuttings. Light rich soil. S. Africa.	1½-3 ft. *Sun*	July, Aug.
"Red"	ALKEKENGI, STRAWBERRY TOMATO, WINTER OR BLADDER CHERRY	Phýsalis Alkekéngi	Striking because of its bright red calyxes surrounding the edible fruit which is showy and lasts a long time. Not hardy in the North. Winter protection necessary. Prop. by seed and division. Good garden soil. S. Europe.	1-2 ft. *Sun*	Fruit July to late Oct.
"Red"	CHINESE LANTERN PLANT	*Phýsalis Franchétti *P. Alkekéngi var. Franchétti*	Interesting and excellent hardy plant, grown for its handsome fruit. The fruit and calyxes are larger than in P. Alkekengi, are very showy and last a long time in good condition. Prop. by seed and division. Japan.	2 ft. *Sun*	"
"Red" 35	DARK BLOOD-RED SILVERY CINQUEFOIL	Potentílla argyrophýlla var. atrosanguínea *P. atrosanguínea*	See page 221.		June, July
"Red"	HYBRID CINQUEFOIL DOUBLE VARS.	Potentílla hỳbrida vars.	See page 221.		June to Sept.
"Scarlet" bet. 12 & 15	RUSSELL'S HYBRID CINQUEFOIL	*Potentílla Russellìana	See page 221.		Early June to Aug.

Color	English Name	Botanical Name and *Synonyms*	Description	Height and *Situation*	Time of Bloom
"Pinkish red" near 33	COBWEB OR SPIDER-WEB HOUSE-LEEK	*Sempervìvum arachnoìdeum	Pretty starry blossoms rise above pale almost white pulpy leaves clustered in tiny cactus-like rosettes which are studded closely together. Cultivated more for its peculiar foliage than for its flowers. Charming carpeting for rock-garden or stone-wall. Prop. by division. Dry sandy well-drained soil. Mts., S. Europe.	3-5 in. *Sun*	July
"Pale red" near 22	ATLANTIC HOUSE-LEEK	*Sempervìvum Atlánticum	See page 221.		June, July
"Red" 24, centre 33	HOUSE-LEEK, OLD-MAN-AND-WOMAN	*Sempervìvum tectòrum	See page 221.		"
"Crimson" 20 brighter & lighter	FIRE PINK	*Silène Virgínica	See page 221		"
"Deep red" 20, 4 inside	PINK ROOT, WORM GRASS	*Spigèlia Marylándica	See page 222.		Late June to early Aug.
"Scarlet" 28	RED HEDGE NETTLE	Stàchys coccínea	Rather a pretty plant. The small flowers are in irregular spikes. Border. Protect slightly in winter. Prop. by seed and division. Any soil. Tex. to Ariz. and Mexico.	1-2 ft. *Half shade*	July
"Bright red"	RED SPIDER-WORT	Tradescántia Virginiàna var. coccínea	See page 117.		Late May to late Aug.
"Orange scarlet" often 16 pinker	CROCUS-FLOWERED BLAZING STAR	**Tritònia crocosmæflòra *Montbrètia crocosmæflòra*	The commonest species. Lovely slender branching plant with long spike-like racemes of flowers, 2 in. across. Tall stiff narrow foliage springs from near the bulb. Protection of mulch or indoor wintering is necessary. Very gay and attractive for the border. Bulbous. Plant in Apr. or May and prop. by scales and offsets. Rich well-drained soil. There are many yellow and orange vars. See Plate, page 327. Var. **Transcendant;* bright red outside, golden yellow within. Hort.	3-4 ft. *Sun*	July to Oct.
"Bright red" sometimes 29	REDDISH BLAZING STAR	**Tritònia ròsea *Montbrètia ròsea*	Spreading tubular flowers spotted with yellow. Short stiff narrow leaves. Good winter protection of mulch is necessary; it is safer to winter indoors. For cultivation, etc., see T. crocosmæflora. S. Africa.	1 ft. *Sun*	July, Aug.

CROCUS-FLOWERED BLAZING STAR. *Tritonia crocosmæflora.*

CROWN VETCH. *Coronilla varia.*

Color	English Name	Botanical Name and *Synonyms*	Description	Height and *Situation*	Time of Bloom
"Pink" 29 lighter	NODDING WILD ONION	Állium cérnuum	Flowers, sometimes white, in loose slightly drooping umbels. Grassy foliage. Easily cultivated in rock or wild garden. Prop. by seed and offsets. Well-drained soil. W. Alleghanies.	12-18 in. *Sun*	July
"Light pink" 22 pale	CHINESE GOAT'S BEARD	**Astílbe Chinénsis	Graceful species. Flowers in compact panicles rising above the handsome foliage. Effective in masses and in the border. Prop. by division. Any good garden soil. China.	1½-2 ft. *Half shade*	July, early Aug.
"Rose" 29 lighter & duller	EVANS' BEGONIA	Begònia Evansiàna *B. díscolor, B. grándis*	Hardy with slight protection. Profusion of large flowers. Foliage green above, red beneath. Prop. by cuttings. Light rich soil. Japan; China.	2 ft. *Sun*	July, Aug.
"Pink" 37 pinker	HEATHER, LING, HEATH	*Callùna vulgàris *Erìca vulgàris*	Evergreen shrub. Small flowers sometimes white, in dense terminal racemes, lasting long when cut. Pretty in margin of shrubbery or in detached clumps. Prop. by division, cuttings or layers. Sandy peaty soil preferable. Europe; Asia Minor.	½-3 ft. *Sun*	"
"Purple pink" 40 duller	GRASS PINK ORCHIS	Calopogon pulchéllus *Limodòrum tuberòsum*	See page 223.		June, July
"Purplish pink" 27	HOARY CEDRONELLA	*Cedronélla càna	See page 223.		June to Sept.
"Deep pink" 31	WHITENED KNAPWEED	Centaurèa dealbàta	See page 223.		Late June to early Aug.
"Rose" 40 duller	ROSY MOUNTAIN BLUET	**Centaurèa montàna var. ròsea	See page 223.		Late May to early July
"Purplish rose" 36 deep	SMALL ROSE OR PINK TICKSEED	*Coreópsis ròsea	See page 223.		June to Sept.
"Pink" 39	CROWN VETCH	Coronílla vària	See page 223 and Plate, page 328.		"
"Pink" 29	CROSSWORT, FŒTID CRUCIANELLA	*Crucianélla stylòsa	See page 224.		"
"Pink etc."	DAHLIA	**Dáhlia vars.	See color "various," page 359.		July to late Oct.
"Deep rose" 29 duller & pale	ALPINE PINK	*Diánthus alpìnus	See page 224.		June, July

JULY

Color	English Name	Botanical Name and *Synonyms*	Description	Height and *Situation*	Time of Bloom
"Delicate pink" 27 light	CHEDDAR PINK	*Diánthus cæsius	See page 121.		Late May to early July
"Magenta pink" 23	CINNAMON PINK	**Diánthus cinnabarìnus	Flowers in tight round clusters. Rigid grass-like leaves. Charming in border or rock-garden. Prop. best by division. Rich soil. Greece.	1 ft. *Sun*	July, Aug.
"Pink" 25, 27, 28, 33, 34 & 35	BROAD-LEAVED PINK	**Diánthus latifòlius	See page 224.		June to Sept.
"Rose" 36 light	ROCK PINK	*Diánthus petræus	See page 224.		Mid. June to Aug.
"Purplish rose" 31	SEGUIER'S PINK	**Diánthus Seguierii	See page 227.		Late June, July
"Deep rose" 25	WILD BLEEDING HEART	**Dicéntra eximía Diélytrae	See page 227		Early June to Aug.
"Rose" 30	BLEEDING HEART	**Dicéntra spectàbilis *Dièlytra s.*	See page 40		Late Apr. to mid. July
"Purplish pink" 32	RED GAS PLANT OR FRAXINELLA	**Dictámnus álbus var. rùbra *D. Fraxinélla var. rùbra*	See page 227.		June, July
"Purplish pink" 29	PALE PURPLE CONE-FLOWER	*Echinàcea angustifòlia *Braunèria pállida*	Rudbeckia-like plant. Flowers smaller than in E. purpurea, of brighter color, varying from pale pink to purple, continuous in bloom. Foliage hairy. Good border plant. Prop. by seed and division. Any soil, preferably rich and sandy. Central U. S. A.	1-3 ft. *Sun*	July, Aug.
"Purplish pink" bet. 32 & 39, centre 13 & 20	PURPLE CONE-FLOWER, BLACK SAMPSON	**Echinàcea purpùrea *Rudbéckia purpùrea*	A compact bushy plant with large showy rayed flowers having drooping petals and dark centres. Blossoms last a long time. Splendid border plant. Prop. by seed and division. Requires good rich soil. U. S. A.	2-3½ ft. *Sun*	"
"Shell-pink" bet. 24 & 17	HEDGE-HOG THISTLE, SIMPSON'S CACTUS	Echinocáctus Símpsoni	See page 227.		June to Sept.
"Magenta" 40	GREAT WILLOW HERB, FIRE WEED, FRENCH WILLOW	Epilòbium angustifòlium *Camænèrion angustifòlium*	See page 228.		June to early Aug.

JULY

Color	English Name	Botanical Name and *Synonyms*	Description	Height and *Situation*	Time of Bloom
"Peach-colored" 38	ROBUST EREMURUS	*Eremùrus robústus	See page 228.		June, July
"Pink" 24	ARMENIAN CRANES-BILL	Geranium Armènum *G. Back-housiànum*	See page 228.		"
"Light pink" veined with 26	LANCASTER CRANES-BILL	*Gerànium sanguíneum var. Lancas-triénse	See page 228.		"
"Pink"	SWORD LILY	**Gladìolus vars.	See color "various," page 365.		July to Oct.
"Pale pink" 36	CREEPING CHALK PLANT	*Gypsóphila repens *G. prostràta*	See page 228.		Early June to mid. July
"Pink" 22 light	MANY-PAIRED FRENCH HONEY-SUCKLE	Hedýsarum multijugum	See page 228.		June to early Aug.
"Rose"	CHANGE-ABLE ROCK ROSE	Heliánthe-mum vulgàre var. mutàbile	See page 228.		Early June to Aug.
"Pur-plish rose" bet. 32 & 40	DELAVAY'S INCAR-VILLEA	*Incarvíllea Delavàyi	See page 229.		Early June to mid. July
"Pink"	JAPANESE LILY	*Lílium Japónicum	A choice species which resembles L. Brownii. 1-5 fragrant funnel-form flowers sometimes purple out-side, always white within. Leafy stems. Pleasing among low shrubs or in border. Not very enduring. Bul-bous. Prop. by offsets, scales, or very slowly by seed. Light rich soil. Avoid direct contact with manure. China; Japan. Var. ***roseum* (*L. Krameri*)*;* (color no. 43 pinker), more graceful than the type, with soft rose colored flowers varying to deep pink. A lovely Lily.	1-3 ft. *Sun or half shade*	July, Aug.
"Pale pink" 36 or 43	TWIN FLOWER	Linnæa boreàlis	See page 229.		June, July
"Pink" 27 light or 29 pale	RAGGED ROBIN, CUCKOO FLOWER	*Lýchnis Flós-cuculi	See page 229.		"
"Bright rose" 31	JUPITER'S FLOWER, UMBELLED LYCHNIS	*Lýchnis Flós-Jovis *Agrostemma Flós-Jovis*	See page 229.		"

Color	English Name	Botanical Name and *Synonyms*	Description	Height and *Situation*	Time of Bloom
"Deep rose" 29	ALCEA MALLOW	*Málva Alcèa	See page 229.		June to Sept.
"Rose" 38 more violet	MUSK MALLOW	*Málva moschàta	A showy plant. Large single fragrant flowers 2 in. across, in clusters. Foliage sweet-smelling. Good in border. Prop. by seed and division. Any soil. Europe.	1-2 ft. *Sun or shade*	June to early Sept.
"Purplish pink" 31	MEXICAN PRIMROSE	*Œnothèra ròsea	See page 229.		June, July
"Light pink"	SAINFOIN, HOLY CLOVER	Onóbrychis satìva *O. viciæfòlia, Hedýsarum o.*	See page 230.		"
"Pink" 32	THORNY REST HARROW	Onònis spinòsus *O. arvênsis var. spinòsus*	See page 230.		"
"Purplish pink" bet. 37 & 38	WILD OR POT MARJORAM	Oríganum vulgàre	See page 230.		"
"Pale pink" 22	ORIENTAL POPPY BLUSH QUEEN	**Papàver orientàle "Blush Queen"	See page 230.		Early June to early July
"Pink or carmine" 32 varying	BEARDED PENTSTEMON	*Pentstèmon barbàtus *Chelòne barbàta*	See page 230.		Early June to mid. July
"Rose" 38 to 27	BELL-FLOWERED PENTSTEMON	*Pentstèmon campanulàtus *P. angustifòlius, P. atropurpùreus*	Branching plant, somewhat shrubby in appearance. Long narrow one-sided spikes of flowers in varying shades of rose, resembling Foxgloves. Desirable border plant. Prop. by seed or division. Any garden soil, not too dry. Mexico.	1½-2 ft. *Sun*	July, Aug.
"Purplish rose" 32	TUBEROUS JERUSALEM SAGE	Phlòmis tuberòsa	See page 230.		June, July
"Deep rose" 40 lighter	CAROLINA PHLOX	*Phlóx Carolìna *P. ovàta*	See page 231.		"
"Pink" 40	SMOOTH PHLOX	*Phlóx glabérrima	Erect growing plant. Flowers in flat clusters, varying to white. Bright glossy foliage. Pretty in border. Prop. by seed, division or cuttings. Rich moist soil. U. S. A.	1-2½ ft. *Sun*	July
"Rose" often 45 lighter	SHRUBBY SMOOTH-LEAVED PHLOX	*Phlóx glabérrima var. suffruticòsa *P. s., P. nìtida*	See page 231.		June to early Aug.

Color	English Name	Botanical Name and *Synonyms*	Description	Height and *Situation*	Time of Bloom
"Reddish pink" 31 brilliant	MOUNTAIN PHLOX	*Phlóx ovàta *P. triflòra*	See page 231.		June, July
"Pink"	PERENNIAL PHLOX	**Phlóx paniculàta vars. *P. decussàta*	See color "various," page 370.		July to Oct.
"Purplish rose" 37 pinker	FALSE DRAGON HEAD, OBEDIENT PLANT, LION'S HEART	*Physostègia Virginiàna *P. Virginica*	Grows in clumps forming an erect bushy plant. Flowers flesh colored, purple or white, in dense spike-like racemes. Transplant and divide frequently. Associate with bold plants in the border. Prop. by division in the spring. Strong rich rather moist soil. N. Amer. Var. ***denticulata;* (*P. denticulata, Dracocephalum denticulatum*), a lower-growing and more slender form. Flowers less showy. N. Y. to Va. Var. **speciosa;* (*P. imbricata, Dracocephalum speciosum*), taller and of more slender habit than the type, with larger darker flowers and broader leaves. Tex.	1-3 ft. *Sun*	Early July to Aug
"Rose" 26	HYBRID CINQUE-FOIL	*Potentílla Hapwoodiàna	See page 231.		Early June to Aug.
"Rose" 38 more violet	PRAIRIE SABBATIA	Sabbàtia campéstris	See page 231.		Late June, July
"Rose" about 25	PINK MEADOW SAGE	*Sálvia praténsis var. ròsea	See page 232.		June, early July
"Pink" 38 paler & more violet	ROCK SOAPWORT	*Saponària ocymoìdes	See page 126.		Late May to Aug.
"Light pink" 38 & paler	DOUBLE BOUNCING BET	Saponària officinàlis var. flòre-plèno	Rambling rapidly spreading plant. Clusters of double flowers varying to to white are borne on leafy stems. Good for naturalizing and in border when kept from spreading. Prop. by seed and division. Any soil. Europe.	$1\frac{1}{2}$-$2\frac{1}{2}$ ft. *Sun or shade*	July, Aug.
"Rose" near 27	AUTUMN SQUILL, STARRY HYACINTH	*Scílla autumnàlis	Charming bulb. Small starry flowers in racemes. Leaves in tufts from the root. Good in rock-garden, wild garden or border. Enrich soil occasionally with top-dressing of manure. Great Britain; N. Africa.	6-9 in. *Sun or shade*	Late July to Oct.
"Purplish pink" 32	PURPLE STONECROP	*Sèdum stoloníferum *S. spùrium*	Trailing plant. Most desirable of the Sedums. Flowers, sometimes white, in clusters 2 in. wide. Flat succulent leaves. Good for carpeting rock-garden or edge of border. Caucasus.	6 in. *Sun*	Mid. July to early Aug.

JULY

Color	English Name	Botanical Name and *Synonyms*	Description	Height and *Situation*	Time of Bloom
"Pink" 23 dull	ORPINE, LIVE-FOREVER	*Sèdum Teléphium	Sturdy plant. Flowers vary to rosy purple and sometimes to white, in wide clusters, not abundant, but lasting long when cut. Foliage succulent. Pretty in dry border or rock-garden. Prop. by seed or offsets. Any soil, preferably sandy. Europe; N. Asia. Var. *purpurascens;* (color no. 32), flowers also purplish, in dense heads on strong stems, height 15-20 in. France. Var. *purpureum;* foliage dark purple. As edging or in rock-garden. N. Asia.	12-18 in. *Sun*	July, Aug.
"Pale" pink" 38	AUVERGNE HOUSELEEK	*Sempervìvum Arvernénse	See page 232.		June, July
"Salmon rose" 22 brighter	VERLOTI'S HOUSELEEK	*Sempervìvum Verloti	Starry flowers on leafy stalks rise well above rather grayish pulpy foliage clustered in small cactus-like rosettes. Pretty carpeting for rock-garden. Prop. by seed. Sandy well-drained soil. French Alps.	6-9 in. *Sun*	July
"Pink" 29 lighter	PINK BEAUTY	Sidálcea malvæflòra var. Lísteri	See page 232.		June, July
"Rose pink" 38	AUTUMN CATCHFLY	Silène Scháfta	See page 233.		June to Sept.
"Deep pink" 24 brilliant	QUEEN OF THE PRAIRIE	*Spiræa lobàta *S. palmàta Filipéndula l. Ulmària rùbra*	See page 233.		Early June to Aug.
"Carmine" 26 deep	PALMATE-LEAVED MEADOW SWEET	**Spiræa palmàta *Filipéndula purpùrea, Ulmària p.*	See page 233.		June, July
"Pink" 22	ELEGANT PALMATE-LEAVED MEADOW SWEET	**Spiræa palmàta var. élegans *Ulmària rùbra var. e.*	See page 233.		Late June, July
"Rose"	HILL-LOVING SEA LAVENDER	Státice collìna *S. Besseriana*	See page 233.		June, July
"Purplish rose" 37	AMERICAN GERMANDER, WOOD SAGE	Teùcrium Canadénse *T. Virgínicum*	Tubular flowers varying to cream white in long spikes. Fragrant foliage, soft and whitish. Wild garden or border. Prop. by division. Moist soil. Eastern U. S. A.	1-3 ft. *Sun*	Late July to late Aug.
"Pink" 29 lighter	SAXIFRAGE-LIKE TUNICA	*Tùnica saxífraga	See page 233.		June, July

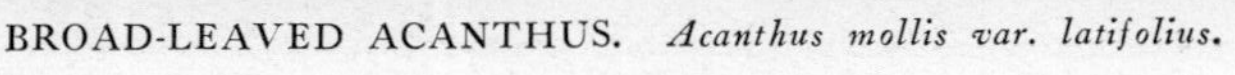

BROAD-LEAVED ACANTHUS. *Acanthus mollis var. latifolius.*

 WILD MONKSHOOD. *Aconitum uncinatum.*

Color	English Name	Botanical Name and *Synonyms*	Description	Height and *Situation*	Time of Bloom
"Pink" near 45	PINK LONG-LEAVED SPEEDWELL	Verónica longifòlia var. ròsea	See page 233.		June, July
"Pink" 29 light	PINK SPIKE-FLOWERED SPEEDWELL	*Verónica spicàta var. ròsea	See page 233.		"
"Purple"	BROAD-LEAVED ACANTHUS	**Acánthus móllis var. latifòlius *A. latifòlius, A. Lusitánicus*	A large and hardy variety. Flowers in loose spikes which rise above the large and striking foliage. Protect in winter. Effective for border or rock-garden. Prop. by seed and division in spring and fall. Light soil. S. Europe. See Plate, page 335.	4 ft. *Sun*	July, Aug.
"Deep purple" 49 darker	WILD MONKSHOOD	**Aconìtum uncinàtum	See page 233 and Plate, page 336.		Mid. June to Sept.
"Purple" 47	GLAND BELL-FLOWER	*Adenóphora Lamárckii	See page 234.		June to Sept.
"Lilac" 26 purpler	LARGE-FLOWERED LEBANON CANDYTUFT	*Æthionèma grandiflòrum	See page 127.		May to Aug.
"Purple"	JAPANESE STAR GRASS	Áletris Japónica	Small flowers in spikes on slender stems which rise much above the glossy clump of grass-like foliage. Prop. by seed and division. Rather moist soil is preferable. Japan.	2-3 ft. *Sun*	July, Aug.
"Pale purple" near 53	LARGE-LEAVED ANEMONOPSIS	*Anemonópsis macrophýlla *A. Califórnica*	Similar to Anemone Japonica, but smaller. Nodding half-closed flowers, pale purple or lilac, in loose racemes. Border. Prop. by seed or division in early spring or late autumn. Rich soil, deep and well-drained. Japan.	2-3 ft. *Half shade*	July
"Violet" 47	RUSH ASTER	*Áster júnceus	Slender plant with slightly branched stems. Flowers in panicles, varying to nearly white. Good for planting along the margins of ponds or in any moist place. Prop. by seed and division. Eastern U. S. A.	1-3 ft. *Sun*	July to early Sept.
"Violet" bet. 46 & 47	SAVORY-LEAVED ASTER	*Áster linariifòlius *Diplopappus linariifòlius*	A neat tufted plant with very showy flowers. Very desirable for naturalizing in wild garden, planted in high dry position in any light soil. Eastern U. S. A.	6-20 in. *Sun*	July to Oct.
"Pale violet" 50 lighter	ARROW-LEAVED ASTER	Áster sagittifòlius	An erect plant with hairy leaves. The flowers are showy and pleasing. Naturalize in wild garden in any ordinary soil. N. Amer.	2-3 ft. *Sun*	July, Aug.
"Pinkish lavender" 43 pinker	BROAD-SCALED BOLTONIA	**Boltònia latisquàma	A delightful plant resembling the Aster, but of more feathery effect with larger flowers than B. asteroides. Profuse bloomer, good for cutting. Attractive in rough places and in the border. Prop. by division. Kans.; Mo.	4 ft. *Sun*	Late July to Oct.

Color	English Name	Botanical Name and *Synonyms*	Description	Height and *Situation*	Time of Bloom
"Purple" 47 dull	LARGE SELF-HEAL	Brunélla grandiflòra *B. Pyrenaìca*	See page 234.		June, July
"Lilac purple" 23 dull	TOM THUMB CALAMINT	Calamíntha glabélla	See page 237.		"
"Pale purple" 44	CUT-LEAVED CALIMERIS	*Calímeris incìsa *Áster i., Boltònia i.*	Aster-like flowers with yellow centres. Erect habit like our native Boltonias. Good in border. Fairly rich soil. Siberia.	1-2 ft. *Sun*	July, Aug.
"Violet" 47 shading darker	CARPATHIAN HAIRBELL	**Campánula Carpática	See page 237.		Late June to late Aug.
"Lilac" 44 bluer & deeper	BRITTLE HAIRBELL	*Campánula frágilis *C. Barrelierii*	Plant of trailing habit. Starry flowers of a delicate blue or lilac, white in the centre, solitary or two together. Pretty in border or rock-garden. Protect slightly in winter. Prop. by cuttings in spring. Rich loam, well-drained. S. Italy.	4-6 in. *Sun*	July
"Pale blue violet" 44 deep	GARGANO HAIRBELL	*Campánula Gargánica	See page 128.		May to Sept.
"Purple" 48	DANE'S BLOOD, CLUSTERED BELL-FLOWER	**Campánula glomeràta	See page 237.		June, July
"Deep purple" 48	LARGE-BLOSSOMED BELL-FLOWER	**Campánula latifòlia var. macrántha *C. macrántha*	See page 238.		Early June to Aug.
"Violet" 47 shading darker	PEACH-LEAVED BELL-FLOWER, PEACH BELLS	**Campánula persicifòlia	See page 238.		Early June to mid. July
"Purple" 39 & 47	RAINER'S BELL-FLOWER	*Campánula Raìnerii	See page 238.		Mid. June to late July
"Violet" 47 lighter or darker	CREEPING OR EUROPEAN BELL-FLOWER	*Campánula rapunculoides	Common in old gardens and spread to roadsides. Long spikes of drooping funnel-form flowers. A continuous bloomer if cut. Foliage good. Extremely pretty to naturalize or for the border. Prop. by seed, division or cuttings. Rich loam well drained. Europe; Siberia; naturalized in Eastern U. S. A.	2-4 ft. *Sun or shade*	July, Aug.

JULY

Color	English Name	Botanical Name and *Synonyms*	Description	Height and *Situation*	Time of Bloom
"Violet" 47 shading to 49	ENGLISH HAIRBELL, BLUE BELLS OF SCOTLAND	**Campánula rotundifòlia	See page 238.		June to late Aug.
"Light purple" 48	COVENTRY BELLS, BLUE FOXGLOVE	*Campánula Trachèlium *C. ucticifòlia*	A common and very hardy species. Sturdy hairy plant with racemes of bell-shaped somewhat drooping flowers bearded within, in clusters of two or three. Rough leaves. Apt to crowd out other species but is a good border plant. Prop. by seed, division or cuttings. Rich well-drained loam. Europe.	2-3 ft. *Sun or half shade*	Mid. July to mid. Aug.
"Violet" 40	VAN HOUTTE'S BELL-FLOWER	**Campánula Van Hoùttei	See page 238.		Early June to mid. July
"Violet" 47	BLUE SUCCORY, BLUE CUPIDONE	*Catanánche cærùlea	See page 239.		June to Sept.
"Purple" 46 shading to 39	MOUNTAIN BLUET OR KNAPWEED	**Centaurèa montàna	See page 239.		"
"Purplish" effect 43	RAM'S HEAD LADY'S SLIPPER	*Cypripèdium arietìnum	See page 131.		Late May to Aug.
"Purplish"	DAHLIA	**Dáhlia vars.	See color "various," page 359.		July to late Oct.
"Deep purple" 49 deep	KASHMIR LARKSPUR	*Delphínium Cashmeriànum	See page 239.		June, July
"Lilac" 43 darker	FRINGED PINK	*Diánthus supérbus	Solitary fragrant flowers with fringed petals. Small bright green leaves. Attractive in border or rock-garden. Prop. best by division. Rich soil. Europe; Japan.	1-2 ft. *Sun*	July, early Aug.
"Light purple" bet. 43 & 44	VERY ROUGH ERIGERON	*Erígeron glabéllus var. ásper	See page 239.		Early June to mid. July
"Purplish lilac" 47 lighter	SHOWY FLEABANE	**Erígeron speciòsus *Stenáctis speciòsa*	See page 240.		June, July
"Light purple"	GLANDULAR STORK'S OR HERON'S BILL	Eròdium macradènium *E. glandulòsum*	See page 240.		June to Sept.
"Purplish"	MANESCAUT'S STORK'S OR HERON'S BILL	Eròdium Manescàvi	See page 240.		June, July

JULY

Color	English Name	Botanical Name and *Synonyms*	Description	Height and *Situation*	Time of Bloom
"Pale lilac" 50	**FORTUNE'S DAY OR PLANTAIN LILY**	***Fúnkia Fórtunei**	Funnel-shaped flowers in racemes well above the gray-green leathery foliage. Effective for margin of border or shrubbery. Prop. by division; sometimes by seed just ripe. Rich soil. Japan.	1-2 ft. *Half shade best*	July
"Pale lilac" 47 pale	**LANCE-LEAVED DAY OR PLANTAIN LILY**	****Fúnkia lancifòlia** *F. Japónica*	Tubular flowers in loose racemes rising above the clumps of foliage. Excellent in the border or the margin of shrubbery. Prop. by division; sometimes by seed just ripe. Rich soil. Japan. *Var. *albo-marginata, (F. albo-marginata);* variegated foliage. Hort. Var. *undulata; (F. undulata)*, variegated and wavy margined leaves. Hort. These variegated vars. are not so good as the type.	1-2 ft. *Half shade best*	Late July to early Sept.
"Lavender" 45 bluer	**OVAL-LEAVED DAY OR PLANTAIN LILY**	***Fúnkia ovàta** *F. cærùlea, F. lanceolàta*	See page 240.		June, early July
"Pale lilac" 43 lighter	**SIEBOLD'S DAY OR PLANTAIN LILY**	***Fúnkia Sieboldiàna** *F. cucullàta, F. gigantèa, F. glaùca, F. Sieboldii, F. Sinênsis*	See page 240		June, July
"Lilac purple" 43	**GOAT'S RUE**	***Galèga officinàlis**	See page 240.		June to Sept.
"Purple" bet. 39 & 47	**HILL-LOVING CRANES-BILL**	**Gerànium collìnum** *G. Londessi*	See page 241.		June, July
"Purple" 42	**WALLICH'S CRANES-BILL**	**Gerànium Wallichiànum**	See page 241.		June to Sept.
"Purple" 40 to 47	**ROCKET, SWEET ROCKET, DAME OR DAMASK VIOLET**	***Hésperis matronàlis**	See page 241.		June, July
"Bluish purple" 49	**HYSSOP**	***Hýssopus officinàlis**	See page 241.		Mid. June to mid. Aug.
"Violet purple" 49	**ENGLISH IRIS**	****Ìris xiphioìdes** *I. Ánglica*	See page 242		Late June, July
"Purplish" 45	**VARIEGATED NETTLE**	**Làmium maculàtum** *L. purpùreum* (Hort.)	See page 136.		Mid. May to late July

JULY

Color	English Name	Botanical Name and *Synonyms*	Description	Height and *Situation*	Time of Bloom
"Purple" near 37 & 40	BLACK PEA OR BITTER VETCH	Láthyrus nìger *Órobus niger*	See page 242.		June, July
"Violet" bet. 47 & 49	TRUE LAVENDER	Lavándula vèra	Fragrant grayish sub-shrub. Flowers occasionally white, in spikes. An old garden favorite, used for oil of lavender. Slight protection needed. Flowers if dried will long retain fragrance. Prop. by young wood cuttings. Light rich open soil. S. Europe.	1-3 ft. *Sun*	Mid. July to Sept.
"Purple" 39 bright & clear	HANDSOME BLAZING STAR	*Liàtris élegans *Lacinària élegans*	Bright flowers in dense slender spikes; buds deeper, color no. 40. Narrow leaves. Striking in border or margin of shrubbery against background. Protect slightly in winter. Prop. by fall-sown seed, division or offsets. Rich or poor soil. Southern U. S. A.	2-3 ft. *Sun or half shade*	Early July to mid. Aug.
"Purple" near 46	LOOSE-FLOWERED BUTTON SNAKEROOT	*Liàtris graminifòlia *Lacinària graminifòlia*	Flowers rise in long spikes above narrow leaves. Good in border or wild garden. Prop. by fall-sown seed or offsets. Rich or poor soil. S. Eastern U. S. A.	2-3 ft. *Sun or half shade*	July, to Oct.
"Deep purple" near 46	DENSE BUTTON SNAKE-ROOT, GAY FEATHER	**Liàtris spicàta *Lacinària spicàta*	Handsome neat species. Flowers, small, in close spikes on unbranched erect stems. Foliage slender. Plant in masses in the border against good background. Prop. by seed and division. Light rich soil is best. Eastern U. S. A. *Var. *montana* (*L. pumila*); lower-growing, 1-1½ ft. high, leaves broader, flower spike shorter. Mts., N. C. and Va.	2-5 ft. *Sun or half shade*	Mid. July to early Sept.
"Lilac" 47 light	KENIL-WORTH IVY, MOTHER OF THOUSANDS	Linària Cymbalària	See page 242.		June to Sept.
"Bluish purple" 44 duller	HIMA-LAYAN LUNGWORT	*Lindelòfia spectàbilis	Showy plant. Pendent clusters of tubular flowers tinged with purple or pink. Sheltered places in the border. Prop. by seed and division. Any well-drained soil. N. India.	1-1½ ft. *Sun*	Early July to early Aug.
"Purple" bet. 47 & 48	TALL BLUE-FLOWERED PERENNIAL LUPINE	**Lupìnus polyphýllus *L. grandi-flòrus*	See page 242.		June, July
"Purple" near 29 duller	SLENDER-BRANCHED PURPLE LOOSE-STRIFE	Lýthrum virgàtum	See page 245.		"

JULY

Color	English Name	Botanical Name and *Synonyms*	Description	Height and *Situation*	Time of Bloom
"Light purple" 44	REQUIEN'S PENNY-ROYAL	Méntha Requieni	Tiny creeping plant. Flowers in scanty whorls. Small leaves. Rock-garden or carpeting for rose-beds. Protect slightly in winter. Prop. by running root-stock. Prefers moist soil. Corsica.	2-3 in. *Sun*	July, Aug.
"Pur-plish" 39	SPEARMINT	Méntha spicàta *M. víridis*	Flowers in slender spikes. Foliage with attractive taste and fragrance. Used for flavoring. Any ordinary soil. Naturalized from Europe; Asia.	1-2 ft. *Sun*	"
"Purple" 37 or 41	WILD BERGAMOT	*Monárda fistulòsa	Striking plant of somewhat rough habit. Large showy open flowers purple in the type with vars. ranging from white to purple. Effective in masses and for naturalization. Prop. by division. Any soil. N. Amer. Var. *media;* (*M. media, M. purpurea*), (color no. 42 lighter or 55). Good var. Flowers deep purple. Hort.	1½-3½ ft. *Sun*	July
"Violet" 47 tinged with 46	MUSSIN'S CATMINT	Népeta Mussíni *N. longifòlia*	Branching plant with rather weak stems, but its many flowers make a pretty color effect in a roomy border. Prop. by seed or division. Any light garden soil. Caucasus; Persia.	2 ft. *Sun*	July, Aug.
"Purple" 47 deep	SCOTCH THISTLE	Onopórdon Acánthium	Vigorous biennial plant of bold habit. Showy flowers about 2 in. wide. Foliage large, silvery and prickly. Effective against dark shrubs. Seeds itself freely. Prop. by seed. Europe.	5-9 ft. *Sun*	"
"Lilac" 43	SNAKE'S-BEARD	Ophiopògon Jabùran	Similar in habit to O. Japonicus but more vigorous. Flowers varying to white in dense racemes. Leathery grass-like leaves. Protect in winter. Sheltered position. Prop. in spring by division. Sandy soil. Japan. Var. *variegatus,* variegated foliage. Japan.	6-24 in. *Sun or half shade*	"
"Pale lilac" 43	GIANT BELL-FLOWER	*Ostròwskia magnífica	Very large bell-shaped flowers, 3-6 in a cluster, borne on stems which branch only above, thus forming with the whorls of foliage very handsome clumps. Mulch in winter. Prop. by seed or by cuttings of the root or young growths in spring. Well-worked and drained sandy soil. Turkestan.	4-5 ft. *Sun*	"
"Violet" 44 pinker	SHARP-LEAVED BEARD-TONGUE	*Pentstèmon acuminàtus	See page 245.		June, early July
"Light purple" 47	DIFFUSE PENTSTE-MON OR BEARD-TONGUE	**Pentstèmon diffùsus	See page 245.		Early June to early July

JULY

Color	English Name	Botanical Name and *Synonyms*	Description	Height and *Situation*	Time of Bloom
"Lilac purple" 45	SLENDER BEARD-TONGUE	*Pentstèmon grácilis	See page 136.		Late May to early July
"Pale violet" 47, edge white	DOWNY PENTSTEMON	*Pentstèmon pubéscens	See page 136.		Late May to mid. July
"Lavender" 44	ONE-SIDED PENTSTEMON OR BEARD-TONGUE	**Pentstèmon secundiflòrus	See page 245.		June, July
"Deep violet"	VIOLET PRAIRIE CLOVER	Petalostèmon violàceus *Kuhnìstera purpùrea*	See page 245.		June until frost
"Lilac," "purple"	PERENNIAL PHLOX	**Phlóx paniculàta vars. *P. decussàta*	See color "various," page 370.		July to Oct.
"Bluish violet" bet. 46 & 47	BELL-FLOWERED HORNED RAMPION	*Phyteùma campanu-loìdes	Ragged looking flowers in spikes. The largest and strongest growing species of the genus. Well grown clumps suitable for rock-garden. Prop. by seed or division in spring. Any garden soil. Caucasus.	1-2 ft. *Sun*	July, Aug.
"Purple" near 30 deeper to white	JAPANESE PRIMROSE	**Prímula Japónica	See page 139.		Late May to Aug.
"Purple" 44 bluer	ROSEMARY, OLD MAN	*Rosmarìnus officinàlis	See page 246.		June, July
"Purple" near 49	SYLVAN SAGE	*Sálvia sylvéstris	Flowers in showy spikes. Oblong leaves. Pretty in the border. Prop. by seed. Any soil. Europe; Asia.	3-3½ ft. *Sun*	July
"Blue purple" 44 deep	VERVAIN SAGE	Sálvia Verbenàca *S. spelmìna, S. Spièmanni*	See page 246.		June, July
"Bluish purple" near 50	SMALL OR LILAC-FLOWERED SCABIOUS	Scabiòsa Columbària	See page 246.		June to Oct.
"Purple" 49	MOUNTAIN SKULLCAP	*Scutellària alpìna	Charming creeping plant. Ornamental tubular flowers, sometimes with yellow lips, in racemes, form rounded tufts of bloom. Leaves oval. Excellent for edge of border or rock-garden. Prop. by seed and division. Any soil. Europe.	8-10 in. *Sun*	Early July to late Sept.
"Purple" 43 deep	WIDOW'S CROSS, BEAUTIFUL STONECROP	Sèdum pulchéllum	See page 246.		June, July

Color	English Name	Botanical Name and *Synonyms*	Description	Height and *Situation*	Time of Bloom
"Rosy purple" 40	PRETTY GROUNDSEL	*Senècio púlcher	Showy plant. Large rayed flowers with yellow centres on almost leafless stems. Leaves arrow-shaped. Sheltered situation in border. Protect slightly. Prop. by root-cuttings. Any well-drained soil. Uruguay; Buenos Ayres.	2-4 ft. *Sun*	July, Aug.
"Pinkish purple" 43	CUSHION PINK, MOSS CAMPION	*Silène acaùlis	See page 246.		June, July
"Light violet" 44 deep	TORREY'S NIGHT-SHADE	Solànum Tórreyi	Plant of erect habit. Few large flowers, 2 in. across, in flat-topped clusters. Bright round yellow berries. Prickly foliage. Border. Prop. by seed or cuttings. Rich loam. Southern U. S. A.	1 ft. *Sun*	July, Aug.
"Red purple" 41	WOOD BETONY	Stàchys Betónica *Betónica officinàlis*	See page 249.		Late June, July
"Violet" often 45	LARGE-FLOWERED WOUND-WORT	*Stàchys grandiflòra *Betónica ròsea*	See page 249.		Mid. June to late July
"Purple" near 37	WOOLLY WOUND-WORT	Stàchys lanàta	See page 249.		"
"Dull violet" 47	TALL SEA LAVENDER	Státice elàta	Spikes of tiny blossoms form a mound well above a striking tuft of large leaves. Good in rock-garden or border. Prop. in spring by seed or in autumn by division. Loose soil preferable. S. Russia.	2 ft. *Sun*	Mid. July to early Sept.
"Blue violet" 44	GMELIN'S SEA LAVENDER	*Státice Gmélini	Delicate plant. Numerous flowers in dense erect paincles. Good for cutting. Tufted root-leaves. Pretty in rock-garden or open border. Prop. by seed. Does best in loose soil of good depth. E. Europe.	1-2 ft. *Sun*	Late July to early Sept.
"Bluish purple" 47	BROAD-LEAVED SEA LAVENDER	**Státice latifòlia	Strikingly effective plant. Clouds of tiny flowers in large spreading spikes. Good for cutting. Bold luxuriant foliage, springing from the root. Excellent in border or margin of shrubbery. Prop. by seed in spring. Leave undisturbed. Deep soil. S. Russia.	1½-2 ft. *Sun*	"
"Purple' near 46	SHOWY SEA LAVENDER	*Státice speciòsa	Blue or white flowers in dense spikes crowning numerous branchlets. Grayish foliage. Good in rock-garden or border. Prop. by seed and division. Sandy well-drained deep soil. Siberia.	1 ft. *Sun*	July

JULY

Color	English Name	Botanical Name and *Synonyms*	Description	Height and *Situation*	Time of Bloom
"Purple" 39	WALL GERMANDER	Teùcrium Chamædrys	Branching plant with clusters of pretty tubular flowers, spotted with red and white. Glossy fragrant foliage, whitish beneath. Border or rocky bank. Prop. by division. Light soil. Europe.	1-2 ft. *Sun*	Mid. July to mid. Aug.
"Lilac" 43	MOTHER OF THYME, CREEPING THYME	Thỳmus Serpýllum	See page 249.		Mid. June to mid. Aug.
"Pale lilac" 37	COMMON GARDEN THYME	Thỳmus vulgàris	See page 249.		June, July
"Pur-plish" 44, 48 or 49	COMMON SPIDER-WORT	*Tradescántia Virginiàna *T. Virgínica*	See page 140.		Late May to late Aug.
"Lilac" near 53	LONG-LEAVED SPEEDWELL	**Verónica longifòlia	Effective sturdy plant of dense growth. Flowers numerous in tall spike-like racemes. Grayish green foliage. Excellent in the border. Prop. by seed and division. Any good garden soil. Central and E. Europe; N. Asia. *Var. *villosa,* (color no. 46). An attractive form. Flowers of deeper color and foliage larger and narrower than that of the type. Siberia.	$2\frac{1}{2}$ ft. *Sun*	Mid. July to Aug.
"Violet" 47 or 49	HORNED VIOLET, BEDDING PANSY	**Vìola cornùta	See page 48.		Late Apr. until frost
"Deep blue" 56	AUTUMN ACONITE, MONKS-HOOD OR WOLFS-BANE	**Aconìtum autumnàle	Large drooping helmet-shaped flowers in spikes. Foliage dark green and deeply cut. The roots are poisonous. Very pretty for the border or rock-garden. Prop. by division. Prefers a rich soil. N. China.	3-5 ft. *Sun or shade*	Mid. July to mid. Sept.
"Blue"	STORK'S PURPLE WOLFBANE	*Aconìtum Cammàrum var. Storkiànum *A. Storkiànum, A. intermèdium*	Dwarf var. of A. Cammarum. Closed helmet-shaped flowers, not very numerous, in spikes. Roots poisonous. Border. Prop. by division. Rich soil preferable. European Alps.	3-4 ft. *Sun or shade*	July, Aug.
"Deep purple blue" near 49	TRUE MONKS-HOOD, OFFICINAL ACONITE	**Aconìtum Napéllus *A. pyramidàle, A. Taùricum*	Ornamental plant. Large and showy helmet-shaped flowers, growing in racemes on erect stems. Leaves deeply cut. Roots and flowers poisonous. Pretty for borders and rough places. Prop. by division. Rich soil preferable. Europe; Asia; N. Amer.	3-4 ft. *Sun or shade*	Mid. July to early Sept.
"Blue" 61	GLAND BELL-FLOWER	*Adenóphora commùnis *A. Fischeri, A. liliflòra, A. liliifòlia*	Rather large drooping bell-shaped flowers in panicles which rise above the leaves. Border. Prop. by seed and cuttings. Rich loam. Temp. Asia; W. Europe.	2-$2\frac{1}{2}$ ft. *Sun*	July to mid. Aug.

Color	English Name	Botanical Name and *Synonyms*	Description	Height and *Situation*	Time of Bloom
"Light blue" 61 duller	POTANNINI'S GLAND BELL-FLOWER	*Adenóphora Potannìni	Bushy plant. Spikes of nodding bell-shaped flowers, similar to Campanula. Foliage springing mostly from the root. Good for mixed border. Should not be disturbed when established. Prop. by seed and cuttings in spring. Rich loam. Turkestan.	2-3 ft. *Sun*	July, Aug.
"Sky blue" 51	AZURE ONIONWORT	Állium azùreum	See page 250.		June, July
"Deep blue" 49	LEAD PLANT	*Amórpha canéscens	Beautiful shrub of the Pea family with clustered spikes of flowers. Foliage very hoary and small. Excellent for shrubbery, rock-garden or border. Prop. by seed, cuttings and layers. Well-drained soil. Southern U. S. A.	1-3 ft. *Sun*	Early July to early Aug.
"Blue" 54	CAPE ALKANET	*Anchùsa Capénsis	See page 142.		Late May to early July
"Blue" 54, buds 41	ITALIAN ALKANET	*Anchùsa Itálica	See page 142.		Late May to mid. July
"Pale blue" 62 lighter	TUFTED HAIRBELL	*Campánula cæspitòsa	See page 145.		May to Aug.
"Pale blue" 46	HAIRY GARGANO HAIRBELL	*Campánula Gargánica var. hirsùta	Tufted trailing plant, hairier than the type. Numerous flowers, shading to white in the centre, in loose racemes on pendent stems. Pretty when hanging over rocky ledges. Prop. by cuttings in spring or by division. Rich well-drained loam. Italy.	3-6 in. *Half shade*	July
"Purplish blue" 46	GREAT BELL-FLOWER	*Campánula latifòlia	See page 253.		June to early Aug.
"Purplish blue" 48	CHIMNEY CAMPANULA, STEEPLE BELL-FLOWER	**Campánula pyramidàlis	Striking plant. Numerous open bell-shaped flowers close to the stem, in long spikes. Best treated as biennial. Effective in border or in isolated clumps. Protect in winter. Prop. by seed, division or cuttings. Rich loam well-drained. Australia.	4-6 ft. *Sun*	July, Aug.
"Light blue" bet. 46 & 52	SARMATIAN BELL-FLOWER	*Campánula Sarmática *C. gummífera*	Masses of velvety nodding light blue flowers in clusters. Gray leaves, wrinkled and hairy. Border. Prop. easily by seed. Rich loam, well-drained. Caucasus.	1-2 ft. *Sun*	July
"Blue" 53	FRÉMONT'S CLEMATIS	*Clématis Frèmonti	Plant of erect habit. Profusion of flowers which are generally drooping. Pretty in border or rock-garden. Winter mulching desirable. Prop. by seed, cuttings, grafting or layers. Rich deep soil. Western U. S. A.	1-2 ft. *Sun*	July, Aug.

Color	English Name	Botanical Name and *Synonyms*	Description	Height and *Situation*	Time of Bloom
"Blue" 60	UNDIVIDED-LEAVED VIRGIN'S BOWER	*Clématis integrifòlia	See page 253.		Mid. June to Aug.
"Blue" 61	MUSK LARKSPUR	*Delphínium Brunoniànum	See page 253.		June, July
"Blue" 62	CAROLINA LARKSPUR	*Delphínium Caroliniànum *D. azùreum, D. virèscens*	Racemes of flowers varying to white. Good for cutting. Second bloom possible if first flowers are cut off. Deeply divided foliage. Attractive in border. Prop. in spring or autumn by seed, division or cuttings; transplant every 3 or 4 years. Deep rich soil, sandy and loamy. U. S. A.	1½-2½ ft. *Sun*	July
"Blue" 60	CAUCASIAN LARKSPUR	*Delphínium Caucásicum	See page 253.		June, July
"Blue" 61	BEE LARKSPUR	**Delphínium elàtum *D. alpìnum, D. pyramidàle*	See page 253.		June to Sept.
"Blue" usually 54	ORIENTAL LARKSPUR	**Delphínium formòsum	See page 253.		June, July
"Blue" 62	GREAT-FLOWERED LARKSPUR	**Delphínium grandiflòrum *D. Sinénse*	Beautiful compact racemes of large flowers good for cutting. Deeply divided foliage. Very attractive in border. Prop. in spring or autumn by seed, division or cuttings. Deep rich soil, sandy and loamy. Siberia. Var. *flore-pleno,* (var. *hybridum flore-pleno*); double flowered. Hort.	1-2 ft. *Sun*	July, Aug.
"Blue" 62 or 54	HYBRID LARKSPUR	**Delphínium hỳbridum	See page 254.		June, July
"Blue" 54 & 47	MAACK'S LARKSPUR	*Delphínium Maackiànum	Flowers blue and deep violet in loose panicles. Divided foliage. If the plant is cut back it will flower again. Effective in the border. Prop. by seed, division or cuttings. A friable or sandy soil, well enriched. Siberia.	3 ft. *Sun or half shade*	July
"Deep blue" effect 46	LARGE-FLOWERED DRAGON'S-HEAD	*Dracocéphalum grandiflòrum *D. Altaiênse*	See page 254.		June, July
"Blue" 46	RUYSCH'S DRAGON'S-HEAD	*Dracocéphalum Ruyschiàna	Belongs to the Sage family. Flowers in whorls. Foliage lance-shaped. Neat border plant, also good for rock-garden. Prop. by seed and division. Rich sandy loam. Enjoys moisture. Siberia.	2 ft. *Half shade*	July
"Blue" 50 & cream	LOFTY GLOBE THISTLE	*Echìnops exaltàtus	Coarse thistle-like biennial differing from other species in its erect stem. Simple flowers in globe-shaped heads. Handsomely cut foliage and silvery stems. Good to naturalize and to plant among shrubs. Prop. by division or cuttings. Any soil. Russia.	5-7 ft. *Sun*	July, Aug.

JULY

Color	English Name	Botanical Name and *Synonyms*	Description	Height and *Situation*	Time of Bloom
"Steel blue" 53 colder	RITRO GLOBE THISTLE	**Echìnops Rìtro	Curious plant somewhat resembling a Thistle with flowers in globe-shaped heads and silvery and bluish foliage spiny and downy. Effective in wild garden or among shrubs, also with other bold plants in the border. Prop. by seed and division. Ordinary soil. S. Europe. See Plate, page 349. **Var. *tenuifolius*, (*E. Ruthĕmcus*); (color no. 53 to 56), the best Globe Thistle. S. Russia.	2-3 ft. *Sun*	July, Aug.
"Pale blue" 52 greenish	ROUND-HEADED GLOBE THISTLE	*Echìnops sphærocéphalus	A fine tall species. Woolly thistle-like plant. Flowers in globe-shaped heads, sometimes white, on white stems. Handsome silvery spiny foliage. Plant in wild garden and shrubbery with Bocconias, Eryngiums, etc., or naturalize. Prop. by division and cuttings. Ordinary soil. Europe.	5-7 ft. *Sun*	"
"Bright blue" 53	ALPINE SEA HOLLY	*Erýngium alpìnum	Odd thistle-like plant. Especially decorative from its steel blue stems and involucres. One of the most beautiful of the genus. Oblong flower-heads, 3 in. across, from spreading blue involucres. Leathery spiny leaves. Excellent for subtropical effect and in the border. Prop. by seed and division. Light soil preferable. Pastures, Swiss Alps.	1½-3 ft. *Sun*	"
"Amethyst blue" 63 lighter	AMETHYST SEA HOLLY	**Erýngium amethýstinum	See page 254 and Plate, page 349.		June to early Sept.
"Blue" 63	BOURGAT'S ERYNGO	*Erýngium Bourgati	Bluish foliage decorative and pungent. Oval flower-heads with large involucres. Spiny leaves deeply divided. Pretty in the border or massed in rock-garden. Prop. by seed and division. Soil, light and sandy. Pyrenees.	1-2 ft. *Sun*	Mid. July to early Sept.
"Blue" 63 lighter	DANEWEED, HUNDRED THISTLE	Erýngium campéstre	Foliage ornamental and peculiar. Roundish flower-heads in panicles. Deeply divided spiny leaves. Pretty for subtropical effect in the border. Prop. by seed and division. Light sandy soil. Europe.	1-2 ft. *Sun*	"
"Blue"	FLAT-LEAVED ERYNGO	**Erýngium plànum	Roundish flower-heads on steel blue stems. Large spiny thistle-like leaves, divided or whole. Excellent for subtropical effect in the border. Prop. by seed and division. Soil light and sandy. E. Europe; N. Asia.	1-3 ft. *Sun*	July, Aug.
"Purplish blue" 46	ORIENTAL GOAT'S RUE	*Galèga orientàlis	Graceful free-growing shrubby plant. Pea-shaped blossoms in dense racemes. Good for cutting. Wild garden. Prop. by seed and division. Any soil, preferably rich loam. Caucasus.	2-4 ft. *Sun*	July

AMETHYST SEA HOLLY. *Eryngium amethystinum.*

RITRO GLOBE THISTLE. *Echinops Ritro.*

 LARGE SMOOTH BEARD-TONGUE. *Pentstemon glaber.*

Color	English Name	Botanical Name and *Synonyms*	Description	Height and *Situation*	Time of Bloom
"Deep blue" 59 lighter	WILLOW GENTIAN	*Gentiàna asclepiadèa	Flowers in racemes covering nearly the whole length of the stems. Border or rock-garden. Prop. by division or very slowly by seed. Leave undisturbed. Rich moist soil. S. Europe.	6-18 in. *Half shade or shade*	July to early Sept.
"Violet" blue" 46 light	IBERIAN CRANES-BILL	Geranium Ibèricum	See page 254.		June to Sept.
"Dark blue"	DUSKY-FLOWERED CRANES-BILL, MOURNING WIDOW	Geranium phæum	See page 254.		Early June to mid. July
"Blue" 46 bluer	MEADOW CRANES-BILL	*Geranium praténse	See page 257.		June to Sept.
"Light blue" 52 deep & dull	HAIR-FLOWERED GLOBE DAISY	*Globulària tricosántha	See page 146.		Late May to Aug.
"Blue" 53	GLOBE DAISY	*Globulària vulgàris	See page 257.		June to Sept.
"Blue"	EASTERN SIBERIAN IRIS	**Ìris Sibírica var. orientàlis *I. S. var. sanguínea, I. S. var. hæmatophýlla, I. h., I. s.*	See page 257.		June, early July
"Light blue" 52 dull	SHEP-HERD'S OR SHEEP SCABIOUS, SHEEP'S BIT	*Jasiòne perénnis	Plant of compact habit. Globe-shaped flowers in profusion. Foliage in tufts. Border, rock-garden or as edging. Easily cultivated. Prop. by seed and division. Any well-drained garden soil. Mts., Central and S. Europe.	1 ft. *Sun or half shade*	Early July to early Aug.
"Blue" 61	AUSTRIAN FLAX	*Lìnum Austrìacum *L. perênne var. A.*	See page 258.		June to Sept.
"Blue" 54 lighter	PERENNIAL FLAX	**Lìnum perénne	See page 149.		Mid. May to Aug.
"Blue" 62 greener	NOOTKA LUPINE	*Lupìnus Nootkaténsis	See page 150.		Late May to early July
"Light blue" bet. 47 & 52	COMMON WILD LUPINE	*Lupìnus perénnis	See page 258.		June, July

Color	English Name	Botanical Name and *Synonyms*	Description	Height and *Situation*	Time of Bloom
"Blue" 57	**EVER-FLOWERING FORGET-ME-NOT**	***Myosòtis palústris var. sempérflorens**	See page 153.		May to Sept.
"Blue" 53	**LARGE-FLOWERED CATMINT**	**Népeta Macrántha**	See page 261.		Late June to early Sept.
"Purplish blue" 46	**LARGE SMOOTH BEARD-TONGUE**	***Pentstèmon glàber** *P. Górdoni, P. speciòsus*	See page 261 and Plate, page 350.		June, early July
"Lilac blue" 46	**LARGE-FLOWERED BEARD-TONGUE**	**Pentstèmon grandiflòrus**	See page 261.		Early June to early July
"Purplish blue" 46 & 41	**SHOWY PENTSTEMON OR BEARD-TONGUE**	****Pentstèmon spectàbilis**	See page 261.		Early June to mid. July
"Purplish blue" 46 dull	**MICHELI'S HORNED RAMPION**	***Phyteùma Michélii**	See page 261.		Late June, July
"Blue" often 56	**BALLOON FLOWER, JAPANESE BELL-FLOWER**	****Platycòdon grandiflòrum** *Campánula grandiflòra, Wahlenbérgia grandiflòra*	A shrubby plant with large showy open bell-shaped flowers, 2-3 in. across, very numerous at the summit of erect leafy stalk. Very attractive for the border or rock-garden. There are white and variegated vars. Prop. in the spring by seed or early division. Well-drained loamy soil. China; Japan. Var. *flore-pleno;* double form, July and August. Var. *Japonicum,* (*P. Japonicum*); more numerous flowers and more bushy habit. Var. *Mariesi;* (color no. 49, 53 or 56 and white), denser. Stronger habit than the type, about 1 ft. high; flowers equally large, purple, lavender, blue or white. Japan.	1-3 ft. *Sun or shade*	Early July to Oct.
"Pale blue" 52	**DWARF JACOB'S LADDER**	****Polemònium hùmile** *P. Richardsonii*	See page 261.		June, July
"Azure blue"	**HIMALAYAN VALERIAN**	***Polemònium réptans var. Himalayànum** *P. grandiflòrum, P. cærùleum var. g.*	See page 261.		Late May to early July

Color	English Name	Botanical Name and *Synonyms*	Description	Height and *Situation*	Time of Bloom
"Blue" 47	**HAIRY RUELLIA**	**Ruéllia ciliòsa**	Hairy or downy plant sometimes erect, sometimes straggling. Flowers 1-2 in. long, single or in clusters. Good for wild garden. Prop. by seed and division. Light soil. U. S. A.	1½ ft. *Sun*	Early July to late Aug.
"Blue" 46 lighter	**MEALY SAGE**	***Sálvia farinàcea**	Dense growing flowers of two distinct shades, branching out from the stalks in pairs forming elongated whorls. Pretty in border. Protect in winter. Prop. by seed, flowering early the first season. Any soil. Texas.	2-3 ft. *Sun*	July, Aug.
"Deep violet blue" near 47	**MEADOW SAGE**	****Sálvia praténsis**	See page 262.		June, early July
"Blue" 53 light	**PINCUSHION FLOWER**	****Scabiòsa Caucásica**	See page 262.		June, July
"Pale blue" often bet. 43 & 44	**GRASS-LEAVED SCABIOUS**	***Scabiòsa graminifòlia**	See page 262.		June to Oct.
"Bluish" near 39 paler	**WOODLAND SCABIOUS**	***Scabiòsa sylvática**	See page 262.		Early June to late Sept.
"Purplish blue" near 49	**BAICAL'S SKULLCAP**	***Scutellària Baicalénsis** *S. macrántha*	Alpine plant of neat half-erect habit, with an abundance of velvety flowers in long racemes. Desirable for rock-garden or border. Prop. by division. Siberia.	9-12 in. *Sun*	July, Aug.
"Deep blue" 46	**COMMON SEA LAVENDER, MARSH ROSEMARY**	****Státice Limònium** *S. marítima*	An effective plant bearing clusters of tiny flowers in numerous spikelets. Large leathery leaves spring from the root. Excellent in rock-garden and border and for cutting. Prop. by seed and division. Deep rich soil. Europe; N. Africa.	1½ ft. *Sun*	"
"Blue" 50	**MARSH FELWORT**	**Swértia perénnis**	Small star-like flowers varying to white, in spikes well above the low tuft of leaves. Bog or rock-garden. Prop. by seed and division. Moist soil. Europe; Asia; N. Western U. S. A.	6-12 in. *Shade*	July
"Blue" 32, turns 61	**PRICKLY COMFREY**	**Sýmphytum aspérrimum**	See page 154.		Late May to mid. July
"Blue" 53	**AUSTRIAN SPEEDWELL**	***Verónica Austrìaca**	Plant of erect vigorous habit. Flowers in showy racemes. Pretty in border. Any good garden soil. Prop. by seed and division. S. Eastern Europe; Asia.	1½-2 ft. *Sun*	Early July to early Aug.
"Blue" 46	**HOARY SPEEDWELL**	****Verónica incàna** *V. cándida, V. neglecta*	See page 262.		Mid. June to late July

Color	English Name	Botanical Name and *Synonyms*	Description	Height and *Situation*	Time of Bloom
"Light blue" 50	COMMON SPEEDWELL	Verónica officinàlis	See page 154.		May to Aug.
"Bright blue" 61 duller	SPIKE-FLOWERED SPEEDWELL	**Verónica spicàta	See page 262.		Early June, July
"Blue" near 47	BROAD-LEAVED HUNGARIAN SPEEDWELL	Verónicum Teùcrium var. latifòlia	See page 263.		June, July
Parti-colored	FRENCH OR CROZY CANNA	**Cánna vars.	See color "various," page 357.		July to late Sept.
Parti-colored	ITALIAN CANNA	**Cánna vars.	See color "various," page 358.		"
Parti-colored	DAHLIA	**Dáhlia vars.	See color "various," page 359.		July to late Oct.
6 shading from 19 to 14	GREAT-FLOWERED GAILLARDIA	**Gaillárdia aristàta *G. grandiflòra*	See page 263.		June to Nov.
Parti-colored	SWORD LILY	**Gladìolus vars.	See color "various," page 365.		July to Oct.
Outside 27 deeper	BROWN'S LILY	**Lílium Brówni *L. Japónicum var. Brównii*	Hardy and vigorous species recommended to beginners. A beautiful Lily with 2 fragrant partly drooping trumpet-shaped flowers with recurved petals, 7-8 in. long, white within and violet-purple without. Deep green foliage. Excellent in border or among low shrubs. Lift and replant every few years. Bulbous. Prop. by offsets or scales. Rich well-drained peaty soil. Avoid direct contact with manure. Japan. See Plate, page 355.	3-4 ft. *Half shade*	July, Aug.
44 & 11	ALPINE TOAD-FLAX	Linària alpìna	Dense spreading plant forming a silvery tuft, covered with small snapdragon-like flowers of violet with orange on the lower lip. Pretty for rock-garden. Prop. by seed and division. Of easy culture in any light soil. Alps.	4-6 in. *Sun*	"
Parti-colored	DOUBLE HYBRID CINQUE-FOIL	*Potentílla hỳbrida vars.	See page 263.		June, July
Various often 21	LONG-LEAVED BEAR'S BREECH	*Acánthus longifòlius	See page 264.		"
Various	BEAR'S BREECH, CUTBER-DILL SEDOCKE	**Acánthus móllis	Flowers, white to purple, in loose spikes. Grown for the large shapely foliage in clump at base of plant. Ornamental for border and rock-garden. Prop. by seed and division. Light well-drained soil. S. Europe.	3 ft. *Sun*	July, Aug.

BROWN'S LILY. *Lilium Browni.*

 HOLLYHOCK. *Althæa rosea.*

Color	English Name	Botanical Name and *Synonyms*	Description	Height and *Situation*	Time of Bloom
Various	VERY PRICKLY BEAR'S BREECH	**Acánthus spinosissímus	Flowers, white to purple, in loose spikes. Scanty dark green and prickly foliage. Protect in winter. Effective for border and rock-garden. Prop. by seed, or division in spring or autumn. Light soil well-drained and rich. S. Europe.	3-4 ft. *Sun*	July, Aug,
Various	PRICKLY BEAR'S BREECH	**Acánthus spinòsus	Flowers, white to purple, in compact spikes. Deeply-cut prickly leaves. See A. spinosissimus. S. Europe.	3-4 ft. *Sun*	"
2, 34, 35, 22, 27, white, etc.	HOLLY-HOCK	**Althæa ròsea	A well known and stately biennial. Large flowers, white, pale yellow, pink, or red, in long spikes which require staking. Foliage mostly in a clump at base of plant. Deep cultivation, much manuring, and watering in dry weather ensure the best results. Subject to fungus disease; Bordeaux mixture should be used in the earliest stages. Cover with manure in winter. There is no plant more effective in rows against houses and garden walls or in clumps at back of border. Prop. by seed, eyes, cuttings or division. Rich loam well drained. China. See Plate, page 356.	5-8 ft. *Sun*	"
Usually 21 or 2	WILD GINGER, CANADA SNAKEROOT	Ásarum Canadénse	See page 264.		Late June to early Aug.
36, 39, 47, 48, 43,45,etc.	CANTER-BURY BELLS	**Campánula Mèdium	See page 264.		Late June, July
48, or white spotted with 45	NOBLE BELL-FLOWER	**Campánula nóbilis	See page 264.		Mid. June to Aug.
Various	FRENCH OR CROZY CANNA	**Cánna vars.	A large-flowered group. Vigorous plants of dwarf compact growth, with large spikes of brilliant flowers, continue in bloom, sometimes from the last of June if started indoors. Foliage in many shades of color and of tropical effect in groups and masses. Plant bulb in spring when danger of frost is over. When the foliage withers dig up the bulb, dry in open air, and store in dry cellar. Prop. by division of rootstock. Light soil, rich, deep and moist. The following are some of the best vars.: **Red vars.**—*Admiral Dewey* or *Tarrytown;* (color no. bet. 11 & 12 deeper), beautiful for bedding; large spikes of deep red flowers, green foliage. *Antoine Crozy;* (color nos. 18 brilliant	1½-4 ft. *Sun*	July to late Sept.

Color	English Name	Botanical Name and *Synonyms*	Description	Height and *Situation*	Time of Bloom
			& 5), flowers rich carmine, leaves green. *President McKinley;* flaming crimson scarlet; large spikes. *President Cleveland;* (color no. 19 more orange), beautiful crimson. *Sir Thomas Lipton;* very large flower spikes; rich crimson. **Parti-colored vars.**—*Charles Henderson;* (color no. 18 redder), crimson and gold flowers. *E. G. Hill;* (color no. 17 brilliant), "scarlet mottled with carmine." *Florence Vaughan;* yellow flowers beautifully spotted; very desirable. *Madame Crozy;* (color nos. 18 & 5), brilliant crimson-scarlet margined with golden yellow. *Queen Charlotte;* (color nos. 18 duller & pinker & 5), scarlet, edged with yellow. *Roslindale;* (color no. 5 & center 12), large spikes of golden flowers variegated with carmine. Very desirable. *Souvenir Antoine Crozy;* (color no. 17 slightly redder), large flowers, vermilion margined with deep golden yellow, very effective. *Yellow Bird;* (color no. 6 light), deep yellow, lower petals slightly mottled; profuse bloomer.		
Various	**ITALIAN CANNA**	****Cánna vars.**	Vigorous tropical foliage plant, unbranched, taller than the French Canna and having longer bloom. Short-lived flowers in irregular racemes terminating stout stalks. Occasionally begins to blossom in June if started indoors. Largely grown for its handsome foliage. Plant in formal garden or group among shrubs. After frost dig up the roots and replant in spring, when there is no longer danger of excessive cold. Prop. by division of root stock. Light rich moist soil. Among the best vars. are: **Yellow vars.**—*Golden Sceptre;* (color no. 5), deep rich yellow. **Red vars.**—*America;* (color no. 13 lighter and yellower), large brilliant red flowers, dark red leaves. *La France;* (color no. 16 deeper), glowing orange-scarlet, glossy dark foliage. *Pluto;* (color no. 18 yellower), large flowers, deep scarlet. **Parti-colored vars.**—*Alemannia;* (color no. 11 & 3 edge), large scarlet flowers margined with yellow; broad leaves. *Aphrodite;* deep yellow flowers spotted with salmon-pink; green foliage. *Edouard André;* (color no. 17 mottled with 11), flowers flaming red with orange spots. *H. Wendland;*	3-4½ ft. *Sun*	July to late Sept.

Color	English Name	Botanical Name and *Synonyms*	Description	Height and *Situation*	Time of Bloom
			(color no. 12 & 3 edge), very large flowers, outer petals scarlet edged with gold, inner petals brilliant red with yellow centres; wide green leaves. *Oceanus;* (color no. 13 brilliant & 4 edge), outer petals fiery red, margined with golden yellow, inner petals scarlet; leaves green. *Pandora;* (color no. 18 with 4), rich red flowers bordered and blotched with gold.		
36, 31, 27, 32	**RED CHRYSANTHEMUM**	****Chrysánthemum coccíneum** *Pyrèthrum hŷbridum, P. ròseum*	See page 264.		June, July
Various	**CACTUS DAHLIA**	****Dáhlia vars.**	One of the most strikingly beautiful of tall plants. The flowers with twisted petals are sometimes single but usually double. The opalescent coloring of those shading from yellowish to pinkish tints is especially fine. Excellent planted in single or double rows. Prop. best by root division or cuttings. Plant tubers in late May or June, about 3 feet apart in sheltered spots in any soil, not too clayey, though best rich and deep. House in winter. Dahlia Juarezi is the parent of the Cactus Dahlia. Mexico. See Plate, page 361. **White vars.**—**Eva;* 4 ft. *Keynes' White;* "ivory white." 5 ft. **Lord Roberts; Miss Webster;* "pure white." 4 ft. *Mrs. A. Pearl;* "creamy white." 4 ft. **Yellow vars.**—*Artus;* "orange buff." 4 ft. *Florence;* "yellowish orange." *H. F. Robertson;* "deep pure yellow." 4 ft. *Mrs. De Lucca;* (color no. 3 to 9), orange yellow. *Mrs. Freeman Thomas;* "clear yellow shading to light orange." 3 ft. *Mrs. H. J. Allcroft;* "soft orange buff." 4 ft. **Pink and Red vars.**—*Ajax;* (color no. 16 brilliant & 17), "orange suffused with salmon and buff." 4 ft. **Britannia;* (color no. 24 pale), salmon pink, free bloomer. 4 ft. ***Clara G. Stredwick;* (color no. bet. 15 & 29), clear bright salmon shading to yellow at base of petals. Large flowers with extremely narrow petals of great length. 3 ft. **Countess of Lonsdale;* (color no. 24 richer and deeper), a blending of salmon and amber shades. 3 ft. *Exquisite;* (color no. bet. 16 & 17), bright scarlet tinted with salmon. 3 ft. *Fire Brand;* (color no. 19 red-	3-6 ft. *Sun*	Late July to late Oct.

Color	English Name	Botanical Name and *Synonyms*	Description	Height and *Situation*	Time of Bloom
			der to 20), rich vermilion shading deeper. *Galliard;* (color no. 19 redder), a true cardinal. 3 ft. **Kriembilde;* (white with color no. 30), "soft Apple-blossom pink with white centre." 3 ft. *Lady Ed. Talbot;* terra cotta shaded with salmon. **Mary Service;* (color no. 24 deep to 31 lighter), apricot shading into orange, margin purplish rose. 4 ft. *Monarch;* (color no. 24 deep, also 26 more orange), orange-red tipped with magenta. 3 ft. *Queen Wilhelmina;* (color no. 33 & deeper), deep pink. *Viscountess Sherbrook;* (centre color no. 11 shading from 15 to 18), "bright terra cotta suffused with apricot." 3 ft. **Dark Red or Maroon vars.**—*Aunt Chloe;* 3 ft. **J. H. Jackson;* (color no. 28 dark), flowers with long narrow petals on long stems. 4 ft. *King of Siam;* 4 ft. **Matchless;* (color no. 28 dark), 4 ft. *Mr. Moore. Night. Uncle Tom;* (color no. 28 dark). 4 ft. **Magenta vars.**—*Austin Cannel;* (color no. 33 brilliant), rose-crimson shading lighter at margin. 4 ft. **Purplish vars.**—*Emperor;* (color no. 26 bluish), indescribable Plum color shading at base of petal to yellow. 3½ ft. *Island Queen;* (color no. 40 to 47), purplish mauve. 3½ ft. **Parti-colored vars.**—*Alpha;* (white with markings color no. 48), white speckled and striped with purple-crimson and lilac. 4 ft. *Columbia;* (white dashed with color no. 18), pure white tipped with vermillion. 4 ft.		
Various	**COLLERETTE DAHLIA**	***Dáhlia vars.**	Collerette Dahlias are single and distinguished by the circle of short petals around the disk. Prop. best by root division. Plant the tubers 1½-3 ft. apart, in sheltered spots in any rich soil not too clayey. See Cactus Dahlia. Var. *Joseph Goujon,* deep scarlet, clear yellow collar. *President Viger;* (color no. 20 deeper), rich claret with lighter edges, and white collar. 2½ ft.	4-6 ft. *Sun*	Late July to late Oct.
Various	**DECORATIVE DAHLIA**	****Dáhlia vars.**	Decorative Dahlias resemble the earlier form of Cactus Dahlia, with broader flatter petals which do not roll backward. Plant 1½ to 3 ft. apart. For prop. and cultivation see Cactus Dahlia. See Plate, page 362. **Vars.**—*Admiral Dewey;* royal purple. *Black Beauty;* deep maroon.	4-6 ft. *Sun*	"

CACTUS DAHLIA

DECORATIVE DAHLIA

QUILLED DAHLIA

Color	English Name	Botanical Name and *Synonyms*	Description	Height and *Situation*	Time of Bloom
			Bronze Beauty; large red-bronze. *Lemon Giant;* lemon-yellow. 5 ft. *Lyndhurst;* (color no. 20 lighter), bright scarlet. 5 ft. *Mephisto;* vermilion. *Nymphæa;* (white to 38 pale), shrimp-pink, shaded darker. 5 ft. *Perle;* white. 3 ft. *Sundew;* orange-scarlet. *Wm. Agnew;* (color no. 19 brighter), bright red. *Wilhelm Miller;* purple, free blooming. 4 ft. *Zulu;* (color no. 28 deeper), blackish red. 5 ft.		
Various	**FANCY DAHLIA**	****Dáhlia vars.**	Fancy, Show and Pompon Dahlias belong to the same category, differing chiefly in size. The flowers are quilled and form symmetrical balls. Prop. most easily by root division or cuttings. Plant the tubers 1½ to 3 ft. apart, in sheltered spots in any rich soil not too clayey. Dahlia rosea is the parent of these Dahlia. **Large-flowered vars.**—*Admiral Schley,* crimson, shaded maroon, each petal striped with white; 3 ft. *Buffalo Bill;* deep golden orange, striped red; 4 ft. *Frank Smith;* (color no. 35 very deep, tipped with pale pink), rich deep purplish maroon, almost black, tipped with pale pink; 5 ft. *Judah;* pale yellow shading to old gold, marked with deep crimson. *Eloise;* (white to color no. 36 with petals tipped 33), white shading to blush-pink, with red margins; 4 ft. *Olympia;* (color no. 33 brighter), very large bright pink flowers, marked with crimson; 4 ft. *Lucy Fawcett;* (color no. 2 faintly marked 33 light), sulphur-yellow, marked with pinkish red; 5 ft. *Lottie Eckford;* (marked with color no. 43 warmer and 33 lighter), white, marked with pink and crimson; 3 ft. *Penelope;* white tinged with lavender near the margin. *Striped Banner,* rich bright red, white striped.	3-6 ft. *Sun*	Late July to late Oct.
Various	**POMPON DAHLIA**	***Dáhlia vars.**	Prop. best by root division. Plant the tubers 1½ to 3 ft. apart in any rich soil, not too clayey. The Dahlia rosea is parent. **Vars.**—*Catherine;* (color no. 3), yellow; 3 ft. *Elegante;* (bet. color nos. 30 & 27, tinged deeper), pink shaded deeper and lighter; 2½ ft. *Le Petit Jean;* plum color. *Little Beauty;* (32 brighter), shrimp-pink; excellent cut flower. *Little Naiad;* rose-lake tinged with crimson. *Little*	3-5 ft. *Sun*	"

Color	English Name	Botanical Name and *Synonyms*	Description	Height and *Situation*	Time of Bloom
			Prince; (26 deeper and white), red with white tips; 3 ft. *Lou Kramer;* yellow and pink, red tipped; 3 ft. *Snowclad;* white, very small; 2½ ft. *Sunshine;* bright scarlet; 3 ft.		
Various	**QUILLED DAHLIA**	****Dáhlia vars.**	Prop. most easily by root division or cuttings. Plant in sheltered spots in any rich soil not too clayey. Plant the tubers 1½ to 3 ft. apart. Quilled are often given under the head of both Decorative and Show Dahlias. See Plate, page 362. **Vars.**—*A. D. Livoni;* (color nos. 32 to 38), clear pink; 4 ft. *Grand Duke Alexis;* ivory white tinged with white near the edges; 4½ ft. *Kaiser Wilhelm;* (5 lighter and softer), old gold with scarlet tips. *Queen Victoria;* (color no. 3), deep yellow; 4 ft. *Ruth;* white faintly tinged with pink near the centre.	4-6 ft. *Sun*	Late July to late Oct.
Various	**SHOW DAHLIA**	****Dáhlia vars.**	Large compact flowers of two or more colors, double to the centre. Prop. best by division. Plant the tubers 1½ to 3 ft. apart in sheltered spots in any rich soil not too clayey. When frost destroys the stems lift the plant and keep over winter in dry cellar. Plant in May when there is no danger of frost. Dahlia rosea is the parent. See Plate, page 362. **Vars.**—*Arabella;* (color no. 36 with 2 centre), primrose, tinted with old rose and lavender; 4 ft. *John Bennett;* (color nos. 18 & 8), golden yellow tipped with red; 3 ft. *John Walker;* (color no. 2 pale), white. *La Phare;* rich scarlet. *Mrs. Dexter;* (color no. 16 pinker), deep salmon pink. *Miss May Loomis;* white tinted with rose. *Black Diamond;* black changing to black maroon. *Pink Dandy;* pink. *Red Hussar;* cardinal-red. *Queen of the Belgians;* (white to 36), delicate pale pink. *Queen of Yellows;* (color no. 3), clear yellow. *Thos. White;* (color no. 28 darker), deep maroon.	3-6 ft. *Sun*	"
Various	**SINGLE DAHLIA**	****Dáhlia vars.**	Prop. most easily by root division or cuttings. Plant the tubers 1½ to 3 ft. apart in sheltered spots in any rich soil not too clayey. Dahlia rosea is parent. **Vars.**—*Anemone;* white. *Annie Hughes;* (color no. 26 with 3 light centre), carmine with yellow centre. *Gold Standard;* very large deep yellow. *Juno;* white tipped with rose,	4-6 ft.	"

Color	English Name	Botanical Name and *Synonyms*	Description	Height and *Situation*	Time of Bloom
			yellow centred. *Mrs. Bowman;* (color no. 3 to white), solferino. *Polly Eccles;* fawn colored, red centred. *Sunset;* (color no. 4 warmer tinged 18), yellowish. *Black Bird;* maroon-black each petal spotted with red. *Danish Cross;* carmine each petal having a white central band. *Gaillardia;* golden yellow, red near the centre. *Lustre;* rose shaded lighter near the centre. *Record;* yellow, tinged with scarlet.		
Often 33 dark	**SCARLET DAHLIA**	**Dáhlia coccínea**	Bushy plant with erect rather slender grayish stems bearing large single flowers having yellow centres and rays recurved at the margin. Vars. yellow and orange. They should be staked. Dahlias are one of the most effective garden plants and are handsome in rows. Store roots in a dry cellar after frost kills the plant; set out preferably in June or July. They take six weeks to come to flower. Prop. by division; each tuber must have an eye, therefore start growth before dividing; cuttings, seed and grafting. Good sandy soil not too rich, for strength may go to foliage. Mexico.	2-5 ft. *Sun*	July to early Oct.
Various	**VON MERCK'S DAHLIA**	**Dáhlia Mérckii** *D. glabràta*	Plant of low spreading habit. Branching flower stems bearing single flowers, often lilac, with yellow centres and short recurved rays which rise 2 or 3 ft. above the foliage, which is particularly finely cut and handsome. Stake the plants. See D. coccinea for further information. Mexico.	2-3 ft. *Sun*	"
Often 17 dark	**COMMON DAHLIA**	**Dáhlia variábilis** *D. ròsea*	Color variable, flowers single and double. Vars. white and yellow. Hairy or smooth grayish or green stems. Leaves entire or in two divisions. Parent of most garden kinds. See D. coccinea for further information. Mexico.	2-5 ft. *Sun*	"
62, 57	**CHINESE LARKSPUR**	****Delphínium grandiflòrum var. Chinénse**	See page 267.		June, July
33, 34, 35, etc.	**SWEET WILLIAM**	****Diánthus barbàtus**	See page 267.		"
Usually bet. 32 & 39	**COMMON FOXGLOVE**	****Digitàlis purpùrea** *D. tomentòsa*	See page 267.		June, early July
Various	**SWORD LILY**	****Gladìolus vars.**	Conspicuous spikes of lily-like flowers which last long when cut. Sword-shaped leaves. Plant at intervals from early spring until July to secure succession of bloom. Striking and very desirable in border. When the	3-4½ ft. *Sun*	July to Oct.

Color	English Name	Botanical Name and *Synonyms*	Description	Height and *Situation*	Time of Bloom
			foliage dies, dig up the bulb, dry in open air and store in dry cellar. Prop. by seed or by small bulb at base of parent bulb. Any garden soil, preferably rich moist sandy loam. Cape Colony; Natal. See Plate, page 367. **Named hybrids. White vars.**—*Angèle;* very effective. *Snow-White;* beautiful pure white, with a faint line of rose on the lower petal. **Pink vars.**—*Madame Monneret;* delicate rose pink. *Pyramide;* large well-opened flowers, bright orange-rose. *Surprise,* purplish rose, very late blooming. **Red vars.**—**Brenchleyensis;* brilliant vermilion-scarlet. *Flamboyant;* large flowers, flaming scarlet. **Parti-colored vars.**—*Agatha;* dashed with lilac-carmine, spotted with clear yellow. *Apollon;* rose-lilac, with large rose spot, and centre striped with white. *Baucis;* large rose flowers, slightly tinted with salmon, blotched with purplish red. *Calypso;* flesh-colored rose, dashed with rose and mottled with carmine. *Ceres;* snow-white and mottled with purplish rose. *Crépuscule;* large spikes of lilac-rose flowers, suffused with carmine and edged with violet. *Eldorado;* yellow, lower petals marked with red. *Fatima;* cream-white streaked with rose-salmon and blotched with violet. *Grandesse;* large well expanded flowers, pinkish white, touched with lilac, blotched with carmine. *La France;* (24 deeper & 3), cream-white with carmine edge and purplish blotch. *Leviathan;* large flowers, delicate rose, streaked with carmine and blotched with purple. *May;* large spikes of pure white flowers, spotted with rose-crimson. *Mr. Jansen;* fine spike of large rose-carmine flowers, edged and slightly streaked with purplish red, and blotched with cream color. *Neron;* tall spike of round flowers, deep crimson, touched with deep blood-red and violet. *Ophir;* dark yellow blotched with purple. *Pepita;* creamy yellow streaked with pink. *Phœbus;* (18 brilliant), vivid red strikingly blotched with white. *Schiller;* sulphur-yellow, with large carmine blotch. *Titania;* beautiful spike, creamy salmon dashed with cherry-red. *Van Dael;* very large flowers, light salmon-pink near		

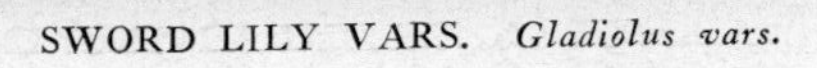

SWORD LILY VARS. *Gladiolus vars.*

JAPANESE IRIS. *Iris lævigata.*

Color	English Name	Botanical Name and *Synonyms*	Description	Height and *Situation*	Time of Bloom
			the centre, streaked blood-red at the petal edges. **Gladiolus Childsi. Red vars.**—*Cardinal;* (color bet. nos. 19 & 20), orange red. **Deborah;* (color nos. 18 & 19 brighter), scarlet-vermilion, with large clear white centre. **Gov. McCormick;* (color no. 18), scarlet vermilion with a white and cardinal red centre, (color no. 20). **Nezinscott;* (effect color no. 18), scarlet-vermilion. *Saratoga;* (effect color no. 19), orange scarlet. **Pink vars.**—*America;* (color no. 36), pale pink, centre deeper (color no. 33 brighter). *Harlequin;* (color no. 36), pale pink dashed with magenta (color no. 31). **Irene;* (color nos. 36 to 38 lighter), pale pink with dark centre (color no. 27). **Little Blush;* (effect color no. 36), pale pink, outer edge dark (color no. 32), with yellow centre (color no. 2). *Lydia;* (color nos. 36 & 37 to 27 brighter, effect 36 spotted with 27), pink with reddish spots. *Mrs. R. A. Goldsmith;* (effect color no. 23 deeper), rose-pink with white centre. **Siboney;* (color nos. 36 to 39), pink varying to heliotrope, with crimson edge (color no. 33). **Splendor;* (color no. bet. 24 & 26), deep pink, crimson centre (color no. 33). **Gladiolus Gandavensis. White vars.**—*Augusta;* (color no. 32), purplish white with small purplish pink centre. **Yellow vars.**—*Canary Bird;* (color nos. 2 greenish & 1), greenish yellow. *Sunshine;* (color nos. 2 greenish & 1), lemon-yellow. **Red vars.**—**Contrast;* (color no. 18 darker), scarlet-vermilion with large clear white centre. **Pink vars.**—*Ajax;* (color no. 36), light pink edged with old-rose (color no. 25). *Chaumont;* (color no. 36 to 27 deeper on edge), pink. *Eugène Scribe;* (color no. 36), light pink edged with rose-red (color no. 27). **Octoroon;* (effect color no. 22), light salmon-pink, spotted slightly with a deeper shade (color no. 30) with centre turning to pale yellow (color no. 1) dashed with crimson (color no. 33).		
Various	**JAPANESE IRIS**	****Ìris lævigàta** *I. Kæmpferi*	See page 267 and Plate, page 368.		June, July

Color	English Name	Botanical Name and *Synonyms*	Description	Height and *Situation*	Time of Bloom
Bet. 38 & 44 or white	BITTER-ROOT, SPATULUM	Lewísia rediviva	See page 268.		June, July
1 spotted brown, 11 to 14	WILD YELLOW OR CANADA LILY	**Lílium Canadénse	See page 268.		"
Often 33 brilliant	MULLEIN PINK, DUSTY MILLER, ROSE CAMPION	*Lýchnis Coronària *Agrostémma C.* *C. tomentòsa*	See page 268.		"
White, 2, 44	COBÆA BEARD-TONGUE	Pentstèmon Cobæa	Few-flowered panicle of large flowers with narrow tube and broad flat top, varying in shades from white to purple. Thick leaves. Suitable for border, rock-garden and margin of shrubbery. Prop. by seed and division. Requires protection. Any soil. S. Western U. S. A.	15-18 in. *Sun*	"
Various	LARGE OR JAPANESE PETASITES	Petasìtes Japónica var. gigantèa	See page 272.		"
Various	EARLY SUMMER FLOWERING PHLOX	*Phlóx glabérrima var. suffruticòsa	See page 272.		"
Various	WILD SWEET WILLIAM	*Phlox maculàta	See page 272.		"
Various	PERENNIAL PHLOX	**Phlóx paniculàta vars. *P. decussàta*	Sturdy erect plant forming clumps 2-3 ft. across. Dense panicles of flowers often 1 ft. long; garden vars. range through all colors except yellow. One of the best plants for brilliant effects forming a striking mass of color in border or edge of shrubbery. The flame-colored kinds should be detached from all except white flowers. Prop. by division; divide every 3 years in late autumn. Moist rich soil. U. S. A. See Plate, page 373. The following are some of the best garden vars.:— White vars.—*Amazon;* about 1½ ft. *Candeur;* dwarf, large flowers. *Fiancée;* pure white, very large flowers. *Independence;* very tall plant; large flowers in large spike. *Ixion;* (eye, color no. 31), small crimson eye, large spike. *Le Cygne;* pure white large flower and very large pyramidal spike. *Mrs. Huerlin;* pure white. *Saisons Lierval;* (eye, color no. 31), white, crimson eye. *Tapis blanc;* pure white.	2-3½ ft. *Sun*	July to Oct.

Color	English Name	Botanical Name and *Synonyms*	Description	Height and *Situation*	Time of Bloom
			Red vars.—*Coccinea;* (color no. 27 deeper & brilliant), small bright flowers in large clusters. **Coquelicot;* (color no. 18 redder), intense orange-scarlet, dark crimson centre. **Etna;* bright, salmon-red, dark-centred. *Flambeau;* salmon-red. *Montagnard;* (color no. 38 brilliant), deep crimson, dark-centred. *Ornament;* (color no. 26 slightly redder), very bright light crimson. *Victor;* (color no. 27), rose-red. **Pink vars.**—*Bicolor;* salmon-pink, crimson-centred. **Charles Darwin;* (color no. 30 yellower, centre 31), bright salmon-pink, crimson centre, very large pyramidal spike. *Daybreak;* (color no. 29 lighter), light pink. *Evenement;* (color no. 30 yellower & brighter), salmon-pink, purple-centred. *Gen. Chanzy;* (bet. color nos. 18 & 26), scarlet-pink, salmon tinted. *James Bennett;* (color no. 29 pale, eye 27), salmon-pink crimson eye. *Le Soleil;* (color no. 31), crimson-pink. *Lothian;* rich salmon-crimson eye. Large spreading spike. Excellent. *Mars le Tours;* (color no. 29 pale, centre 31), pink and white. *Pantheon;* (color no. 30), pink. *Pink Beauty;* (color no. 38), pale pink, enormous spike. **R. P. Struthers;* (color no. 26 bright), crimson-pink, orange tinted, dark centred; very good. **Magenta vars.**—*Eclaireur;* (color no. 33 very deep & brilliant), deep magenta, large flowers. **Lilac vars.**—*Amphitryon;* large, lilac suffused with white. *Crêpuscule;* (color no. 39 lighter), very large, lilac, with lighter margins, crimson centred. *Cross of Honor;* lilac, white margined. *Esperance;* (color no. 40, bluer towards 41), large and beautiful, magenta-lilac shading to white. **Purple vars.**—**Blue Hills;* (color no. 48 deeper to 55), intense violet. *Huxley;* (color bet. nos. 40 & 41, centre white), deep purple-magenta with white centre. *Iris;* deep magenta purple, crimson centred. *Le Mahdi;* (color bet. nos. 40 & 41), bright deep velvety purple flowers in large clusters. *Lord Rayleigh;* (color no. 47), effect blue, dark violet, centre purple. *Wm. Ramsey;* (color no. 40 very deep) velvety flowers, deep purple-magenta, dark centre; fine spike.		

JULY

Color	English Name	Botanical Name and *Synonyms*	Description	Height and *Situation*	Time of Bloom
Often bet. 46 & 47	SAGE	Sálvia officinàlis	See page 273.		June, early July
Often 39 or 38	EGYPTIAN OR GYPSIES' ROSE	*Scabiòsa arvénsis	See page 273.		Early June to mid. Aug.
25, 42, lighter, white	PURPLE MULLEIN	*Verbáscum phœníceum *V. ferrigíneum*	See page 273.		June, July
Various	PANSY, HEART'S-EASE	**Viòla trícolor	See page 273.		Mid. Apr. to mid. Sept.

PERENNIAL PHLOX. *Phlox paniculata.*

"A SHADY WALK IS PLEASANT IN AUGUST"

AUGUST

WHITE TO GREENISH

Color	English Name	Botanical Name and *Synonyms*	Description	Height and *Situation*	Time of Bloom
"White"	SNEEZE-WORT	Achillèa Ptármica	See page 167.		June to mid. Sept.
"White"	DOUBLE SNEEZE-WORT	**Achillèa Ptármica var. "The Pearl"	See page 167.		June to Oct.
"White"	SIBERIAN MILFOIL OR YARROW	*Achillèa Sibírica *A. Mongòlica, A. ptarmicoìdes*	See page 274.		July to Oct.
"Nearly white"	WHITE MONKS-HOOD OR OFFICINAL ACONITE	**Aconìtum Napéllus var. álbum *A. pyramidàle var. a., A. Taúricum var. a*	See page 274.		Mid. July to early Sept.
"White"	WHITE POTANIN-NI'S GLAND BELL-FLOWER	*Adenóphora Potaninni var. álba	See page 274.		July, Aug.
"White"	ST. BERNARD'S LILY	*Anthéricum Liliàgo	See page 274.		Mid. July to early Aug.
"White"	BOUQUET STAR-FLOWER, YARROW-LEAVED STARWORT	*Áster ptarmicoìdes *Chrysópsis álba*	See page 277.		July, Aug.
"White"	TALL FLAT-TOP WHITE ASTER	*Áster umbellàtus *Doellingèria umbellàta*	Showy plant. Flowers large. Foliage long and narrow. Good for naturalizing in half shady places. U.S.A.	7 ft. *Half shade*	Late Aug. to early Oct.
"Pinkish white"	PLUME POPPY	**Boccònia cordàta *B. Japónica*	See page 277.		Early July to early Aug.
"White"	ASTER-LIKE BOLTONIA	*Boltònia asteroìdes *B. glastifòlia*	Plant resembling the Aster in growth and flower; effect, however, is more feathery. Covered in blooming season with flowers varying to purple. Attractive in rough places or in mixed border, associated with Asters, etc. Prop. by division. Any soil. Central U. S. A. See Plate, page 377.	2-8 ft. *Sun*	Late Aug. to mid. Oct.

AUGUST

Color	English Name	Botanical Name and *Synonyms*	Description	Height and *Situation*	Time of Bloom
"White"	WOAD-LEAVED BOLTONIA	**Boltònia glastifòlia	Blossoms about one month earlier than B. asteroides and is smaller and weaker. Flowers with yellow centres. Pale green foliage. Prop. by division. Any soil. N. Amer.	4-5 ft. *Sun*	Aug., Sept.
"White"	WHITE CARPATHIAN HAIRBELL	**Campánula Carpática var. álba	See page 171.		Late June to late Aug.
"White"	WHITE TURBAN BELL-FLOWER	**Campánula Carpática var. turbinàta álba *C. t. var. a.*	See page 278.		July to Sept.
"White"	MILK-WHITE BELL-FLOWER	*Campánula lactiflòra	See page 278.		July, Aug.
"White"	WHITE CHIMNEY-PLANT OR STEEPLE BELLS	*Campánula pyramidàlis var. álba	See page 278.		"
"White"	WHITE BLUE BELLS OF SCOTLAND	*Campánula rotundifòlia var. álba	See page 171.		June to Sept.
"White"	WOLVES'-THISTLE	Carlìna acaùlis	Low dwarf plant. Flower-heads, sometimes 6 in. wide, surrounded by petal-like scales, scarcely borne above the thistle-like leaves. Good rock-garden plant. Prop. by seed in spring, or cuttings. Any garden soil. Europe.	3-9 in. *Sun*	Aug.
"Cream white"	TARTARIAN CEPHALARIA	*Cephalària Tatárica	See page 278.		Early July, Aug.
"Cream white"	WHITE SMOOTH TURTLE HEAD	*Chelòne glàbra *C. oblìqua var. álba*	See page 281.		July, Aug.
"White"	SHASTA DAISY	*Chrysánthemum "Shasta Daisy"	See page 173.		June to Sept.
"White"	GIANT DAISY	**Chrysánthemum uliginòsum *Pyrèthrum uliginòsum*	A fine plant forming a dense upright bush branching above with a profusion of large, single, rayed and yellow-centred flowers, about 2½ in. across, in lax clusters. Leaves pale green. Good for cutting. Needs plenty of water. Blooms the first year from seed; prop. also by division, cuttings or suckers. Rich loam, moist and heavy. Hungary.	4-5 ft. *Sun*	Aug., Sept.

ASTER-LIKE BOLTONIA. *Boltonia asteroides.*

 WHITE SNAKE-ROOT. *Eupatorium ageratoides.*

AUGUST

Color	English Name	Botanical Name and *Synonyms*	Description	Height and *Situation*	Time of Bloom
"White"	AMERICAN BUGBANE	Cimicífuga Americàna *Actæa prodocárpa*	Flowers of disagreeable odor in long terminal racemes. Handsome foliage. Plant against dark background in some shaded spot. Prop. by seed and division. Any good soil. Alleghany Mts.	2-4 ft. *Sun or half shade*	Aug.
"Creamy white"	BLACK SNAKEROOT	*Cimicífuga racemòsa *C. serpentària*	See page 281.		July, early Aug.
"White"	DAHLIA	**Dáhlia vars.	See page 359.		Late July to late Oct.
"White"	WHITE LARGE-FLOWERED LARKSPUR	**Delphínium grandiflòrum var. álbum	See page 282.		July, Aug.
"White"	CYCLAMEN POPPY	Eomècon chionántha	See page 282.		"
"White"	WHITE GREAT WILLOW HERB OR FIRE WEED	*Epilòbium angustifòlium var. álbum *E. spicàtum var. a.*	See page 177.		June to early Aug.
"White"	HORSE-WEED, BUTTER-WEED	Erígeron Canadénsis	See page 177.		June to Sept.
"White"	RATTLE-SNAKE-MASTER, BUTTON SNAKE-ROOT	Eryngium aquáticum *E. yuccæfò-lium*	See page 177		June to Oct.
"White"	WHITE SEA HOLLY	*Erýngium plànum var. álbum	See page 283.		July to mid. Aug.
"White"	WHITE SNAKE-ROOT	Eupatòrium ageratoìdes	Bushy native plant found in woods. Flower-heads small, but numerous, in flat clusters, covering the plant with masses of bloom. Good for cutting. Border or naturalization. Prop. by division or cuttings. Any ordinary soil. Eastern N. Amer. See Plate, page 378.	3-5 ft. *Sun*	Aug., Sept.
"White"	TALL THOR-OUGHWORT OR BONESET	Eupatòrium altíssimum	See page 283.		July, Aug.
"White"	THOROUGH-WORT, BONESET, INDIAN SAGE	Eupatòrium perfoliàtum	See page 283.		"
"White"	FLOWERING SPURGE	Euphórbia corollàta	See page 283.		"

AUGUST

Color	English Name	Botanical Name and *Synonyms*	Description	Height and *Situation*	Time of Bloom
"Pure white"	SUBCORDATE DAY LILY	*Funkia subcordàta *F. álba, F. cordàta, F. Japónica, F. liliiflòra, F. macrántha*	See page 283		July, Aug.
"White"	LARGE-FLOWERED SUBCORDATE PLANTAIN LILY	**Fúnkia subcordàta var. grandiflòra *F. álba var. g., F. liliiflòra var. g.*	Large fragrant tu ular flowers in spikes project above the mound of large leaves. Effective along walks or in the margin of shrubbery. Prop. by division, sometimes by seed just ripe. Rich moist soil. Japan	1-2 ft *Half shade*	Late Aug., Sept.
"White"	WHITE OR GREAT HEDGE BEDSTRAW, WILD MADDER	Gàlium Mollùgo	See page 178		Early June to late Aug.
"White"	SYLVAN BEDSTRAW	*Gàlium sylváticum	See page 284		July, Aug.
"White"	CAPE OR GIANT SUMMER HYACINTH	**Galtònia cándicans *Hyacinthus cándicans*	See page 284.		"
"White"	WINTERGREEN, CHECKERBERRY, BOXBERRY	Gaulthèria procúmbens	See page 284.		July, Aug.
"White"	WHITE GENTIAN	*Gentiàna álba	Terminal clusters of dull tubular flowers tinted with green or yellow. Cool locations in rock-garden or border. Prop. very slowly by seed, also by division. Leave undisturbed. Light moist soil, rich and sandy. N. Amer.	1-1½ ft. *Half shade*	Aug.
"White"	SWORD LILY	**Gladìolus vars.	See page 365		July to Oct.
"Greenish white"	RATTLESNAKE PLANTAIN	Goodyèra pubéscens	See page 285.		July, Aug.
"Pinkish white"	POINTED-LEAVED CHALK-PLANT	*Gypsóphila acutifòlia	See page 285.		Mid. July to late Aug.
"Pinkish white"	BABY'S BREATH	**Gypsóphila paniculàta	See page 285.		July, Aug.
"Greenish"	HERNIARY, RUPTUREWORT	Herniària glàbra	See page 285.		"
"White"	WHITE CORAL BELLS	Heùchera sanguínea var. álba *H. álba*	See page 179.		June to late Sept.

Color	English Name	Botanical Name and *Synonyms*	Description	Height and *Situation*	Time of Bloom
"Whitish" faintly 32	**HAIRY ALUM ROOT**	**Heùchera villòsa** *H. caulèscens*	See page 179.		Late June to Sept.
"White"	**WHITE SWAMP ROSE OR ROSE MALLOW**	***Hibíscus Moscheùtos var. álbus** *H. palùstris var. àlbus*	One of the best kinds of this vigorous plant. Native along sea-coast. Large open flowers like single Hollyhocks, 4-8 in. across and borne close to the stem. Leaves large and handsome. Winter mulch desirable. Handsome in groups in border or shrubbery. Prop. by seed. Any good loam, preferably sandy. N. Amer.	4 ft. *Sun or half shade*	Aug., Sept.
"White"	**CRIMSON EYE SWAMP ROSE**	***Hibíscus Moscheùtos var. "Crimson Eye"**	A handsome variety. Wide open flowers sometimes 7 in. across, pure white with velvety crimson eye, and bronze tinted foliage. See H. M. var. albus. N. J.	3-5 ft. *Sun or half shade*	"
"White"	**GOLD-BANDED OR JAPAN LILY**	****Lílium auràtum**	See page 286.		Mid. July to mid. Aug.
"White"	**HEART-LEAVED LILY**	***Lílium cordifòlium**	Vigorous plant. Funnel-shaped flowers, 3-5 in. long, marked with purplish brown within, are borne in racemes. Red-tinged leaves. Rather hard to grow. Bulbous. Prop. by scales or offsets. Light well-drained vegetable soil. Avoid direct contact with manure. Japan.	3-4 ft. *Sun or half shade*	Aug., Sept.
"White"	**GIANT LILY**	***Lílium gigantèum**	See page 286.		July, early Aug.
"White"	**COMMON TRUMPET LILY**	***Lílium longiflòrum**	See page 286.		"
"White"	**GREAT WHITE LOBELIA**	****Lobèlia syphilítica var. álba**	Handsome variety. Flowers in long spikes. Leaves almost stemless on the flower stalks. For damp grounds or bogs. Prop. by seed or cuttings. Hort	2-3 ft. *Sun*	Mid. Aug. to Oct.
"White"	**DOUBLE WHITE OR EVENING CAMPION**	***Lýchnis álba var. flòre-plèno** *L. vespertìna var. f.-p.*	See page 289.		Mid. July to mid. Sept.
"White"	**SINGLE AND DOUBLE WHITE MALTESE CROSS**	****Lýchnis Chalcedónica vars. álba & álba plèna**	See page 180.		June to early Aug.
"White"	**WHITE MUSK MALLOW**	***Málva moschàta var. álba**	See page 289.		July to early Sept.

AUGUST

Color	English Name	Botanical Name and *Synonyms*	Description	Height and *Situation*	Time of Bloom
"White"	DOUBLE SCENTLESS CAMOMILE	*Matricària inodòra var. plenissima *M. i. var. ligulòsa, var. múltiplex, M. grandiflòra, Chrysánthemum i. var. flòre-plèno*	See page 183.		June to Sept.
"Yellowish white"	BALM	Melíssa officinàlis	See page 183		June to early Aug.
"White"	SHOWY PRIMROSE	*Œnothèra speciòsa *Hartmánnia speciòsa*	Shrubby plant. Large broad flowers which turn pink, borne freely. Very attractive in rock-garden or border. Prop. by seed, which blooms first year, division or cuttings. Any soil. S. Western U. S. A.	1½-2 ft. *Sun*	Aug., Sept.
"White"	WHITE REST-HARROW	Onònis arvénsis var. álba *O. spinòsa var. álba*	See page 290.		Mid. July to early Aug.
"White"	WHITE ICELAND POPPY	**Papàver nudicaùle var. álbum	See page 184.		Late Apr. to mid. June, late Aug., Sept.
"White"	AMERICAN FEVERFEW, PRAIRIE DOCK	Parthènium integrifòlium	See page 184.		June to Sept.
"White"	PERENNIAL PHLOX	**Phlóx paniculàta vars. *P. decussàta*	See page 370.		July to Oct.
"White"	WHITE FALSE DRAGON-HEAD	Physostègia Virginiàna var. álba *P. Virginica var. álba*	See page 290.		Early July to Sept.
"White"	WHITE BALLOON FLOWER	**Platycòdon grandiflòrum var. álbum *Campánula g. var. a., Wahlenbergia g. var. a.*	See page 187.		June to Oct.

Color	English Name	Botanical Name and *Synonyms*	Description	Height and *Situation*	Time of Bloom
"Greenish white"	**SACALINE**	**Polýgonum Sachalinénse**	Flowers inconspicuous, in drooping clusters. Leaves dull green, often 1 ft. long, and very numerous. Requires plenty of room and may spread too much. Most effective in rough grounds or near water. Prop. by seed, sown in permanent position, and by division. Any garden soil. Island of Sachalian.	8-12 ft. *Sun or shade*	Late Aug., Sept.
"White"	**JAPAN KNOTWEED**	**Polýgonum Sièboldi** *P. cuspidàtum, P. Zuccaírnii*	A smaller species than P. Sachalinense and a freer bloomer. Quick-growing and useful. See P. Sachalinense. Japan.	3-5 ft. *Sun*	Aug., Sept.
"White"	**PEARL-WORT**	**Sagìna subulàta** *S. pilífera, Spérgula p., S. subulàta*	See page 291		July, Aug.
"White"	**WHITE WOODLAND SCABIOUS**	***Scabiòsa sylvática var. albiflora**	See page 189.		Early June to late Sept.
"White"	**WHITE HYACINTH SQUILL**	****Scílla hyacinthoìdes var. álba** *S. parviflòra var. álba*	A free bloomer. Open bell-shaped flowers in many-flowered racemes. Long and narrow foliage. Charming in the border or rock-garden. Bulbous. Prop. by offsets. Light soil enriched with manure. Mediterranean Region.	1-1½ ft. *Sun or half shade*	Aug.
"Cream white"	**WHITE STONECROP**	***Sèdum álbum**	See page 291.		Mid. July to late Aug.
"White"	**WHITE SHOWY SEDUM**	***Sèdum spectábile var. álbum** *S. Fabària var. álbum*	One of the best Sedums. Flowers in large, showy clusters surmount leafy stems. Glaucous grayish foliage. Very hardy. Good for massing in rock-garden. Prop. by division. Ordinary garden soil. Japan.	1½-2 ft. *Sun*	Late Aug. to mid. Sept.
"White"	**WHITE COMMON SEA LAVENDER OR MARSH ROSEMARY**	**Státice Limònium var. álba**	See page 292.		July, Aug.
"White"	**MOUNTAIN WILD THYME**	**Thýmus Serpýllum var. montànus** *T. montànus, T. Chamædrys*	See page 193.		Early June to mid. Aug.
"White"	**WHITE SPIDER-WORT**	***Trades-cántia Virginiàna var. álba**	See page 94.		Late May to late Aug.

AUGUST

Color	English Name	Botanical Name and *Synonyms*	Description	Height and *Situation*	Time of Bloom
"Whitish"	GREAT VIRGINIAN SPEEDWELL, CULVER'S ROOT	*Verónica Virgínica *Leptándra Virgínica*	Strong, upright plant, more or less downy. Numerous flowers in tall spike-like racemes. Effective in border though of somewhat coarse appearance. Rich soil preferable. Eastern U. S. A. Var. *Japonica* resembles the type but blooms a month earlier. Japan.	2-6 ft. *Sun*	Early Aug. to early Sept.
"White"	WHITE HORNED VIOLET OR BEDDING PANSY	**Vìola cornùta var. álba	See page 23.		Late Apr. until frost
"Yellow" 5	WOOLLY-LEAVED MILFOIL	*Achillèa tomentòsa	See page 96.		Late May to mid. Sept.
"Pale yellow" 4	EGYPTIAN MILFOIL OR YARROW	Achillèa Tournefórtii *A. Ægyptìaca*	See page 293.		July to Oct.
"Yellow" 2	PALE YELLOW WOLFSBANE	*Aconìtum Lycóctonum *A. barbàtum, A. ochroleùcum, A. squarròsum*	See page 294		July to early Sept.
"Yellow" 6 very deep	DOTTED PICRADENIA	Actinélla scapòsa *Picradenia scapòsa*	See page 294.		July, Aug.
"Yellow" 6 lighter	CROWN-BEARD	*Actinómeris squarròsa	See page 294.		Mid. July to late Aug.
"Yellow" 6	SILVERY MADWORT	Alýssum argéntium *A. alpéstre*	See page 194.		June to early Aug.
"Yellow" 4	BEAKED MADWORT	*Alýssum rostràtum *A. Wièrzbichii*	See page 194		Early June to early Aug.
"Yellow" 5	GOLDEN MARGUERITE	*Ánthemis Kélwayi *A. tinctòria var. Kélwayi*	See page 194.		Mid. June to Oct.
"Yellow" 2, centre 6	GOLDEN MARGUERITE, ROCK CAMOMILE	*Ánthemis tinctòria	See page 96		Mid. May to Oct.
"Yellow" 3 & 2	GOLDEN-SPURRED COLUMBINE	**Aquilègia chrysántha *A. leptocèros var. c.*	See page 96		Late May to late Aug.
"Orange-yellow" 4	MOUNTAIN TOBACCO OR SNUFF	*Árnica montàna	See page 294.		July, Aug.

Color	English Name	Botanical Name and *Synonyms*	Description	Height and *Situation*	Time of Bloom
"Yellow"	ITALIAN CANNA	**Cánna vars.	See page 357.		July to late Sept.
"Yellow" 6	WILD SENNA	*Cássia Marylándica	See page 295.		July, Aug.
"Yellow" 5	BLUE-LEAVED CENTAURY OR STAR THISTLE	Centaurèa glastifòlia	See page 295.		"
"Yellow" 5	SHOWY CENTAURY OR KNAPWEED	**Centaurèa macrocéphala	See page 295.		Mid. July to Sept.
"Yellow" 6	HAIRY GOLDEN ASTER	Chrysópsis villòsa	Rather coarse plant. Irregular clusters of aster-like flowers terminating leafy stalks. Suitable for wild garden. Prop. by seed and division. Blooms the first year from seed. Any garden soil. N. Amer. Var. *Rutteri;* (color no. 6), larger than the species, 1-3 ft. high; blooms later and is desirable in the border.	1-2 ft. *Sun*	Aug.
"Yellow" 3	ERECT SILKY CLEMATIS	*Clématis ochroleùca *C. sericea*	See page 296.		July, Aug.
"Yellow" 6 paler	LARKSPUR TICKSEED	*Coreópsis delphinifòlia	See page 296.		Mid. July to mid. Sept.
"Yellow" 5	LARGE-FLOWERED TICKSEED	**Coreópsis grandiflòra	See page 196.		June to Sept.
"Yellow" 6	LANCE-LEAVED TICKSEED	**Coreópsis lanceolàta	See page 196.		"
"Yellow" 6 deeper	STAR TICKSEED	*Coreópsis pubéscens *C. auriculàta*	See page 296.		Mid. July to late Sept.
"Clear yellow" 5	TALL TICKSEED	*Coreópsis trípteris	Vigorous branching plant with large, rayed flowers. Broad leaves divided into 3 parts. Showy in the border. Prop. by seed and division. Any soil. Central U. S. A.	4-8 ft. *Sun*	Aug. Sept.
"Yellow" 6 paler	WHORLED TICKSEED	*Coreópsis verticillàta *C. tenuifòlia*	See page 296.		July, Aug.
"Yellow"	DAHLIA	**Dáhlia vars.	See page 359.		July to late Oct.
"Yellow" 4	LADY'S BEDSTRAW	Gàlium vèrum	See page 198.		June to Sept.
"Yellow" 5	YELLOW GENTIAN	Gentiàna lùtea	See page 299.		July, Aug.

Color	English Name	Botanical Name and *Synonyms*	Description	Height and *Situation*	Time of Bloom
"Yellow"	SWORD LILY	**Gladìolus vars.	See page 365.		July to Oct.
"Yellow" 4	BROAD-LEAVED GUM-PLANT	Grindèlia squarròsa	See page 198.		Late June to Sept.
"Yellow" 5 deeper	SNEEZE-WEED, FALSE OR SWAMP SUN-FLOWER	**Helènium autumnàle *H. grandi-flòrum*	Striking plant. Daisy-like flowers with prominent yellow centres and drooping petals with 3-5 teeth. Valuable for back row of border or margin of shrubbery. Prop. by seed, division or cuttings. Moist rich soil, preferable. N. Amer.	2-5 ft. *Sun*	Aug. to late Sept.
"Yellow" bet. 5 & 6	LARGE-FLOWERED SNEEZE-WEED	**Helènium autumnàle var. grandiflòrum	Taller than the type with larger flowers of deeper color; deeper centres. See H. autumnale. N. Amer.	2-6 ft. *Sun*	Aug., Sept.
"Yellow" 6	DWARF SNEEZE-WEED	**Helènium autumnàle var. pùmilum	Dwarf variety, vigorous and free-flowering. Daisy-like flowers with prominent yellow centres and drooping petals. Good for cutting. See H. autumnale. N. Amer.	1-2 ft. *Sun*	"
"Yellow" 5	TALL SNEEZE-WEED	**Helènium autumnàle var. supérbum *H. grandiflò-rum var. supêrbum*	Vigorous plant. Flowers on leafy branching stalks, with bright reflexed petals, and protruding yellow centres. Suitable for shrubbery and mixed border and excellent for cutting. Prop. by seed, division or cuttings. Loam, rich and moist. N. Amer.	2-6 ft. *Sun*	"
"Yellow" 5	BIGELOW'S SNEEZE-WEED	*Helènium Bígelovii	See page 300.		July, Aug.
"Yellow" 6	BOLAND-ER'S SNEEZE-WEED OR SNEEZE-WORT	*Helènium Bolánderi	See page 198.		June to Sept.
"Yellow" 5	PURPLE-HEADED SNEEZE-WEED	*Helènium nudiflòrum *Leptopoda brachypoda*	See page 300.		July to Oct.
"Yellow" 5	NARROW-LEAVED OR SWAMP SUN-FLOWER	Heliánthus angustifòlius	An erect plant, slightly branching. Rayed flowers 2-3 in. broad, generally solitary. Dark green narrow leaves. Good in masses in shrubbery border. Prop. by division; divide every 2 years. Light, moist soil. Eastern States.	2-6 ft. *Sun*	Aug. to Nov.
"Yellow" 5	THIN-LEAVED OR WILD SUN-FLOWER	*Heliánthus decapétalus	A profusion of large, rayed flowers terminating branching stems. Rough foliage. Good in shrubbery, or for naturalization in wild garden. Prop. by division; divided every 2 years. Dry soil, not too heavy. N. Amer.	2-5 ft. *Sun*	Aug. to late Sept.

AUGUST

Color	English Name	Botanical Name and *Synonyms*	Description	Height and *Situation*	Time of Bloom
			Var. *multiflorus;* (*H. multiflorus*), double flowers. Var. *multiflorus var. flore-pleno,* flowers almost completely double. Var. *multiflorus var. maximus,* single flowers, very large with pointed rays. *Soleil d'Or,* quilled petals; striking in border. Hort.		
"Yellow" 5 to 6 darker, centre dull	**SHOWY SUNFLOWER**	**Heliánthus lætiflòrus**	Numerous single flowers 2-4 in. across on wiry stems. Rough leaves. Good for cutting and in clumps in shrubberies. Gross feeder. Prop. by seed, division or cuttings. Loose dry soil. Central U. S. A.	4-8 ft. *Sun*	Aug. Sept.
"Yellow" 5	**SMOOTH SUNFLOWER**	***Heliánthus lævigàtus**	Rather small flowers as produced abundantly on light graceful stems which are clothed with smooth leaves. Plant among shrubs. Prop. by division; divide every 2 years. Dry soil, not too heavy. S. Eastern U. S. A.	2-5 ft. *Sun*	Mid. Aug. to Oct.
"Yellow" 5	**HAIRY SUNFLOWER**	****Heliánthus móllis**	See page 300.		July, Aug.
"Yellow" 5	**STIFF SUNFLOWER**	***Heliánthus rígidus** *H. Missouriênsis*	See page 300.		"
"Yellow" bet. 5 & 6	**STIFF SUNFLOWER MISS MELLISH**	****Heliánthus rígidus var. "Miss Mellish"**	Vigorous leafy plant with large graceful flowers. Desirable, isolated or massed in waste places. Prop. by division; divide every 2 years. Dry soil, not too heavy. Hort.	6 ft. *Sun*	Late Aug. Sept.
"Orange yellow" 6	**PALE-LEAVED WOOD SUNFLOWER**	**Heliánthus strumòsus**	See page 300.		Mid. July to late Sept.
"Yellow" 5	**WOOLLY SUNFLOWER**	**Heliánthus tomentòsus**	Large flowers with dull grayish centres, borne on stout branching stalks. Very large rough leaves. Plant among shrubs. Prop. by division; divide every 2 years. Dry soil, not too heavy. U. S. A.	4-9 ft. *Sun*	Late Aug. to late Sept.
"Yellow" 5	**THROATWORT SUNFLOWER**	**Heliánthus tracheliifòlius**	See page 300.		Mid. July to early Sept.
"Yellow" 6	**OXEYE, FALSE SUNFLOWER**	***Heliópsis lævis** *H. helianthoìdes*	See page 300.		Mid. July to late Sept.
"Yellow" 6, centre 7	**PITCHER'S OXEYE OR FALSE SUNFLOWER**	***Heliópsis lævis var. Pitcheriàna** *H.P.*	See page 303.		"

AUGUST

Color	English Name	Botanical Name and *Synonyms*	Description	Height and *Situation*	Time of Bloom
"Yellow" 5	LESSER YELLOW DAY LILY	*Hemerocállis mìnor *H. gramìnea, H. graminifòlia*	See page 303.		July, Aug.
"Yellow" 1	HOARY MARSH OR ROSE MALLOW	*Hibìscus incànus	Very similar to H. Moscheutos with large crimson-centred flowers like Hollyhocks, and pretty foliage which is whitish beneath. Needs slight protection. Suitable for back of border. Prop. by seed or cuttings. Rich moist soil. S. Eastern U. S. A.	3-5 ft. *Sun*	Aug., Sept.
"Yellow" 5	NARROW-LEAVED HAWK-WEED	Hieràcium umbellàtum	Flowers in umbels on erect stems well clothed with lance-shaped leaves. Border plant. Any ordinary soil. Northern N. Amer.; Kamtschatka; N. Asia; Europe.	1-2 ft. *Sun*	"
"Golden yellow" 5	SHAGGY HAWK-WEED	Hieràcium villòsum	See page 199.		June to mid. Aug.
"Yellow" 5	AARON'S BEARD, ROSE OF SHARON	Hypéricum calycìnum	See page 303.		July, Aug.
"Yellow" 5	KALM'S ST. JOHN'S WORT	**Hypéricum Kalmiànum	Shrub of somewhat twisted growth. Blossoms small in 3-7 flowered clusters. Narrow leaves, bluish tinted. Valuable as a foliage plant. Prop. by seed, cuttings or suckers. Prefers sandy loam. N. Amer.	2-4 ft. *Shade*	Aug.
"Yellow" bet. 5 & 6	GOLD FLOWER, ST. JOHN'S-WORT	**Hypéricum Moseriànum	See page 303.		July, Aug.
"Yellow" 5	SWORD-LEAVED ELECAM-PANE	*Ínula ensifòlia	See page 303.		Mid. July to Sept.
"Deep yellow" 7	GLANDULAR FLEABANE OR INULA	*Ínula glandulòsa	See page 304		July, early Aug.
"Yellow" 6	ELECAM-PANE	*Ínula Helènium	See page 304		Mid. July to late Aug.
"Yellow"	TUCK'S FLAME FLOWER	*Kniphòfia Túckii	See page 200.		June to Sept.
"Yellow" 5	GRAY-HEADED CONE-FLOWER	*Lépachys pinnàta *Ratibida p., Rudbêckia p.*	See page 200.		June to mid. Sept.

AUGUST

Color	English Name	Botanical Name and *Synonyms*	Description	Height and *Situation*	Time of Bloom
"Yellow" 3	DALMATIAN TOADFLAX	*Linària Dalmática	See page 203.		Early June to early July, late July to late Aug.
"Yellow" 6	GOLDILOCKS	Linosỳris vulgàris *Chrysocòma Linosỳris*	Showy plant of the Daisy order Profuse clusters of small flowers. Border or rock-garden. Prop. by seed and division. Any garden soil. Europe.	6-12 in *Sun*	Aug.
"Yellow" 3	BIRD'S-FOOT TREFOIL, BABIES' SLIPPERS	Lòtus corniculàtus	See page 203.		June to Oct.
"Bright yellow" 6 to 7	GOLDEN SPIDER LILY	Lýcoris aùrea *Amarýllis aùrea, Nerìne aùrea*	Large bright flowers in umbels, 3-4 in. in diameter, appearing before the sword-like gray-green foliage and deepening in color when mature. Bulbous. China.	1-3 ft.	Aug., Sept.
"Yellow" 3	LARGE-FLOWERED BIENNIAL EVENING PRIMROSE	*Œnothèra biénnis var. grandiflòra *Œ. Lamarckiàna*	See page 204.		June to Sept.
"Yellow" 3	FRASER'S EVENING PRIMROSE	**Œnothèra glaùca var. Fràseri *Œ. Fràseri*	See page 204.		"
"Lemon yellow" 3	LINEAR-LEAVED EVENING PRIMROSE	Œnothèra lineàris *Œ. fruticòsa var. l., Œ. ripària*	See page 204.		June to early Aug.
"Yellow" 3	MISSOURI PRIMROSE	**Œnothèra Missouriénsis *Œ. macrocárpa, Megaptèrium M.*	See page 204		"
"Yellow" 5	TAURIAN GOLDEN DROP	Onósma stellulàtum var. Taùricum *O. Taùricum*	See page 306.		July, Aug.
"Pale yellow" 6	MANY-SPINED OPUNTIA	Opúntia Missouriénsis *O. fèrox, O. splêndens*	See page 306.		"
"Yellow" 6	WESTERN PRICKLY PEAR	Opúntia Rafinésquii *O. mesacántha*	See page 306.		"

AUGUST

Color	English Name	Botanical Name and *Synonyms*	Description	Height and *Situation*	Time of Bloom
"Dull yellow" 3 duller	**BARBERRY FIG, COMMON PRICKLY PEAR**	**Opúntia vulgàris** *O. Opúntia*	See page 207.		June to Sept.
"Yellow" 5 or 6 brilliant	**ICELAND POPPY**	****Papàver nudicaùle**	See page 35.		Late Apr. to July, late Aug. to Oct.
"Yellow" 3	**SHRUBBY CINQUE-FOIL OR FIVE-FINGER**	**Potentílla fruticòsa**	See page 207.		June to Sept.
"Golden yellow" effect 7	**HYBRID CINQUE-FOIL OR FIVE-FINGER**	***Potentílla "Gloire de Nancy"**	See page 208.		"
"Golden yellow" 5	**BACHEL-OR'S BUTTONS**	**Ranúnculus àcris var. flòre-plèno**	See page 105.		Mid. May to Sept.
"Yellow" 5	**YELLOW OR CALIFORNIA CONE-FLOWER, WHORTLE-BERRY-LEAVED KNOTWEED**	***Rudbéckia Califórnica**	An effective plant and the largest flowered species. Solitary daisy-like flowers 5 in. across, with prominent dark centres, on erect robust unbranching stems. Divide frequently. Prop. by seed and division. Any garden soil. Cal.	2-4 ft. *Sun or half shade*	Aug. to mid. Sept.
"Yellow" 5	**LARGE-FLOWERING CONE-FLOWER**	***Rudbéckia grandiflòra**	Solitary flowers with drooping rays and purplish disks, on erect branching stalks. Long oval pointed leaves. Effective in border. Ark.; S. Central U. S. A.	2-3½ ft. *Sun*	Late Aug. Sept.
"Yellow" 5	**TALL OR GREEN-HEADED CONE-FLOWER**	***Rudbéckia laciniàta**	See page 309.		Mid. July to late Aug.
"Yellow" bet. 5 & 6	**GOLDEN GLOW**	****Rudbéckia laciniàta var. flòre-plèno**	See page 309.		Late July to late Sept.
"Yellow" 5	**LARGE CONE-FLOWER**	***Rudbéckia máxima**	Flowers 4-6 in. across. Long drooping petals. Purple cone-shaped centre, 1-2 in. high. Leaves grayish green. Needs slight protection. Prop. by seed, division or cuttings. Any garden soil. Ark. to Tex.	4-9 ft. *Sun or half shade*	Aug., Sept.
"Yellow" 6, brown centre	**SHOWY CONE-FLOWER**	**Rudbéckia speciòsa** *R. áspera*	See page 309.		July, early Aug.

Color	English Name	Botanical Name and *Synonyms*	Description	Height and *Situation*	Time of Bloom
"Yellow" bet. 5 & 6	SWEET CONE-FLOWER	**Rudbéckia subtomentòsa	Daisy-like flowers, petals fairly numerous, sometimes darker near the dull brown centre. Splendid border plant. Prop. by seed, division or cuttings. Any garden soil. Prairies, U. S. A.	3-5 ft. *Sun or shade*	Mid. Aug. to Oct.
"Deep yellow" 6	THIN-LEAVED CONE-FLOWER	**Rudbéckia tríloba	See page 309.		July, Aug.
"Pale yellow" 2	JUPITER'S DISTAFF	*Sálvia glutinòsa	See page 310.		Mid. July to early Aug.
"Yellow" 3 pale	WEBB'S SCABIOUS OR PIN-CUSHION FLOWER	Scabiòsa ochroleùca *S. Webbiàna*	See page 208.		June to early Sept.
"Yellow" 5	AIZOON STONECROP	*Sèdum Aizóon	See page 208.		Mid. June to mid. Aug.
"Golden yellow" 4	ORANGE STONECROP	*Sèdum Kamtscháticum	See page 310.		July, Aug.
"Yellow" 2 greenish	STONE-HORE, STONE ORPINE, TRIP-MADAM	*Sèdum refléxum	See page 310.		Early July to early Aug.
"Yellow" 4	CRESTED STONE-HORE, STONE ORPINE OR TRIP-MADAM	Sèdum refléxum var. cristàtum *S. monstròsum, S. robústum*	See page 310.		"
"Yellow"	HENS-AND-CHICKENS, HOUSELEEK	*Sempervìvum globíferum *S. sobolíferum*	See page 310		July, Aug.
"Yellow" 2 deep	HEUFFEL'S HOUSELEEK	*Sempervìvum Heúffelii	Evergreen plant. Somewhat bell-shaped flowers. Reddish succulent foliage in rosettes with fringed edges. Excellent foliage plant for rock-garden and walls. Prop. by offsets. Any sandy soil. Transylvania; Greece.	6-8 in. *Sun*	Aug.
"Pale yellow" 2 bright	HAIRY HOUSELEEK	*Sempervìvum hìrtum	See page 310.		July, Aug.

Color	English Name	Botanical Name and *Synonyms*	Description	Height and *Situation*	Time of Bloom
"Yellow" 5	**DORIAN GROUND-SEL OR RAGWEED**	**Senècio Dòria**	See page 313.		July, Aug.
"Yellow" 5	**ROUGH ROSINWEED**	**Sîlphium aspérrimum**	See page 313.		July to early Sept.
"Yellow" 5	**COMPASS PLANT, PILOT WEED**	****Sîlphium laciniàtum**	See page 313.		Mid. July to mid. Sept.
"Yellow" 6 lighter	**CUP PLANT, INDIAN CUP**	***Sîlphium perfoliàtum**	See page 313.		"
"Yellow" 5	**PRAIRIE DOCK**	***Sîlphium terebinthin-àceum**	Rather coarse-growing plant bearing many small blossoms like Sunflowers. Rough leaves about a foot long rise from the base. Effective in wild gardens or shrubbery. Prop. by seed and division. Any garden soil. Prairies, N. Amer. Var. *pinnatifidum* has deeply divided leaves.	6-8 ft. *Sun*	Mid. Aug. to late Sept.
"Yellow" 6 lighter	**WHORLED ROSINWEED**	**Sîlphium trifoliàtum**	See page 313.		Mid. July to mid. Sept.
"Yellow" 5 to 6	**GOLDEN-ROD**	***Solidàgo**	See page 313.		Late July to early Oct.
"Yellow" 2	**FALSE RHUBARB, FEN RUE**	**Thalîctrum flàvum**	See page 313.		July, Aug.
"Yellow"	**CROCUS-FLOWERED BLAZING STAR**	****Tritònia crocosmæ-flòra vars.** *Montbrètia c.*	See page 314.		July to Oct.
"Yellow" 6	**YELLOW HORNED VIOLET OR BEDDING PANSY**	****Vìola cornùta var. lùtea màjor**	See page 35.		Late Apr. until frost
"Yellow orange" 9	**GOLDEN PERUVIAN LILY**	***Alstrœmè-ria aurantìaca** *A. aurea*	See page 317.		July, Aug.
"Red orange" shading 10 to 12	**BUTTERFLY WEED, PLEURISY ROOT**	****Asclèpias tuberòsa**	See page 317.		Early July to early Aug.
"Orange" 13	**BLACK-BERRY OR LEOPARD LILY**	***Belemcánda Chinénsis** *B. punctàta, Pardán-thus C., P. Sinênsis, Íxia Chinênsis*	See page 317		July, Aug.

AUGUST

Color	English Name	Botanical Name and *Synonyms*	Description	Height and *Situation*	Time of Bloom
"Orange" 7	DOUBLE ORANGE DAISY	*Erígeron aurantìacus	See page 317.		July, Aug.
"Orange" 5 & 14	STRIPED SNEEZE-WEED	*Helènium autumnàle var. strìàtum	Daisy-like flowers, with drooping yellow petals marked with deep crimson, and prominent centres, gold and purple. Good for cutting. Back row of border or margin of shrubbery. Prop. by seed, division or cuttings. Moist rich soil preferable. N. Amer.	4-5 ft. *Sun*	Aug., Sept.
"Orange" 6 tinged 20	STRIPED PURPLE-HEADED SNEEZE-WEED	**Helènium nudiflòrum var. grandi-céphalum strìàtum	Larger flowers than the type, 2 in. across; drooping yellow petals blotched with crimson. Distinctive border plant. Prop. by seed, division or cuttings. Moist rich soil preferable. Hort.	3-4 ft. *Sun*	Aug.
"Orange" 10 deeper	ORANGE DAY LILY	*Hemerocállis aurantìaca	See page 317.		July, early Aug.
"Tawny orange" 14 brighter	BROWN DAY LILY, MAHOGANY LILY	Hemerocállis fúlva *H. dìsticha*	See page 318.		"
"Orange" 12	ORANGE HAWK-WEED	Hieràcium aurantìacum	See page 211.		June to Oct.
"Scarlet orange" effect 12 brighter	EVER-BLOOMING FLAME FLOWER	**Kniphòfia Pfítzerii	Handsome plant. Most profuse and continuous in bloom of all the Kniphofias. Spikes, 12 in. long, of rich orange-scarlet flowers salmon-rose at the edge. Good for cutting. Protect in winter. Striking massed among tropical plants against dense background. Prop. by division. Any well-drained soil. Hort.	3-4 ft. *Sun or half shade*	Early Aug. to early Oct.
"Apricot" 10 pinker	SHINING THUNBERGIAN LILY	**Lílium élegans var. fúlgens *L. f., L. Bátemanniæ, L. sanguíneum*	See page 318.		Mid. July to early Aug.
"Salmon orange" bet. 16 & 8	DR. HENRY'S LILY	*Lílium Hénryi	A plant recently introduced and unconventional in habit, somewhat like L. speciosum. Flowers dotted with red-brown, in flat loose clusters. Thick foliage. Effective massed in border or on margin of Rhododendron bed. Bulbous. Prop. by offsets or scales. Light peaty soil. Avoid direct contact with manure. China.	5-6 ft. *Half shade*	Aug., Sept
"Reddish orange" 12 redder	WILD ORANGE-RED OR PHILADELPHIA LILY	*Lílium Philadélphicum	See page 318.		July, Aug.

AUGUST

Color	English Name	Botanical Name and *Synonyms*	Description	Height and *Situation*	Time of Bloom
"Reddish orange" bet. 19 & 20	AMERICAN TURK'S CAP LILY	**Lílium supérbum	See page 321.		Early July to early Aug.
"Or-ange" 12 redder	TIGER LILY	**Lílium tigrìnum	See page 321.		Mid. July to Sept.
"Deep orange" bet. 12 & 17	ORANGE ICELAND POPPY	*Papàver nudicaùle var. aurantìacum	See page 36.		Late Apr. to July, late Aug., Sept.
"Deep orange" bet. 12 & 17	SMALL ICELAND POPPY	*Papàver nudicaùle var. miniàtum	See page 36		Late Apr. to July, mid. Aug. to Oct
"Or-ange" 11	RUSSELL'S CINQUE-FOIL OR FIVE-FINGER	Potentílla Russelliàna	See page 321.		July, Aug.
"Or-ange" 7, centre 21 dark	ORANGE CONE-FLOWER	*Rudbéckia fúlgida	Small daisy-like flowers, rays yellow or orange, centres black. Prop. by seed, division or cuttings. Grows in dry soil. Penn. to the Mississippi and South.	1-2½ ft. *Sun or half shade*	Mid. Aug. to early Oct.
"Or-ange" 6 to 18	POTTS' BLAZING STAR	**Tritònia Póttsii *Montbrètia P.*	See page 321.		July to Oct.
"Red" 27	RED YARROW OR MILFOIL	*Achillèa Millefòlium var. rùbrum	See page 321.		Mid. July to mid. Sept.
"Purple crimson" bet. 41 & 42	WESTERN SILKY OR SILVERY ASTER	*Áster seríceus *A. argênteus*	Pretty dwarf kind. Flowers about 1½ in. across, solitary on the branchlets. Leaves silvery and silky. Does well under trees and in the border. Prop. by division. Well-drained soil. Central U. S. A.; Tex.; N. Amer.	1-2 ft. *Sun or shade*	Late Aug. to early Oct.
"Violet crimson" 27 bright	PURPLE POPPY MALLOW	*Callírrhoe involucrata	See page 212.		June to Sept.
"Red"	FRENCH OR CROZY CANNA	**Cánna vars.	See page 357.		July to late Sept.
"Red"	ITALIAN CANNA	**Cánna vars.	See page 358.		"
"Ma-roon" near 33	DARK PURPLE KNAPWEED	Centaurèa atropurpùrea	See page 212.		June to Sept.

AUGUST

Color	English Name	Botanical Name and *Synonyms*	Description	Height and *Situation*	Time of Bloom
"Purplish red" 33 & 32	**LYON'S TURTLE HEAD**	***Chelòne Lyoni**	Forms thick clumps. Flowers in dense showy spikes terminate stalks clothed with deep green glossy foliage. A profuse bloomer. Excellent border plant. Prop. by seed, division in the spring, or by cuttings. Any garden soil, preferably rich. S. Eastern U. S. A.	2 ft. *Half shade*	Aug.
"Deep rose" 34 lighter	**RED TURTLE HEAD**	***Chelòne oblìqua** *C. purpùrea*	Less vigorous than C. Lyoni, but of a finer color. Flowers in showy spikes terminate leafy stalks. Excellent border plant. There is a good white var. Prop. by seed, division in the spring, or by cuttings. Any rich garden soil. S. Eastern U. S. A.	1½-2 ft. *Half shade*	"
"Red"	**DAHLIA**	****Dáhlia vars.**	See page 359.		July to late Oct.
"Orange scarlet" 18 more orange	**SOUTHERN SCARLET LARKSPUR**	***Delphínium cardinàle**	See page 322.		July, Aug.
"Crimson" 33	**SEROTIN'S PURPLE CONE-FLOWER**	***Echinàcea purpùrea var. serótina** *E. intermèdia*	See page 322.		"
"Orange scarlet"	**CHILOE AVENS**	***Gèum Chiloénse** *G. coccíneum,* (Hort.)	See page 213.		Late June to early Aug.
"Red"	**SWORD LILY**	****Gladìolus vars.**	See page 365.		July to Oct.
"Red" 26 brilliant	**CORAL OR CRIMSON BELLS**	****Heùchera sanguínea**	See page 213.		June to late Sept.
"Bright red" 18	**BULB-BEARING LILY**	***Lílium bulbíferum**	See page 323.		July, Aug.
"Orange red" bet. 19 & 20	**CAROLINA LILY**	**Lílium supérbum var. Caroliniànum** *L. Caroliniànum*	A pretty Lily though not so effective as the type. Dwarf var. with 1-3 delicate broad flat flowers having pointed petals spotted within with black and marked with yellow. Broader foliage than the type. Plant in quantity in the border or edge of shrubbery. Bulbous. Prop. by offsets and scales, or very slowly by seed. Any well-drained soil. Avoid direct contact with manure. S. Eastern U. S. A.	2-3 ft. *Sun or half shade*	Aug.

Color	English Name	Botanical Name and *Synonyms*	Description	Height and *Situation*	Time of Bloom
"Crimson" 20 bright & rich	CARDINAL FLOWER, INDIAN PINK	**Lobèlia cardinàlis	A favorite native of beautiful color. Brilliant flowers in spikes on erect unbranching stems. Leaves narrow mostly on lower part of stalk. Naturalize near water or plant preferably in shaded border. Prop. by seed, division or green-wood cuttings. Resows itself. Rich soil, preferably moist. Wet places in Eastern N. Amer.	2-4 ft. *Sun or shade*	Aug. to mid. Sept.
"Scarlet" 33 warmer	SHINING CARDINAL FLOWER	*Lobèlia fúlgens *L. formòsa* *L. cardinàlis* (Hort.)	See page 323.		July, Aug.
"Red" 11, 17 & 18 brilliant	SHAGGY LYCHNIS	*Lỳchnis Haageana	See page 217.		Early June to early Aug.
"Red" 24 redder	BLOOD-RED AMARYLLIS	Lycòris sanguínea	See page 324.		July, Aug.
"Rosy red" 18	HALL'S AMARYLLIS	*Lycòris squamígera *Amarỳllis Hállii*	See page 324.		"
"Scarlet" 18 duller	SCARLET MONKEY FLOWER	*Mímulus cardinàlis	See page 217.		June to Sept.
"Red" 20 lighter, centre 21 redder	OSWEGO TEA, BEE OR FRAGRANT BALM	**Monárda dídyma	See page 217.		Mid. June to early Sept.
"Scarlet" 18	TORREY'S BEARDED PENTSTEMON	**Pentstèmon barbàtus var. Tórreyi *P. Tórreyi*	See page 325.		Early July to early Aug.
"Purplish red" 41 deeper	HARTWEG'S LARGE-FLOWERED HYBRID PENTSTEMON	Pentstèmon gentianoìdes hýbrida grandiflòra *P. Hártwegi hýbrida g.*	See page 218.		June to Sept.
"Red"	PERENNIAL PHLOX	**Phlóx paniculàta vars. *P. decussàta*	See page 370.		July to Oct.
"Vermilion" 18 redder	CAPE FUCHSIA	*Phygèlius capénsis	See page 325.		July, Aug.

Color	English Name	Botanical Name and *Synonyms*	Description	Height and *Situation*	Time of Bloom
"Red"	ALKE-KENGI STRAW-BERRY, TOMATO, WINTER OR BLADDER CHERRY	Phýsalis Alkekéngi	See page 325		Fruit July to late Oct.
"Red"	CHINESE LANTERN PLANT	*Phýsalis Franchétti *P. Alkekéngi var. Franchétti*	See page 325		"
"Red"	HYBRID CINQUE-FOIL DOUBLE VARS.	Potentílla hỳbrida vars.	See page 221		June to Sept.
"Deep red" 20, 4 inside	PINK ROOT, WORM GRASS	*Spigèlia Marylándica	See page 222.		Late June to early Aug.
"Red" 26 deep	TARTARIAN SEA LAVENDER	*Státice Tatárica *S. Besseriàna, S. incàna var. hỳbrida*	Broadly branching plant bearing dainty spikelets of blossoms good for cutting. From the base of the plant rises a clump of large leathery leaves. Effective in rock-garden or border. Prop. by seed in the spring. Loose soil preferable. Tartary.	1-2 ft. *Sun*	Early Aug. to early Sept.
"Bright red"	RED SPIDER-WORT	Tradescántia Virginiàna var. coccínea	See page 117.		Late May to late Aug.
"Orange scarlet" often 16 pinker	CROCUS-FLOWERED BLAZING STAR	*Tritònia crocosmæflòra *Montbrètia c.*	See page 326.		July to Oct.
"Bright red" some-times 29	REDDISH BLAZING STAR	*Tritònia ròsea *Montbrètia ròsea*	See page 326.		July, Aug.
"Ver-milion" 18 redder	CALIFORNIA FUCHSIA, HUMMING-BIRD'S TRUMPET	*Zauschnèria Califórnica	Rather sprawling plant, requiring close setting for compact growth. Its numerous bell-shaped blossoms are effective in the rock-garden. Give sheltered position and protect in winter. Prop. by seed, division or cuttings. Light sandy loam. Cal. to Wyoming and Mexico.	¾-2 ft. *Sun*	Late Aug., Sept.
"Light pink" 22 pale	CHINESE GOAT'S BEARD	**Astílbe Chinénsis	See page 329.		July, early Aug.
"Rose" 29 lighter & duller	EVAN'S BEGONIA	Begònia Evansiàna *B. discolor, B. grandis*	See page 329.		July, Aug.

AUGUST

Color	English Name	Botanical Name and *Synonyms*	Description	Height and *Situation*	Time of Bloom
"Pink" 37 pinker	HEATHER, LING, HEATH	*Callùna vulgàris *Erica vulgàris*	See page 329.		July, Aug.
"Pur-plish pink" 27	HOARY CEDRO-NELLA	*Cedronélla càna	See page 223.		June to Sept.
"Deep pink" 31	WHITENED KNAPWEED	Centaurèa dealbàta	See page 223.		Late June to early Aug.
"Pur-plish rose" 36 deep	SMALL ROSE OR PINK TICKSEED	*Coreópsis ròsea	See page 223.		June to Sept.
"Pink" 39	CROWN VETCH	Coronílla vària	See page 223.		"
"Pink" 29	CROSS-WORT, FOETID CRUCIAN-ELLA	*Crucianélla stylòsa	See page 224.		"
"Pink etc."	DAHLIA	**Dáhlia vars.	See page 359.		July to late Oct.
"Ma-genta pink" 23	CINNAMON PINK	**Diánthus cinnabarìnus	See page 330.		July, Aug.
"Pink" 25, 27, 28, 33, 34, 35	BROAD-LEAVED PINK	**Diánthus latifòlius	See page 224.		June to Sept.
"Pur-plish" pink 29	PALE PURPLE CONE FLOWER	*Echinàcea augustifòlia *Braunèria pàllida*	See page 330.		July, Aug.
"Pur-plish pink" bet. 32 & 39, centre 13 & 20	PURPLE CONE-FLOWER, BLACK SAMPSON	**Echinàcea purpùrea *Rudbéckia p.*	See page 330.		"
"Shell pink" bet. 24 & 17	HEDGE-HOG THISTLE, SIMPSON'S CACTUS	Echinocáctus Símpsoni	See page 227.		June to Sept.
"Ma-genta" 40	GREAT WILLOW HERB, FIRE WEED, FRENCH WILLOW	Epilòbium angustifòlium *Camænerion angustifòlium*	See page 228.		June to early Aug.
"Pink"	SWORD LILY	**Gladìolus vars.	See page 365.		July to Oct.

SHOWY SEDUM. *Sedum spectabile.*

 HANDSOME RED LILY. *Lilium speciosum var. rubrum.*

AUGUST

Color	English Name	Botanical Name and *Synonyms*	Description	Height and *Situation*	Time of Bloom
"Pink" 22 light	MANY-PAIRED FRENCH HONEY-SUCKLE	Hedýsarum multijugum	See page 228.		June to early Aug.
"Pink" 22 very pale, centre 33	HALBERT-LEAVED ROSE MALLOW	*Hibíscus militàris *H. Virginicus*	Vigorous plant. Flowers, pale pink or white, with purple centres, 4-5 in. wide. Leaves heart-shaped, downy beneath. Border. Prop. by seed. Any garden soil, preferably sandy and rich. S. Eastern U. S. A.	2-6 ft. *Sun or half shade*	Aug., early Sept.
"Rose" 36, centre 33	SWAMP ROSE, ROSE MALLOW	*Hibíscus Moscheùtos *H. palùstris*	Vigorous plant. One of the best kinds. Native along the coast. Flowers, 4-8 in. across, like those of the Hollyhock, usually light rose with purple eye. Also a white var. Large effective foliage. Handsome in groups in the border or along edge of shrubbery. Winter mulch desirable. Any good loam, preferably sandy. Swamps, Eastern U. S. A.	3-5 ft. *Sun or half shade*	Aug., Sept.
"Pink"	JAPANESE LILY	*Lílium Japónicum	See page 331.		July, Aug.
"Pink" 36 spotted 27	SPOTTED LILY	**Lílium speciòsum *L. lancifò-lium*	White flowers in broad panicles with twisted and reflexed petals suffused with pink and spotted with red. Graceful foliage. Effective massed in the border. Bulbous. Prop. by offsets or scales. Light well-drained soil. Avoid direct contact with manure. Japan. Var. *album*, (*L. præcox*), usually has pure white fragrant flowers. Hort.	2-4 ft. *Sun or half shade*	Aug., Sept.
"Pink" white & 23	HANDSOME MELPO-MENE LILY	**Lílium speciòsum var. Melpómene	Bright flowers heavily spotted with crimson-pink and margined with white, droop in broad clusters. Effective massed in the border. Bulbous. Prop. by offsets or scales. Any well-drained soil. Avoid direct contact with manure. Japan.	2-4 ft. *Sun or half shade*	"
"Reddish pink" white & 26 deeper	HANDSOME RED LILY	**Lílium speciòsum var. rùbrum	The best variety, more vigorous than the type. Clusters of drooping flowers charming in shape and beautiful in color. Graceful leaves. Very effective in masses. Bulbous. Prop. by offsets or scales. Any well-drained soil. Avoid direct contact with manure. Japan. See Plate, page 400.	2-4 ft. *Sun or half shade*	"
"Deep rose" 29	ALCEA MALLOW	*Málva Alcèa	See page 229.		June to Sept.
"Rose" 38 more violet	MUSK MALLOW	*Málva moschàta	See page 332		July to early Sept.

Color	English Name	Botanical Name and *Synonyms*	Description	Height and *Situation*	Time of Bloom
"Rose" 38 to 27	BELL-FLOWERED PENTSTEMON	*Pentstèmon campanulàtus *P. angustifólius, P. atropurpúreus*	See page 332.		July, Aug.
"Rose" often 45 lighter	SHRUBBY SMOOTH-LEAVED PHLOX	*Phlóx glabérrima var. suffruticòsa *P.s., P. nìtida*	See page 231.		June to early Aug.
"Pink"	PERENNIAL PHLOX	**Phlóx paniculàta vars. *P. decussàta*	See page 370		July to Oct.
"Light pink" 38 & paler	DOUBLE BOUNCING BET	Saponària officinàlis var, flòre-plèno	See page 333.		July to Sept.
"Rose" 27	AUTUMN SQUILL, STARRY HYACINTH	*Scílla autumnàlis	See page 333.		Late July to Oct.
"Flesh-color" 36	PINK HYACINTH SQUILL	**Scílla hyacinthoìdes var. ròsea *S. parviflòra var. ròsea*	Bell-shaped flowers in panicles, blooming in profusion. Very long and narrow foliage. Charming in border and rock-garden. Bulbous. Soil enriched with manure. Mediterranean Region.	1-1½ ft. *Half shade*	Aug.
"Pink"	LYDIAN STONECROP	Sèdum Lýdium	Small, compact, turf-like evergreen, Flowers very small. Succulent greenish foliage tipped with red. Good for covering bare spots or for edging. Prop. by division. Dry soil. Asia Minor.	3-6 in. *Sun*	Aug., Sept.
"Pink" 38 lighter	SIEBOLD'S STONECROP	*Sèdum Sieběldii	An evergreen with spreading branches. Flowers in dense clusters. Glaucous leaves, bluish, almost gray, with edges slightly pink. Good for the border or rock-garden. Prop. by division or cuttings. Japan. Var. *variegatum,* (*S. variegatum*), grayish variegated foliage. Japan.	9 in. *Sun*	Aug. to mid. Sept.
"Rose" 38 or 32 lighter	SHOWY SEDUM	**Sèdum spectábile *S. Fabaria*	Distinctive and one of the best Sedums. Flowers in large showy clusters, sometimes purplish or whitish, surmounting leafy stems. Foliage grayish. Very hardy. Good in masses. Prop. by division. Ordinary garden soil. Japan. See Plate, page 399. Var. *variegatum,* variegated foliage.	1½-2 ft. *Sun*	Late Aug. to mid. Sept.

Color	English Name	Botanical Name and *Synonyms*	Description	Height and *Situation*	Time of Bloom
"Purplish pink" 32	PURPLE STONECROP	*Sèdum stoloníferum *S. spùrium*	See page 333.		Mid. July to early Aug.
"Pink" 23 dull	ORPINE, LIVE-FOREVER	*Sèdum Teléphium	See page 334.		July, Aug.
"Rose pink" 38	AUTUMN CATCHFLY	Silène Scháfta	See page 233.		June to Sept.
"Purplish rose" 37	AMERICAN GERMANDER, WOOD SAGE	Teùcrium Canadénse *T. Virgínicum*	See page 334.		Late July to late Aug.
"Purple"	BROAD-LEAVED ACANTHUS	**Acánthus móllis var. latifòlius *A. l., A. Lusitánicus*	See page 337.		July, Aug.
"Deep purple" 49 darker	WILD MONKSHOOD	**Aconìtum uncinàtum	See page 233.		Mid. June to Sept.
"Purple" 47	GLAND BELL-FLOWER	*Adenóphora Lamárckii	See page 234.		June to Sept.
"Purple"	JAPAN STAR GRASS	Áletris Japónica	See page 337.		July, Aug.
"Purple" 46	BESSARABIAN ASTER	*Áster Améllus var. Bessarábicus *A. Bessarábicus*	One of the most beautiful of the dwarf Asters. Very graceful. Large flowers in abundance; orange centres. Good for cutting. Var. *elegans*, similar, with darker flowers. Plant in shrubbery, wild garden or border. Prop. by seed and division. Ordinary soil. Europe.	2 ft. *Sun*	Aug., Sept.
"Violet" 47	RUSH ASTER	*Áster júnceus	See page 337.		July to early Sept.
"Violet" bet. 46 & 47	SAVORY-LEAVED ASTER	*Áster linariifòlius *Diplopappus linariifòlius*	See page 337.		July to Oct.
"Violet" 44 pale	LINDLEY'S ASTER	*Áster Lindleyànus	Flowers with brownish centres rather showy and abundant. Stems stout and erect, branching at top. Good for open position in wild garden. Prop. by seed and division. Ordinary soil. Eastern States.	1-3½ ft. *Sun*	Aug. to early Oct.

AUGUST

Color	English Name	Botanical Name and *Synonyms*	Description	Height and *Situation*	Time of Bloom
"Purple" 48 redder	NEW ENGLAND ASTER OR STARWORT	**Áster Nòvæ Ángliæ	Perhaps the most beautiful and conspicuous of the tall Asters. Flowers originally rich violet (color no. 46 lighter), with yellow centres, but varieties include *albus,* white; *roseus* (color no. 47 redder), deep rose-colored; *ruber* (color no. bet. 40 & 41). Striking in the shrubbery or border or to naturalize. Prop. by seed and division. Grows readily in ordinary soil, preferably moist. New England. See Plate, page 405.	3-7 ft. *Sun*	Mid. Aug. to late Sept.
"Pale violet"	ARROW-LEAVED ASTER	Áster sagittifòlius	See page 337.		July, Aug.
"Violet" 47	SIBERIAN ASTER OR STARWORT	*Áster Sibíricus	Pretty and hairy. Single terminal flowers. Foliage rather broad and long. Good for the rock-garden. Prop. by seed and division. Europe; Rocky Mts.	1 ft. or less *Sun*	Aug., Sept.
"Purple" 49	LOW SHOWY OR SEASIDE PURPLE ASTER	*Áster spectábilis	Beautiful flowers with light yellow centres. Long and narrow foliage. Good border plant. Coast, Mass. to Del.	2 ft. *Sun or half shade*	Late Aug. to mid. Sept.
"Pinkish lavender" 43 pinker	BROAD-SCALED BOLTONIA	**Boltònia latisquàma	See page 337.		Late July to Oct.
"Pale purple" 44	CUT-LEAVED CALIMERIS	*Calímeris incìsa *Áster incìsus, Boltònia incìsa*	See page 338.		July, Aug.
"Violet" 47 shading darker	CARPATHIAN HAIRBELL	**Campánula Carpática	See page 235.		Late June to late Aug.
"Pale blue violet" 44 deep	GARGANO HAIRBELL	*Campánula Gargánica	See page 128.		May to Sept.
"Violet" 47 lighter or darker	CREEPING OR EUROPEAN BELL-FLOWER	*Campánula rapunculòides	See page 338.		July, Aug.
"Violet" 47 shading to 49	ENGLISH HAIRBELL, BLUE BELLS OF SCOTLAND	**Campánula rotundifòlia	See page 238.		June to late Aug.
"Purplish" 48	COVENTRY BELLS, BLUE FOXGLOVE	*Campánula Trachèlium *C. uticifòlia*	See page 339.		Mid. July to mid. Aug.

NEW ENGLAND ASTER. *Aster Novæ Angliæ.*

STOKES' ASTER. *Stokesia cyanea.*

Color	English Name	Botanical Name and *Synonyms*	Description	Height and *Situation*	Time of Bloom
"Violet" 47	BLUE SUCCORY OR CUPIDONE	*Catanánche cærùlea	See page 239.		June to Sept.
"Purple" 46 shading to 39	MOUNTAIN BLUET OR KNAPWEED	**Centaurèa montàna	See page 239.		"
"Purple" 37 deeper	MEADOW SAFFRON	**Cólchicum autumnàle	Clusters of crocus-like flowers appear after the large coarse foliage which dies in June. Protect in winter. Plant thickly in Aug. or early Sept. and where the exposed flowers get the support of other foliage or in grass which is not mown early. Do not disturb unless flowers deteriorate; then separate after the leaves die. Bulbous; can also be sown from seed. Rich light soil. White, striped and also double vars. Europe.	3-4 in. *Sun or half shade*	Aug., Sept.
"Purplish" 48	MARYLAND DITTANY	*Cunìla Mariàna	Plant of tufted habit. Profusion of small flowers in clusters. Heart-shaped leaves, 1 in. long. Neat plant for sunny border or wild garden. Middle U. S. A.	1 ft. *Sun*	"
"Purplish"	DAHLIA	**Dáhlia vars.	See page 359.		July to late Oct.
"Lilac" 43 darker	FRINGED PINK	*Diánthus supérbus	See page 339.		July, early Aug.
"Light purple"	GLANDULAR STORK'S OR HERON'S BILL	Eròdium macradènium *E. glandulòsum*	See page 240.		June to Sept.
"Pale lilac" 47 pale	LANCE-LEAVED DAY OR PLANTAIN LILY	**Fúnkia lancifòlia *F. Japónica*	See page 340.		Late July to early Sept.
"Lilac purple" 43	GOAT'S RUE	Galèga officinàlis	See page 240.		June to Sept.
"Purple" 42	WALLICH'S CRANES-BILL	Geranium Wallichiànum	See page 241.		"
"Bluish purple" 49	HYSSOP	Hýssopus officinàlis	See page 241.		Mid. June to mid. Aug.

Color	English Name	Botanical Name and *Synonyms*	Description	Height and *Situation*	Time of Bloom
"Violet" 55	COMMON LAVENDER	**Lavàndula Spìca**	Fragrant woolly sub-shrub, whiter and of lower and more compact habit than L. vera; flowers in shorter, denser spikes. Slight protection necessary. An old favorite on account of its fragrance. Border. Prop. by young wood cuttings. Light, rich, open soil. S. Europe.	1-2 ft. *Sun*	Late Aug. to mid. Sept.
"Violet" bet. 47 & 49	TRUE LAVENDER	**Lavàndula vèra**	See page 341.		Mid. July to Sept.
"Purple" 33 bluer	TWO-FLOWERED BUSH CLOVER	**Lespedèza bícolor** *Desmòdium penduliflòrum*	Graceful shrub. Pea-shaped flowers in racemes which branch out from the main stalk. Small round leaves. Ornamental in border and edges of shrubbery. Prop. by seed and division. Japan; N. China.	6–10 ft. *Sun or half shade*	Early Aug. to early Oct.
"Purple" 32	CYLINDRIC BLAZING STAR	***Liàtris cylindràcea** *Lacinària cylindràcea*	Resembles L. squarrosa though more of a dwarf plant. Wild garden or border. Prop. by seed sown in autumn or by offsets. Rich soil preferable. Western N. Y. and West.	1 ft. *Sun or half shade*	Aug.
"Purple" 39 very bright & clear	HANDSOME BLAZING STAR	***Liàtris élegans** *Lacinària èlegans*	See page 341.		Early July to mid. Aug.
"Purple" 46	LOOSE-FLOWERED BUTTON SNAKE-ROOT	***Liàtris graminifòlia** *Lacinària graminifòlia*	See page 341.		July to Oct.
"Purple" 42 lighter	DENSE-SPIKED BLAZING STAR	****Liàtris pycnostàchya**	Flowers in dense spikes continuing in bloom for a long time. Thick grass-like foliage. Good in masses in the border. Prop. by seed or division. Rich soil is best. Central U. S. A.	3-5 ft. *Half shade*	Aug. to mid. Sept.
"Dark lavender" 47	SCARIOUS BLAZING STAR	****Liàtris scariòsa**	Tubular flowers in large, lax spikes on stout whitish stalks. Foliage lance-shaped. Effective in masses. Prop. by seed and division. Rich soil preferable. Canada; U. S. A.	1-5 ft. *Sun*	Aug. Sept.
"Deep purple" 46	DENSE BUTTON SNAKE-ROOT, GAY FEATHER	****Liàtris spicàta** *Lacinària spicàta*	See page 341.		Mid. July to early Sept.
"Magenta purple" 45 more purple	COMMON BLAZING STAR, COLIC-ROOT	***Liàtris squarròsa** *Lacinària squarròsa*	Flowers showy, in few heads. Foliage somewhat grass-like. Mass in wild garden or border. Prop. by fall-sown seed or by offsets. Easily cultivated in any light, though preferably rich soil. Penn., S. and W.	1-2 ft. *Sun or shade*	Aug., early Sept.

AUGUST

Color	English Name	Botanical Name and *Synonyms*	Description	Height and *Situation*	Time of Bloom
"Lilac" 47 light	KENILWORTH IVY, MOTHER-OF-THOUSANDS	Linària Cymbalària	See page 242.		June to Sept.
"Bluish purple" 44 duller	HIMALAYAN LUNGWORT	*Lindelòfia spectábilis	See page 341.		Early July to early Aug.
"Light purple" 44	REQUIEN'S PENNYROYAL	Méntha Requieni	See page 342.		July, Aug.
"Purplish" 39	SPEARMINT	Méntha spicàta *M. víridis*	See page 342.		"
"Violet" 47 tinged 46	MUSSIN'S CATMINT	Népeta Mussíni *N. longifòlia*	See page 342.		"
"Purple" 47 deep	SCOTCH THISTLE	Onopórdon Acánthium	See page 342.		"
"Lilac" 43	SNAKE'S BEARD	Ophiopògon Jabùran	See page 342.		"
"Pale lilac" 43	GIANT BELL-FLOWER	*Ostrówskia magnífica	See page 342.		"
"Deep violet"	VIOLET PRAIRIE CLOVER	Petalostèmon violàceus *Kuhnístera purpùrea*	See page 245.		June until frost
"Lilac purple"	PERENNIAL PHLOX	**Phlóx paniculàta vars. *P. decussàta*	See page 370.		July to Oct.
"Bluish violet" bet. 46 & 47	BELL-FLOWERED HORNED RAMPION	*Phyteùma campanuloìdes	See page 343.		July, Aug.
"Bluish purple" near 50	SMALL OR LILAC-FLOWERED SCABIOUS	Scabiòsa Columbària	See page 246.		June to Oct.
"Purple" 49	MOUNTAIN SKULLCAP	*Scutellària alpìna	See page 343.		Early July to late Sept.
"Rosy purple" 40	PRETTY GROUNDSEL	*Senècio púlcher	See page 344.		July, Aug.
"Light violet" 44 deep	TORREY'S NIGHTSHADE	Solànum Tórreyi	See page 344.		"
"Dull violet" 47	TALL SEA LAVENDER	Státice elàta	See page 344.		Mid. July to early Sept.

AUGUST

Color	English Name	Botanical Name and *Synonyms*	Description	Height and *Situation*	Time of Bloom
"Rose lilac" bet. 44 & 38	CHOICE SEA LAVENDER	Státice exímia	Flowers in somewhat one-sided clusters. Coarse leaves with wavy margins spring from the root. Plant in isolated clumps or in rock-garden. Prop. by seed and division. Sandy well-drained soil. Songoria.	1-2 ft. *Sun*	Aug.
"Blue violet" 44	GMELIN'S SEA LAVENDER	*Státice Gmélini	See page 344.		Late July to early Sept.
"Bluish purple" 47	BROAD-LEAVED SEA LAVENDER	**Státice latifòlia	See page 344.		"
"Purple" 39	WALL GERMANDER	Teùcrium Chamaedrys	See page 345.		Mid. July to mid. Aug.
"Lilac" 43	MOTHER OF THYME, CREEPING THYME	Thỳmus Serpýllum	See page 249.		Mid. June to mid. Aug.
"Purplish" 44, 48 or 49	COMMON SPIDERWORT	*Tradescántia Virginiàna *T. Virgínica*	See page 140.		Late May to late Aug.
"Deep blue purple" 55	SUBSESSILE LONG-LEAVED VERONICA	**Verónica longifòlia var. subséssilis *V. spicàta*	One of the best Veronicas. Showy vigorous branching plant, covered with long dense spikes of small brilliant flowers. Excellent for the border. Prop. by division. Rich loam preferable. Japan.	2-3 ft. *Sun*	Early Aug. to mid. Sept.
"Violet" 47 or 49	HORNED VIOLET, BEDDING PANSY	**Vìola cornùta	See page 48.		Late Apr. until frost
"Deep blue" 56	AUTUMN ACONITE, MONKSHOOD OR WOLFSBANE	**Aconìtum autumnàle	See page 345.		Mid. July to mid. Sept.
"Blue"	STORK'S PURPLE WOLFSBANE	*Aconìtum Cammàrum var. Storkiànum *A. S., A. intermèdium*	See page 345.		July, Aug.
"Deep purple blue" near 49	TRUE MONKSHOOD, OFFICINAL ACONITE	**Aconìtum Napéllus *A. pyramidàle A. Tauricùm*	See page 345.		Mid. July to early Sept.

AUGUST

Color	English Name	Botanical Name and *Synonyms*	Description	Height and *Situation*	Time of Bloom
"Blue" 61	GLAND BELL-FLOWER	*Adenóphora commùnis *A. Fischeri, A. liliflòra, A. lilifòlia*	See page 345.		July to mid. Aug.
"Light blue" 61 duller	POTANNINI'S GLAND BELL-FLOWER	*Adenóphora Potannìni	See page 346.		July, Aug.
"Deep blue" 49	LEAD PLANT	*Amórpha canéscens	See page 346.		Early July to early Aug.
"Lilac blue" 53 lighter & brighter	MAACK'S ASTER	*Áster Máackii	Dwarf plant. Smooth leaves and stems with large flowers. Good border plant. Prop. by seed and division. Good rich soil. Japan.	1-2 ft. *Sun*	Mid. Aug. to late Sept.
"Purplish blue" 46	GREAT BELL-FLOWER	*Campánula latifòlia	See page 253.		June to early Aug.
"Blue" 48	CHIMNEY CAMPANULA, STEEPLE BELL-FLOWER	*Campánula pyramidàlis	See page 346.		July, Aug.
"Cobalt blue" 61	BLUE-FLOWERED LEADWORT	Ceratostígma plumbaginoìdes *Plumbàgo Lárpentæ, Valoràdia p.*	Beautiful half-shrubby plant which forms neat tufts and is covered with a wealth of flowers that gradually become violet. Excellent for rock-garden or edging. Protect in winter. Prop. by division. Warm light soil. China.	6-12 in. *Sun*	Late Aug., Sept.
"Blue" 53	FREMONT'S CLEMATIS	*Clématis Frèmonti	See page 346.		July, Aug.
"Light blue" 44 bluer	HERACLEUM-LEAVED CLEMATIS	*Clématis heracleæfòlia *C. tubulòsa*	Erect sturdy plant. Hyacinth-shaped flowers in terminal and axillary clusters. Bright green foliage. Border or rock-garden. Prop. by division and cuttings, and with difficulty by seed. Rich deep soil. China. Var. *Davidiana*, (*C. tubulosa var. Davidiana*): (color no. 54 lighter). Best form. Stems, 4 ft. tall, need slight support. Flowers with Orange-blossom fragrance. Leaves very large. Fine border plant. Japan; China.	2-3 ft. *Sun*	Aug., Sept.
"Blue" 61	BEE LARKSPUR	**Delphínium elàtum *D. alpìnum, D. pyramidàle*	See page 253.		June to Sept.
"Blue" 62	GREAT-FLOWERED LARKSPUR	**Delphínium grandiflòrum *D. Sinénse*	See page 347.		July, Aug.

Color	English Name	Botanical Name and *Synonyms*	Description	Height and *Situation*	Time of Bloom
"Blue" 50 & cream	LOFTY GLOBE THISTLE	*Echìnops exaltàtus	See page 347.		July, Aug.
"Steel blue" 53 colder	RITRO GLOBE THISTLE	**Echìnops Rìtro	See page 348.		"
"Pale blue" 52 greenish	ROUND-HEADED GLOBE THISTLE	*Echìnops sphæro-céphalus	See page 348.		"
"Bright blue" 53	ALPINE SEA HOLLY	*Erýngium alpìnum	See page 348.		"
"Ame-thyst blue" 63 lighter	AMETHYST SEA HOLLY	**Erýngium amethýsti-num	See page 254.		June to early Sept.
"Blue" 63	BOURGAT'S ERYNGO	*Erýngium Bourgati	See page 348.		Mid. July to early Sept.
"Blue" 63 lighter	DANE-WEED, HUNDRED THISTLE	Erýngium campéstre	See page 348.		"
"Blue"	FLAT-LEAVED ERYNGO	**Erýngium plànum	See page 348.		July, Aug.
"Deep blue" 56	CLOSED BOTTLE OR BLIND GENTIAN	Gentiàna Andréwsii *G. Càtesbæi*	A common wild herb. Flowers, which never open, stemless and supported on whorls of leaves. Brookside, rock-garden or wild garden. Prop. by division or very slowly by seed. Leave undisturbed. Good moist soil. N. Amer.	2 ft. *Half shade*	Aug., Sept.
"Deep blue" 59 lighter	WILLOW GENTIAN	*Gentiàna asclepiadèa	See page 351.		July to early Sept.
"Blue" 60	WIND FLOWER, HARVEST BELLS	Gentiàna Pneumo-nánthe	Funnel-shaped flowers in racemes terminating leafy stalks. Pretty when planted in grass on the banks of streams. Prop. very slowly by seed, also by division. Leave undisturbed. Rich deep soil mixed with humus Europe: N. Asia.	6 in. *Half shade*	Aug., Sept.
"Light blue" 62 greener	BARREL OR SOAPWORT GENTIAN	*Gentiàna Saponària *G. Càtesbæi*	Closed flowers about 1 in. long, erect in clusters. Wild or rock-garden. Leave undisturbed. Prop. by seed just ripe. Deep loam well-drained. N. Amer.	1-2 ft. *Sun or half shade*	"
"Violet" blue" 46 light	IBERIAN CRANES-BILL	Gerànium Ibèricum	See page 254.		June to Sept.
"Blue" 46 bluer	MEADOW CRANES-BILL	*Gerànium praténse	See page 257.		"

Color	English Name	Botanical Name and *Synonyms*	Description	Height and *Situation*	Time of Bloom
"Blue" 53	GLOBE DAISY	*Globulària vulgàris	See page 257.		June to Sept.
"Light blue" 52 dull	SHEP-HERD'S OR SHEEP SCABIOUS, SHEEP'S BIT	*Jasiòne perénnis	See page 351.		Early July to early Aug.
"Blue" 61	AUSTRIAN FLAX	*Lìnum Austrìacum *L. pérenne var. A.*	See page 258.		June to Sept.
"Blue" 46	GREAT LOBELIA, BLUE CARDINAL FLOWER	**Lobèlia syphilítica	Tubular flowers in long leafy spikes, on slightly hairy stalks. Foliage large, smooth or hairy. Good border plant. Moist soil. Eastern U. S. A.	1-3 ft. *Sun*	Mid. Aug. to late Sept.
"Blue" 57	EVER-FLOWER-ING FOR-GET-ME-NOT	*Myosòtis palústris var. sempérflorens	See page 153.		May to Sept.
"Blue" 53	LARGE-FLOWERED CATMINT	Népeta macrántha	See page 261.		Late June to early Sept.
"Blue" often 56	BALLOON FLOWER, JAPANESE BELL-FLOWER	**Platycòdon grandiflòrum *Wahlenbérgia, grandiflòra, Campánula g.*	See page 352.		Early July to Oct.
"Blue" 47	HAIRY RUELLIA	Ruéllia ciliòsa	See page 353.		Early July to late Aug.
"Blue" 52	PITCHER'S SAGE	*Sálvia azùrea var. grandiflòra *S. Pítcheri*	A sky-colored plant, slightly downy. Flowers blue, varying to white. Pretty in border. Protect in winter. Light sandy soil. S. Central U. S. A.	1-5 ft. *Sun*	Late Aug., Sept.
"Blue" 46 lighter	MEALY SAGE	*Sálvia farinàcea	See page 353.		July, Aug.
"Pale blue" often bet. 43 & 44	GRASS-LEAVED SCABIOUS	*Scabiòsa graminifòlia	See page 262.		June to Oct.
"Bluish" near 39 paler	WOOD-LAND SCABIOUS	*Scabiòsa sylvática	See page 262.		Early June to late Sept.
"Blue" 58	HYACINTH SQUILL	**Scílla hyacinthoìdes *S. parviflòra*	Rather shy bloomer. Small open bell-shaped flowers in many-flowered racemes. Long narrow leaves in a spreading clump at the base. Very pretty in the border or wild garden. Bulbous. Prop. by offsets. Light soil enriched with manure. Mediterranean Region.	1-1½ ft. *Sun or half shade*	Aug

AUGUST

Color	English Name	Botanical Name and *Synonyms*	Description	Height and *Situation*	Time of Bloom
"Blue" 57 deep	BLUE HYACINTH SQUILL	**Scílla hyacinthoìdes var. cærùlea *S. parviflòra var. c.*	Small bell-shaped flowers in racemes. Leaves in a spreading clump at the base of the plant. Charming in the border or wild garden. Bulbous. Light soil enriched with manure. Mediterranean Region.	1-1½ ft. *Sun or shade*	Aug.
"Blue" 49	BAICAL'S SKULLCAP	*Scutellària Baicalénsis *S. macrántha*	See page 353.		July, Aug.
"Deep blue" 46	COMMON SEA LAVENDER, MARSH ROSEMARY	**Státice Limònium *S. marìtima*	See page 353.		"
"Blue" 52 dull	STOKES' ASTER	**Stokèsia cyànea	Somewhat aster-like flowers, few to many on leafy stalks. Useful for cutting. Excellent in border. Protect slightly in winter. Prop. by division. Loamy soil, sandy and well-drained. S. Atlantic States. See Plate, page 406.	1-1½ ft. *Sun*	Aug. to early Oct.
"Blue" 53	AUSTRIAN SPEED-WELL	*Verónica Austrìaca	See page 353.		Early July to early Aug.
Parti-colored white & 34	PARKIN-SON'S CHECK-ERED MEADOW SAFFRON	*Cólchicum Párkinsoni	Distinct species. Crocus-like flowers, white checkered and barred with purple, unsupported by foliage. Small leaves appear in spring. Plant in clumps in rock-garden or border among carpeting plants. Protect in winter. Bulbous. Prop. by offsets. Sandy loam, rich and light. Greek Archipelago.	4-6 in. *Sun or half shade*	Aug., Sept.
Parti-colored	FRENCH OR CROZY CANNA	**Cánna vars.	See page 357.		July to late Sept.
Parti-colored	ITALIAN CANNA	**Cánna vars.	See page 358.		"
Parti-colored	DAHLIA	**Dáhlia vars.	See page 359.		July to late Oct.
6 shading from 19 to 14	GREAT-FLOWERED GAILLARDIA	**Gaillárdia aristàta *G. grandiflòra*	See page 263		June to Nov.
Parti-colored	SWORD LILY	**Gladìolus vars.	See page 365.		July to Oct.
Parti-colored	BROWN'S LILY	**Lílium Brówni *L. Japónicum var. Brównii*	See page 354.		July, Aug.
44 & 11	ALPINE TOADFLAX	Linària alpìna	See page 354.		"

AUGUST

Color	English Name	Botanical Name and *Synonyms*	Description	Height and *Situation*	Time of Bloom
Various	BEAR'S BREECH, CUTBER-DILL SEDOCKE	**Acánthus móllis	See page 354.		July, Aug.
Various	VERY PRICKLY BEAR'S BREECH	**Acánthus spinossímus	See page 357.		"
Various	PRICKLY BEAR'S BREECH	**Acánthus spinòsus	See page 357.		"
2, 34, 35, 22, 27, white etc.	HOLLY-HOCK	**Althæa ròsea	See page 357.		"
Various	FRENCH OR CROZY CANNA	**Canna vars.	See page 357.		July to late Sept.
Various	ITALIAN CANNA	**Canna vars.	See page 358.		"
Various	DAHLIA	**Dáhlia vars.	See page 359.		July to late Oct.
Often 33 dark	SCARLET DAHLIA	Dáhlia coccínea	See page 365.		July to early Oct.
Various	VON MERCK'S DAHLIA	Dáhlia Mérckii *D. Glabràta*	See page 365.		"
Often 17 dark	COMMON DAHLIA	Dáhlia variábilis *D. ròsea*	See page 365.		"
Various	SWORD LILY	**Gladìolus	See page 365.		July to Oct.
White, 2, 44	COBÆA BEARD-TONGUE	*Pentstèmon Cobæa	See page 370.		July, Aug.
Various	PERENNIAL PHLOX	**Phlóx paniculàta vars. *P. decussàta*	See page 370.		July to Oct.
Often 39 or 38	EGYPTIAN OR GYPSIES' ROSE	*Scabiòsa arvénsis *S. vària*	See page 273.		Early June to mid. Aug.
Various	PANSY, HEART'S-EASE	**Vìola trícolor	See page 68.		Mid. Apr. to mid. Sept.

SEPTEMBER

WHITE TO GREENISH

Color	English Name	Botanical Name and *Synonyms*	Description	Height and *Situation*	Time of Bloom
"White"	**SNEEZE-WORT**	**Achillèa Ptármica**	See page 167		June to mid. Sept.
"White"	**DOUBLE SNEEZE-WORT**	****Achillèa Ptármica var. "The Pearl"**	See page 167.		June to Oct.
"White"	**SIBERIAN MILFOIL OR YARROW**	***Achillèa Sibírica** *A. Mongòlica, A. ptarmicoìdes*	See page 274.		July to Oct.
"Nearly white"	**WHITE MONKS-HOOD OR OFFICINAL ACONITE**	****Aconìtum Napéllus var. álbum** *A. pyramidàle var. a. A. Taùricum, var. a.*	See page 274.		Mid. July to early Sept.
"White"	**JAPANESE BANE-BERRY**	**Actæa Japónica**	Dense spikes of flowers, good for cutting and lasting a long time. Effective border plant. Needs light rich soil. Japan.	2½ ft. *Half shade*	Sept., Oct.
"Cream white"	**ANEMONE HONORINE JOBERT**	****Anemòne Japónica var. álba**	One of the best autumn plants. Similar to the type. Large flowers with yellow centres. Handsome foliage. Good for cutting. Plant in masses under trees or in clumps in the border. Leave undisturbed and protect in winter. Prop. by seed and division. Rich soil. Hort. See Plate, page 418. **A. Japonica var. "The Whirlwind;"* a strong vigorous plant like the type, but with semi-double white flowers 2½-3 in. across, blooming more abundantly and lasting longer. Hort.	2-3 ft *Sun or half shade*	Late Sept. to early Nov.
"White"	**TALL FLAT-TOP WHITE ASTER**	**Áster umbellàtus** *Dœllingèria umbellàta*	See page 375.		Late Aug. to early Oct.
"Whit-ish" turns 39	**VARIOUS-COLORED STARWORT**	***Áster versícolor**	Dwarf species with flowers about 1 in. across, changing from white to mauve. Plant in mixed border or rock-garden. Sandy soil Prop. by division. N. Amer.	2-3 ft. *Sun*	Sept., early Oct.

AUTUMN

ANEMONE HONORINE JOBERT. *Anemone Japonica var. alba.*

SEPTEMBER

Color	English Name	Botanical Name and *Synonyms*	Description	Height and *Situation*	Time of Bloom
"White"	ASTER-LIKE BOLTONIA	*Boltònia asteroìdes *B. glastifòlia*	See page 375.		Late Aug. to mid. Oct.
"White"	WOAD-LEAVED BOLTONIA	**Boltònia glastifòlia	See page 376.		Aug., Sept.
"White" centre 5 to 6	NIPPON CHRYSANTHEMUM	*Chrysánthemum Nippónicum	Large glistening daisy-like flowers sometimes 4 in. across, with yellow centres. Thick foliage. Good for border. Thrives in moderately light enriched soil. Japan.	2 ft. *Sun*	Sept., Oct.
"White"	GIANT DAISY	**Chrysánthemum uliginòsum *Pyrèthrum u.*	See page 376.		Aug., Sept.
"White"	DAHLIA	**Dáhlia vars.	See page 359.		Late July to late Oct.
"White"	RATTLE-SNAKE-MASTER, BUTTON SNAKEROOT	Erýngium aquáticum *E. yuccæfòlium*	See page 177.		June to Oct.
"White"	WHITE SNAKEROOT	Eupatòrium ageratoìdes	See page 379.		Aug., Sept.
"White"	LARGE-FLOWERED SUBCORDATE PLANTAIN LILY	**Fúnkia subcordàta var. grandiflòra *F. álba var.g., F. liliiflòra var. g.*	See page 380.		Late Aug., Sept.
"White"	SWORD LILY	**Gladìolus vars.	See page 365.		July to Oct.
"White"	WHITE CORAL BELLS	Heùchera sanguínea var. álba *H. álba*	See page 179.		June to late Sept.
"White"	CALIFORNIA ROSE MALLOW	*Hibíscus Califórnicus	Vigorous shrubby plant. Flowers 4-5½ in. across, sometimes pink, with purple centres. Heart-shaped leaves, gray beneath. Good in border. Protect in winter. Prop. by seed. Any garden soil, preferably sandy and rich. Cal.	5-7 ft. *Sun or half shade*	Sept., Oct.
"White"	WHITE SWAMP ROSE OR ROSE MALLOW	*Hibíscus Moscheùtos var. álbus *H. palústris var. álbus*	See page 381.		Aug., Sept.
"White"	CRIMSON EYE SWAMP ROSE	*Hibíscus Moscheùtos var. "Crimson Eye"	See page 381.		"

SEPTEMBER

Color	English Name	Botanical Name and *Synonyms*	Description	Height and *Situation*	Time of Bloom
"White"	JAPANESE BUSH CLOVER	*Lespedèza Japónica *Desmòdium Japónicum*	Resembles L. Sieboldi, but blooms a fortnight later. Strong vigorous plant with a profusion of pea-shaped flowers drooping in elongated racemes. Dull pale green compound foliage. Ornamental in the border or edge of shrubbery. Prop. by division. Easily cultivated. Japan.	4-6 ft. *Sun*	Sept., Oct.
"White"	AUTUMN SNOW-FLAKE	*Leucòjum autumnàle *Acis autumnàlis*	Not hardy everywhere. Related to Snowdrop. Fragrant drooping bell-shaped flowers, red-tinted, 1-3 on slender stems, followed by slender shrubby foliage. Attractive in border and in shady sheltered part of rock-garden or shrubbery. Bulbous. Prop. by offsets. Any soil. Portugal.	3-9 in. *Half shade*	Sept.
"White"	HEART-LEAVED LILY	*Lílium cordifòlium	See page 381.		Aug., Sept.
"White"	GREAT WHITE LOBELIA	**Lobèlia syphilítica var. álba	See page 381.		Mid. Aug. to Oct.
"White"	DOUBLE WHITE OR EVENING CAMPION	*Lýchnis álba var. flòre-plèno *L. vespertìna var. f.-p.*	See page 289.		Mid. July to mid. Sept.
"White"	WHITE MUSK MALLOW	*Málva moschàta var. álba	See page 289.		July to early Sept.
"White"	SHOWY PRIMROSE	*Œnothèra speciòsa *Hartmánnia s.*	See page 382		Aug., Sept.
"White"	WHITE ICELAND POPPY	**Papàver nudicaùle var. álbum	See page 184.		Late Apr. to mid. June, late Aug., Sept.
"White"	PERENNIAL PHLOX	**Phlóx paniculàta vars. *P. decussàta*	See page 370.		July to Oct.
"White"	WHITE BALLOON FLOWER	**Platycòdon grandiflòrum var. álbum *Campánula g. var. a., Wahlenbérgia g. var. a.*	See page 187.		June to Oct.
"Greenish white"	SACALINE	Polýgonum Sachalinénse	See page 383.		Late Aug., Sept.

SEPTEMBER

Color	English Name	Botanical Name and *Synonyms*	Description	Height and *Situation*	Time of Bloom
"White"	JAPAN KNOTWEED	Polýgonum Sièboldi *P. cuspidàtum. P. Zuccarínii*	See page 383.		Aug., Sept.
"White"	WHITE WOODLAND SCABIOUS	*Scabiòsa sylvática var. albiflòra	See page 189.		Early June to late Sept.
"White"	WHITE SHOWY SEDUM	*Sèdum spectábile var. álbum *S. Fabària var. álbum*	See page 383.		Late Aug. to mid. Sept.
"Whitish"	GREAT VIRGINIAN SPEED-WELL, CULVER'S ROOT	*Verónica Virgínica *Leptàndra Virgínica*	See page 384.		Early Aug. to early Sept.
"White"	WHITE HORNED VIOLET OR BEDDING PANSY	**Vìola cornùta var. álba	See page 23.		Late Apr. until frost
"Yellow" 5	WOOLLY-LEAVED MILFOIL	*Achillèa tomentòsa	See page 96.		Late May to mid. Sept.
"Pale yellow" 4	EGYPTIAN MILFOIL OR YARROW	Achillèa Tournefórtii *A. Ægyptiaca*	See page 293.		July to Oct.
"Yellow" 2	PALE YELLOW WOLFS-BANE	*Aconìtum Lycóctonum *A. barbàtum, A. ochroleùcum, A. squarròsum*	See page 294.		July to early Sept.
"Yellow" 5	GOLDEN MARGUE-RITE	*Ánthemis Kélwayi *A. tinctòria var. Kélwayi*	See page 194.		Mid. June to Oct.
"Yellow" 2, centre 6	GOLDEN MARGUE-RITE, ROCK CAMOMILE	*Ánthemis tinctòria	See page 96.		Mid. May to Oct.
"Yellow"	ITALIAN CANNA	**Cánna vars.	See page 357		July to late Sept.
"Yellow" 6 paler	LARKSPUR TICKSEED	*Coreópsis delphinifòlia	See page 296		Mid. July to mid. Sept.

SEPTEMBER

Color	English Name	Botanical Name and *Synonyms*	Description	Height and *Situation*	Time of Bloom
"Yellow" 6 deeper	STAR TICKSEED	Coreópsis pubéscens *C. auriculàta*	See page 296.		Mid July to late Sept.
"Clear yellow" 5	TALL TICKSEED	*Coreópsis trípteris	See page 385.		Aug., Sept.
"Yellow"	DAHLIA	**Dáhlia vars.	See page 359.		July to late Oct.
"Yellow"	SWORD LILY	**Gladìolus vars.	See page 365.		July to Oct.
"Yellow" 5 deeper	SNEEZE-WEED, FALSE OR SWAMP SUN-FLOWER	**Helènium autumnàle *H. grandi-flòrum*	See page 386.		Aug. to late Sept.
"Yellow" bet. 5 & 6	LARGE-FLOWERED SNEEZE-WEED	**Helènium autumnàle var. grandiflòrum	See page 386.		Aug., Sept.
"Yellow" 6	DWARF SNEEZE-WEED	**Helènium autumnàle var. pùmilum	See page 386.		"
"Yellow" 5	TALL SNEEZE-WEED	**Helènium autumnàle var. supérbum *H. grandiflò-rum var. s.*	See page 386.		"
"Yellow" 5	PURPLE-HEADED SNEEZE-WEED	*Helènium nudiflòrum *Leptopoda brachypoda*	See page 300.		July to Oct.
"Yellow" 5	NARROW-LEAVED OR SWAMP SUN-FLOWER	Heliánthus angustifòlius	See page 386.		Aug. to Nov.
"Yellow" 5	THIN-LEAVED OR WILD SUN-FLOWER	*Heliánthus decapétalus	See page 386		Aug. to late Sept.
"Yellow" 6	WILD SUN-FLOWER, INDIAN POTATO	Heliánthus gigantèus	Striking plant. Flowers about 2½ in. across, rays 1 in. long, borne on strong hairy stalks. Rough foliage. Effective in shrubbery. Prop. by division; divide every 2 years. Dry or moist soil, not too heavy. N. Amer. See Plate, page 423.	3-12 ft. *Sun*	Early Sept. to mid. Oct.
"Yellow" 5 to 6 darker, centre dull	SHOWY SUN-FLOWER	Heliánthus lætiflòrus	See page 387.		Aug., Sept.

WILD SUNFLOWER. *Helianthus giganteus.*

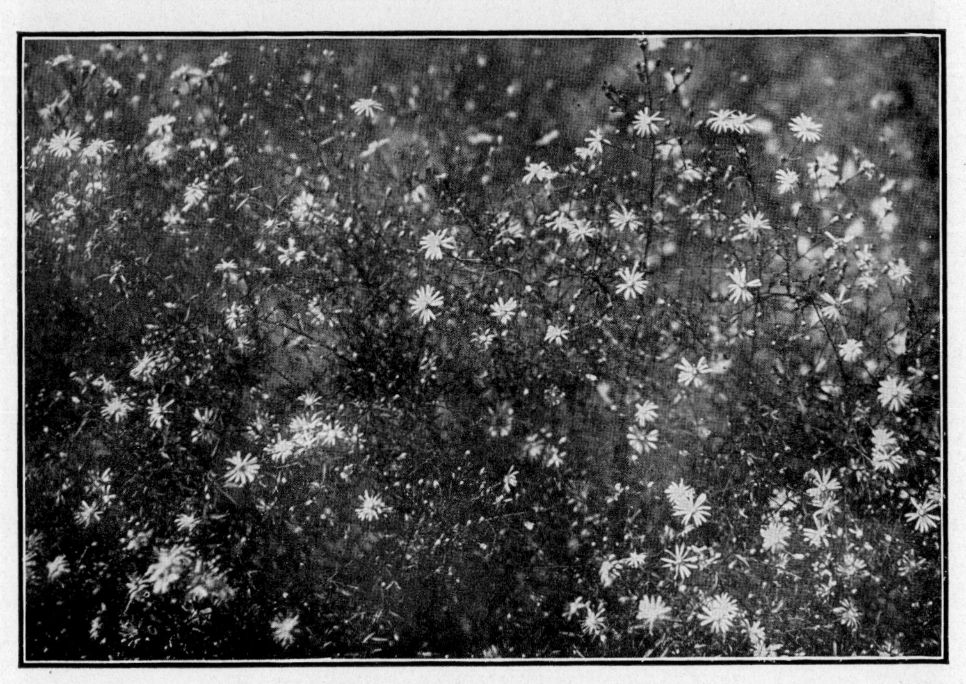

PRAIRIE ASTER. *Aster turbinellus.*

LINEAR-LEAVED SUNFLOWER. *Helianthus orgyalis.*

SEPTEMBER

Color	English Name	Botanical Name and *Synonyms*	Description	Height and *Situation*	Time of Bloom
"Yellow" 5	SMOOTH SUN-FLOWER	*Heliánthus lævigàtus	See page 387.		Mid. Aug. to Oct.
"Yellow" 5 richer	MAXI-MILIAN'S SUN-FLOWER	*Heliánthus Maximíliani	One of the most desirable Sunflowers, graceful and stately; flowers very late. Very large single blossoms and willow-like foliage. Excellent for cutting. Good among shrubs. Prop. by division; divide every 2 years. Dry soil, not too heavy. Western U. S. A.	2-8 ft. *Sun*	Early Sept. to Nov.
"Yellow" 5 richer	LINEAR-LEAVED SUN-FLOWER	*Heliánthus orgyàlis *H. giganteùs var. crinítus*	Graceful and decorative plant. Long spikes of small flowers terminate unbranched stalks densely sheathed with drooping grass-like foliage. Plant among shrubs. Prop by division; divide every 2 years. Dry soil, not too heavy. S. Western U. S. A. See Plate, page 424.	8-10 ft. *Sun*	Sept., early Oct.
"Yellow" bet. 5 & 6	STIFF SUN-FLOWER MISS MELLISH	**Heliánthus rígidus var. "Miss Mellish"	See page 387.		Late Aug., Sept.
"Orange yellow" 6	PALE-LEAVED WOOD SUN-FLOWER	Heliánthus strumòsus	See page 300.		Mid. July to late Sept.
"Yellow" 5	WOOLLY SUN-FLOWER	Heliánthus tomentòsus	See page 387.		Late Aug. to late Sept.
"Yellow" 5 richer	JERUSALEM ARTICHOKE	Heliánthus tuberòsus	Plant often grown as vegetable. Numerous flowers 2-3 in. in diameter. Foliage rough. Effective in shrubbery or wild places but apt to become a bad weed. Prop. by seed, division or cuttings. Loose dry loam.	5-12 ft. *Sun*	Sept.
"Yellow" 5	THROAT-WORT SUN-FLOWER	Heliánthus trachelii-fòlius	See page 300.		Mid. July to early Sept.
"Yellow" 6	OXEYE, FALSE SUN-FLOWER	*Heliópsis lævis *H. helian-thoìdes*	See page 300.		Mid. July to late Sept.
"Yellow" 6, centre 7	PITCHER'S OXEYE OR FALSE SUN-FLOWER	*Heliópsis lævis var. Pitcheriàna *H. P.*	See page 303.		"
"Yellow" 1	HOARY MARSH OR ROSE MALLOW	*Hibíscus incànus	See page 388.		Aug., Sept.

Color	English Name	Botanical Name and *Synonyms*	Description	Height and *Situation*	Time of Bloom
"Yellow" 5	NARROW-LEAVED HAWK-WEED	Hieràcium umbellàtum	See page 388.		Aug., Sept.
"Yellow" 5	GRAY-HEADED CONE-FLOWER	*Lépachys pinnàta *Ratíbida p.*, *Rudbéckia p.*	See page 200.		June to mid. Sept.
"Yellow" 3	BIRD'S-FOOT TREFOIL, BABIES' SLIPPERS	Lòtus corniculàtus	See page 203.		June to Oct.
"Bright yellow" 6 to 7	GOLDEN SPIDER LILY	Lycòris aùrea *Amarýllis a.*, *Nerìne a.*	See page 389.		Aug., Sept.
"Yellow" 5 or 6 brilliant	ICELAND POPPY	**Papàver nudicaùle	See page 35.		Late Apr. to July, late Aug. to Oct.
"Yellow" 5	YELLOW OR CALIFORNIA CONE-FLOWER, WHORTLE-BERRY-LEAVED KNOT-FLOWER	*Rudbéckia Califórnica	See page 390		Aug. to mid. Sept.
"Yellow" 5	LARGE-FLOWERING CONE-FLOWER	*Rudbéckia grandiflòra	See page 390.		Late Aug., Sept.
"Yellow" bet. 5 & 6	GOLDEN GLOW	**Rudbéckia laciniàta var. flòre-plèno	See page 309.		Late July to late Sept.
"Yellow" 5	LARGE CONE-FLOWER	*Rudbéckia máxima	See page 390.		Aug., Sept.
"Yellow" 5	SHINING CONE-FLOWER	*Rudbéckia nítida	Closely related to R. maxima, differing principally in height and foliage. Leaves bright green. Prop. by seed, division or cuttings. Any garden soil. Southern U. S. A.	2-4 ft. *Sun*	Sept.
"Yellow" bet. 5 & 6	SWEET CONE-FLOWER	**Rudbéckia subtomentòsa	See page 391.		Mid. Aug. to Oct.
"Yellow" 3 pale	WEBB'S SCABIOUS OR PIN-CUSHION FLOWER	Scabiòsa ochroleùca *S. Webbiàna*	See page 208.		June to early Sept.

Color	English Name	Botanical Name and *Synonyms*	Description	Height and *Situation*	Time of Bloom
"Yellow" 5	ROUGH ROSINWEED	Sílphium aspérrimum	See page 313		July to early Sept.
"Yellow" 5	COMPASS PLANT, PILOT WEED	**Sílphium laciniàtum	See page 313.		Mid. July to mid. Sept.
"Yellow" 6 lighter	CUP PLANT, INDIAN CUP	*Sílphium perfoliàtum	See page 313.		"
"Yellow" 5	PRAIRIE DOCK	*Sílphium terebinthinàceum	See page 392.		Mid. Aug. to late Sept.
"Yellow" 6 lighter	WHORLED ROSINWEED	Sílphium trifoliàtum	See page 313.		Mid. July to mid. Sept.
"Yellow" 5 to 6	GOLDEN-ROD	*Solidàgo	See page 313.		Late July to early Oct.
"Yellow"	CROCUS-FLOWERED BLAZING STAR	**Tritònia crocosmæflòra vars. *Montbrètia c.*	See page 314.		July to Oct.
"Yellow" 6	YELLOW HORNED VIOLET OR BEDDING PANSY	**Vìola cornùta var. lùtea màjor	See page 35.		Late Apr. until frost
"Orange" 5 & 14	STRIPED SNEEZE-WEED	*Helènium autumnàle var. striàtum	See page 393.		Aug., Sept.
"Orange" 12	ORANGE HAWK-WEED	Hieràcium aurantìacum	See page 211.		June to Oct.
"Scarlet orange" 13	MACOWAN'S FLAME FLOWER	*Kniphòfia Macówani	A very neat dwarf species. Spikes of flowers rise well above grass-like foliage. Pretty in the border. Take up and store in winter or protect with leaves. Prop. by division. Any well-drained soil. S. Africa.	1-2 ft. *Half shade*	Sept., early Oct.
"Scarlet orange" effect 12 brighter	EVER-BLOOMING FLAME FLOWER	**Kniphòfia Pfítzerii	See page 393.		Early Aug. to early Oct.
"Salmon orange" bet. 16 & 8	DR. HENRY'S LILY	*Lílium Hényri	See page 393.		Aug., Sept.
"Deep orange" bet. 12 & 17	ORANGE ICELAND POPPY	*Papàver nudicaùle var. aurantìacum	See page 36.		Late Apr. to July, late Aug., Sept.

Color	English Name	Botanical Name and *Synonyms*	Description	Height and *Situation*	Time of Bloom
"Deep orange" bet. 12 & 17	SMALL ICELAND POPPY	*Papàver nudicaùle var. miniàtum	See page 36.		Late Apr. to July, mid. Aug. to Oct.
"Orange" 7, centre 21 dark	ORANGE CONE-FLOWER	*Rudbéckia fúlgida	See page 394.		Mid. Aug. to early Oct.
"Orange" 6 to 18	POTTS' BLAZING STAR	*Tritònia Póttsii *Montbrètia P.*	See page 321.		July to Oct.
"Red" 27	RED YARROW OR MILFOIL	*Achillèa Millefòlium var. rùbrum	See page 321.		Mid. July to mid. Sept.
"Purple crimson" bet. 41 & 42	WESTERN SILKY OR SILVERY ASTER	*Áster serîceus *A. argênteus*	See page 394.		Late Aug. to early Oct.
"Red"	FRENCH OR CROZY CANNA	**Cánna vars.	See page 357.		July to late Sept.
"Red"	ITALIAN CANNA	**Cánna vars.	See page 358.		"
"Red"	DAHLIA	**Dáhlia vars.	See page 359.		July to late Oct.
"Red"	SWORD LILY	**Gladìolus vars.	See page 365.		July to Oct.
"Red" 26 brilliant	CORAL OR CRIMSON BELLS	**Heùchera sanguínea	See page 213.		June to late Sept.
"Scarlet" 9 to 18	FLAME FLOWER, TORCH LILY, RED-HOT POKER PLANT	*Kniphòfia aloìdes *K. Uvària, Tritoma Uvària*	Striking and tropical looking plant. Pyramidal spikes of flaming flowers, lower blossoms yellow, turning to red at the top, rise from a clump of drooping grass-like leaves. Effective massed in border against background or in shrubbery. Take up and store in winter or protect with leaves. Prop. by division. Any well-drained soil. Africa. See Plate, page 429. Var. *grandiflora;* (color no. bet. 9 & 16), spikes and flowers larger than the type. Hort. Var. *glaucescens;* (color no. bet. 9 & 16), "vermilion-scarlet flowers changing to a more orange color, one of the freest bloomers." Foliage gray-green. S. Africa. Var. *nobilis,* (*T. nobilis*); (color no. 17), a very fine robust form, 6-7 ft. high. Flowers, with prominent anthers, 1½ in. long varying to orange-scarlet.	3 ft. *Half shade*	Sept., early Oct.

RED-HOT POKER PLANT. *Kniphofia aloides.*

JAPANESE WINDFLOWER. *Anemone Japonica.*

SEPTEMBER

Color	English Name	Botanical Name and *Synonyms*	Description	Height and *Situation*	Time of Bloom
"Coral red" 16 brilliant	CORAL-RED FLAME FLOWER	*Kniphòfia corallìna	A cross between K. Macowani and K. Uvaria. Free-flowering dwarf species. Oval spikes of flowers shading to rose. Good for cutting. See K. aloides. Hort.	1½-2 ft. *Half shade*	Sept.
"Crimson" 20 bright & rich	CARDINAL FLOWER, INDIAN PINK	**Lobèlia cardinàlis	See page 396.		Aug. to mid. Sept.
"Red" 20 lighter centre 21 redder	OSWEGO TEA, BEE OR FRAGRANT BALM	**Monárda dídyma	See page 217.		Mid. June to early Sept.
"Red"	PERENNIAL PHLOX	**Phlóx paniculàta vars. *P. decussàta*	See page 370.		July to Oct.
"Red"	ALKEKENGI, STRAWBERRY, TOMATO, WINTER OR BLADDER CHERRY	Phýsalis Alkekéngi	See page 325.		Fruit July to late Oct.
"Red"	CHINESE LANTERN PLANT	*Phýsalis Franchétti *P. Alkekéngi var. Franchètti*	See page 325.		"
"Red" 26 deep	TARTARIAN SEA LAVENDER	*Státice Tatárica *S. Besseriàna, S. incàna var. hỳbrida*	See page 397.		Early Aug. to early Sept.
"Orange scarlet" often 16 pinker	CROCUS-FLOWERED BLAZING STAR	*Tritònia crocosmæflòra *Montbrètia c.*	See page 326.		July to Oct.
"Vermilion" 18 redder	CALIFORNIA FUCHSIA, HUMMING-BIRD'S TRUMPET	*Zauschnèria Califórnica	See page 397.		Late Aug., Sept.
"Rose" 36, darker markings	JAPANESE WIND-FLOWER	**Anemòne Japónica	One of the best autumn plants. Beautiful large flowers with yellow centres. Leaves handsome and mostly in clumps at base. Good for cutting. Lovely under trees in masses, or in clumps in the border. Leave undisturbed and protect in winter. Prop. by seed and division. Rich soil. China; Japan. See Plate, page 430. *"Queen Charlotte,"* (color no. 40 duller), a fine distinct form having all the good qualities of the type; semi-double flowers. Hort.	2-4½ ft. *Sun or half shade*	Late Sept. to early Nov.

Color	English Name	Botanical Name and *Synonyms*	Description	Height and *Situation*	Time of Bloom
"Rosy pink" 31 & 32	PINK JAPANESE ANEMONE	**Anemòne Japónica var. rùbra	Similar to A. Japonica; flowers and massive foliage of wax-like texture. Good for cutting and attractive for the border and in masses. Treat like A. Japonica. Prop. by seed and division. Rich soil. Hort.	4-5 ft. *Sun or half shade*	Late Sept. to early Nov.
"Lilac rose" near 43	HYBRID ITALIAN STARWORT	*Áster Améllus var. hỳbridus	Beautiful and graceful dwarf Aster, bearing an abundance of large flowers with orange centres. Good for planting in the border. Prop. by seed and division. Europe.	2 ft. *Sun*	Sept.
"Rose" effect 43	VARIEGATED MEADOW SAFFRON	*Cólchicum variegàtum	Rather large crocus-like flowers, checkered with purple, appearing before the foliage. Plant in clumps in grassy places or among foliage in rock-garden and border. Bulbous. Sandy loamy soil, rich and light. Asia Minor; Greece.	3-6 in. *Sun or half shade*	Early Sept. to early Oct.
"Pink etc."	DAHLIA	**Dáhlia vars.	See page 359.		July to late Oct.
"Pink"	SWORD LILY	**Gladìolus vars.	See page 365.		July to Oct.
"Pink" 22 very pale centre 33	HALBERT-LEAVED ROSE MALLOW	*Hibíscus militàris *H. Virgínicus*	See page 401.		Aug., early Sept.
"Rose" 36, centre 33	SWAMP ROSE, ROSE MALLOW	*Hibíscus Moscheùtos *H. palústris*	See page 401.		Aug., Sept.
"Pink" 36, spotted 27	SPOTTED LILY	**Lílium speciòsum *L. lancifòlium*	See page 401.		"
"Pink" white & 23	HANDSOME MELPOMENE LILY	**Lílium speciòsum var. Melpómene	See page 401.		"
"Reddish pink" white & 26 deeper	HANDSOME RED LILY	**Lílium speciòsum var. rùbrum	See page 401.		"
"Rose" 38 more violet	MUSK MALLOW	*Málva moschàta	See page 332.		July to early Sept.
"Pink"	PERENNIAL PHLOX	**Phlóx paniculàta vars. *P. decussàta*	See page 370.		July to Oct.
"Rose" 27	AUTUMN SQUILL, STARRY HYACINTH	*Scílla autumnàlis	See page 333.		Late July to Oct.

CORAL BELLS. *Heuchera sanguinea.*

NEW ENGLAND ASTER. *Aster Nova Anglia.*

Color	English Name	Botanical Name and *Synonyms*	Description	Height and *Situation*	Time of Bloom
"Purplish pink" 27 lighter	EWER'S TURKESTAN STONECROP	*Sèdum Èwersii var. Turkestánicum	More or less trailing plant. Flowers in close globular clusters. Broad glaucous leaves. "Hardy in Mass." Good for carpeting. Prop. preferably by division. Sandy soil best. Turkestan.	4-6 in. *Sun*	Sept., early Oct.
"Pink"	LYDIAN STONECROP	Sèdum Lýdium	See page 402.		Aug., Sept.
"Pinkish" dull 22 in effect	LARGEST STONECROP	*Sèdum máximum	Vigorous bushy plant, bold and stately. Flowers, spotted with red, in lax panicles. Broad pulpy leaves. Good for massing in the border or rock-garden. Prop. preferably by division. Easily grown in any poor gravelly soil. Europe; Asia. Var. *purpureum*, purple leaves.	1-2 ft. *Sun*	Mid. Sept. to early Oct.
"Pink" 38 lighter	SIEBOLD'S STONECROP	*Sèdum Siebòldii	See page 402.		Aug. to mid. Sept.
"Rose" 38 or 32 lighter	SHOWY SEDUM	**Sèdum spectábile *S. Fabària*	See page 402.		Late Aug. to mid. Sept.
"Pinkish" 39 lighter	JAPANESE TOAD LILY	*Tricýrtis hírta *T. Japónica*	An interesting plant. Lily-like purple-spotted flowers on erect leafy stems. Blooms so late that it is apt to be injured by frost. Associate with Trilliums, etc., in rock-garden. Prop. by offsets. Light sandy loam with leaf-mold. Japan. Var. *nigra,* desirable as it flowers a few weeks earlier. Hort.	1-3 ft. *Half shade*	Late Sept., Oct.
"Purple" 46	BESSARABIAN ASTER	*Áster Améllus var. Bessarábicus *A. Bessarábicus*	See page 403.		Aug., Sept.
Violet" 47	RUSH ASTER	*Áster júnceus	See page 337.		July to early Sept.
"Violet" bet. 46 & 47	SAVORY-LEAVED ASTER	*Áster linariifòlius *Diplopappus linariifòlius*	See page 337.		July to Oct.
"Violet" 44 pale	LINDLEY'S ASTER	*Áster Lindleyànus	See page 403.		Aug. to early Oct.
"Purple" 48 redder	NEW ENGLAND ASTER OR STARWORT	**Áster Nòvæ Ángliæ	See page 404 and Plate, page 434.		Mid. Aug. to late Sept.
"Blue violet" 44	NEW YORK STARWORT	*Áster Nòvi Bélgii	Vigorous plant with numerous flowers. Shrubbery and border. Prop. by seed and division. Prefers moist soil. N. Amer.	3-5 ft. *Sun*	Sept., early Oct.

SEPTEMBER

Color	English Name	Botanical Name and *Synonyms*	Description	Height and *Situation*	Time of Bloom
"Bluish purple" 41 light	LATE PURPLE ASTER	*Áster pàtens	Low spreading plant with slender stems. Solitary flowers with yellow centres. Foliage rather rough. One of the weaker and shorter lived species. Border plant. There is a pink var. Prop. by division. Any soil. N. Amer.	1-2 ft. *Sun or half shade*	Sept.
"Blue violet" 44 pinker	SHORT'S ASTER	**Áster Shórtii	A tall pretty species. Flowers abundant in long clusters in late fall. Good for the border and in groups. Prop. by division. N. Amer.	2-4 ft. *Sun or shade*	Late Sept. to late Oct.
"Violet" 47	SIBERIAN ASTER OR STARWORT	*Áster Sibíricus	See page 404.		Aug., Sept.
"Purple" 49	LOW SHOWY OR SEASIDE PURPLE ASTER	*Áster spectábilis	See page 404.		Late Aug. to mid. Sept.
"Purple" 46	TARTARIAN ASTER	**Áster Tatáricus	Late blooming species with large flowers. Lance-shaped foliage sometimes 2 ft. long. Effective in the shrubbery or rear of the border. Prop. by seed and division. Siberia.	7 ft. or less *Sun or half shade*	Late Sept., Oct.
"Mauve" 47 light & pinker	PRAIRIE ASTER	**Áster turbinéllus	Graceful plant, valuable for its late flowers, which grow in clusters. Excellent for the border or for naturalization. Prop. by division. Easy of cultivation. N. Amer. See Plate, page 423.	2-4 ft. *Sun*	Mid. Sept. to Nov.
"Pinkish lavender" 43 pinker	BROAD-SCALED BOLTONIA	**Boltònia latisquàma	See page 337.		Late July to Oct.
"Purple" 37 deeper	MEADOW SAFFRON	**Cólchicum autumnàle	See page 407.		Aug., Sept.
"Violet" 37 deeper	HANDSOME MEADOW SAFFRON	**Cólchicum speciòsum	Very large crocus-like flowers, varying to pink, appear earlier than the leaves. Plant in clumps in grass or foliage in rock-garden or border. Bulbous. Sandy loamy soil, rich and light. Caucasus.	6-12 in. *Sun or half shade*	Early Sept. to early Oct.
"Lilac" bet. 44 & 45	SAFFRON CROCUS	Cròcus satìvus	The commonest fall-blooming species. Fragrant funnel-form flowers marked with deeper and lighter shades of the same color; sometimes with white. Border or rock-garden. Bulbous. Prop. by offsets. Deep well-drained soil. S. Europe; Asia Minor.	6 in. *Sun*	Sept., early Oct.
"Bright lilac" 44 or 46	HANDSOME AUTUMN CROCUS	*Cròcus speciòsus	The handsomest kind. Large funnel-form flowers, with prominent orange stigmas, and striped with darker lilac, appear earlier than the long narrow leaves. Charming grouped in the border or rock-garden. Bulbous. Prop. by offsets. Deep well-drained soil. S. Europe; Asia Minor.	6-8 in. *Sun*	"

SEPTEMBER

Color	English Name	Botanical Name and *Synonyms*	Description	Height and *Situation*	Time of Bloom
"Purplish" 48	MARYLAND DITTANY	*Cunìla Marìana	See page 407.		Aug., Sept.
"Purplish"	DAHLIA	**Dáhlia vars.	See page 359.		July to late Oct.
"Pale lilac" 47 pale	LANCE-LEAVED DAY OR PLANTAIN LILY	**Fúnkia lancifòlia *F. Japónica*	See page 340.		Late July to early Sept.
"Violet" 55	COMMON LAVENDER	Lavándula Spìca	See page 408.		Late Aug. to mid. Sept.
"Purple" 33 bluer	TWO-FLOWERED BUSH CLOVER	Lespedèza bícolor *Desmòdium p.*	See page 408.		Early Aug., early Oct.
"Rose purple" 39 & 40	VON SIEBOLD'S BUSH CLOVER	**Lespedèza Siebòldi *L. formòsa, L. racemòsa, Desmòdium penduliflòrum*	Splendid vigorous plant. Drooping pea-like flowers in long graceful clusters. Profusion of dull green compound leaves. Ornamental in the border or edge of shrubbery. Prop. by division or green cuttings. Easily cultivated. Japan.	4-6 ft. *Sun*	Early Sept. to mid. Oct.
"Purple" 46	LOOSE-FLOWERED BUTTON SNAKEROOT	*Liàtris graminifòlia *Lacinària graminifòlia*	See page 341.		July to Oct.
"Purple" 42 lighter	DENSE-SPIKED BLAZING STAR	**Liàtris pycnostàchya	See page 408.		Aug. to mid. Sept.
"Dark lavender" 47	SCARIOUS BLAZING STAR	**Liàtris scariòsa	See page 408.		Aug., Sept.
"Deep purple" 46	DENSE BUTTON SNAKE-ROOT, GAY FEATHER	**Liàtris spicàta *Lacinària spicàta*	See page 341		Mid. July to early Sept.
"Magenta purple" 45 more purple	COMMON BLAZING STAR, COLIC ROOT	*Liàtris squarròsa	See page 408.		Aug., early Sept.
"Deep purple" 39	WING-ANGLED LOOSE-STRIFE	Lýthrum alàtum	Erect bushy plant. Solitary stemless axillary flowers. Leaves small and numerous, alternate on the flower stalks. Shrubbery or water side. Prop. by division. Any moist soil. N. Amer.	2-5 ft. *Sun*	Sept., early Oct.
"Deep violet"	VIOLET PRAIRIE CLOVER	Petalostèmon violàceus *Kuhnìstera purpùrea*	See page 245.		June to frost

Color	English Name	Botanical Name and *Synonyms*	Description	Height and *Situation*	Time of Bloom
"Purple, lilac"	PERENNIAL PHLOX	**Phlóx paniculàta vars. *P. decussàta*	See page 370.		July to Oct.
"Bluish purple" near 50	SMALL OR LILAC-FLOWERED SCABIOUS	Scabiòsa Columbària	See page 246.		June to Oct.
"Purple" 49	MOUNTAIN SKULLCAP	*Scutellària alpìna	See page 343.		Early July to late Sept.
"Dull violet" 47	TALL SEA LAVENDER	Státice elàta	See page 344.		Mid. July to early Sept.
"Blue violet" 44	GMELIN'S SEA LAVENDER	Státice Gmélini	See page 344.		Late July to early Sept.
"Bluish purple" 47	BROAD-LEAVED SEA LAVENDER	**Státice latifòlia	See page 344.		"
"Purple" 41 richer	GREAT IRONWEED	*Vernònia Arkansàna *V. crinìta*	Rough, vigorous and strong growing plant. Flower-heads in large flat-topped clusters. Striking for use among shrubs or in the back of border. Prop. by division. Good soil necessary. Mo.; Kan. to Texas; Plains, S. Western U. S. A.	8-12 ft. *Sun*	Mid. Sept. to early Oct.
"Deep blue purple" 55	SUBSESSILE LONG-LEAVED VERONICA	**Verónica longifòlia var. subséssilis *V. spicàta*	See page 410.		Early Aug. to mid. Sept.
"Violet" 47 or 49	HORNED VIOLET, BEDDING PANSY	**Vìola cornùta	See page 48.		Late Apr. until frost
"Deep blue" 56	AUTUMN ACONITE, MONKS-HOOD OR WOLFSBANE	**Aconìtum autumnàle	See page 345.		Mid. July to mid. Sept.
"Deep purple blue" near 49	TRUE MONKS-HOOD OR OFFICINAL ACONITE	**Aconìtum Napéllus *A. pyramidàle, A. Taùricum*	See page 345.		Mid. July to early Sept.
"Light blue" 44 bluer	SMOOTH ASTER	**Áster læ̀vis	Beautiful species with smooth foliage. Flowers with yellow centres in dense panicles. Good for bold effects in the border or shrubbery. Prop. by seed and division. Any good soil. N. Amer. Var. *formosissima;* (color no. bet. 44 & 46), late Aug. to late Sept.	4 ft. *Sun or half shade*	Sept., early Oct.

Color	English Name	Botanical Name and *Synonyms*	Description	Height and *Situation*	Time of Bloom
"Lilac blue" 53 lighter & brighter	MAACK'S ASTER	**Åster Máackii	See page 411.		Mid. Aug. to late Sept.
"Cobalt blue" 61	BLUE-FLOWERED LEADWORT	Ceratostígma plumbaginoìdes *Plumbàgo Lárpentæ, Valoràdia p.*	See page 411.		Late Aug., Sept.
"Light blue" 44 bluer	HERACLEUM-LEAVED CLEMATIS	*Clématis heracleæfòlia *C. tubulòsa*	See page 411.		Aug., Sept.
"Light blue" 46	ERECT HERACLEUM-LEAVED CLEMATIS	**Clématis heracleæfòlia var. stáns *C. stáns*	Robust plant of erect habit. Tube-shaped flowers in clusters. Excellent foliage. Charming in the border or rock-garden. Prop. by division. Rich deep soil. Japan.	4-5 ft. *Sun*	Sept., early Oct.
"Blue" 51	MIST FLOWER	*Conoclínium cœlestìnum *Eupatòrium cœlestìnum*	Flowers in tight flat-topped clusters on leafy stems. Useful for the border. Protect slightly in winter. Prop. by cuttings. Eastern U. S. A.	1-2 ft. *Sun*	"
"Amethyst blue" 63 lighter	AMETHYST SEA HOLLY	**Erýngium amethýstinum	See page 254.		June to early Sept.
"Blue" 63	BOURGAT'S ERYNGO	*Erýngium Bourgati	See page 348.		Mid. July to early Sept.
"Blue" 63 lighter	DANEWEED, HUNDRED THISTLE	Erýngium campéstre	See page 348.		"
"Deep blue" 56	CLOSED BOTTLE OR BLIND GENTIAN	Gentiàna Andréwsii *G. Càtesbæi*	See page 412.		Aug., Sept.
"Deep blue" 59 lighter	WILLOW GENTIAN	*Gentiàna asclepiadèa	See page 351.		July to early Sept.
"Blue" 53 lighter	FRINGED GENTIAN	*Gentiàna crinìta	A charming native biennial found in damp meadows. Erect with lovely terminal fringed flowers, which close in dull weather. Wild or rock-garden. Prop. very slowly by seed, also by division. Leave undisturbed. Good moist soil. N. Amer. See Plate, page 441.	1-2 ft. *Half shade*	Sept., Oct.
"Blue" 60	WIND FLOWER, HARVEST BELLS	Gentiàna Pneumonánthe	See page 412.		Aug., Sept.
"Light blue" 62 greener	BARREL OR SOAPWORT GENTIAN	*Gentiàna Saponària *G. Càtesbæi*	See page 412.		Aug., Sept.

Color	English Name	Botanical Name and *Synonyms*	Description	Height and *Situation*	Time of Bloom
"Blue" 46	GREAT LOBELIA, BLUE CARDINAL FLOWER	**Lobèlia syphilítica	See page 413.		Mid. Aug. to late Sept.
"Blue" often 56	BALLOON FLOWER, JAPANESE BELL-FLOWER	**Platycòdon grandiflòrum *Wahlenbérgia grandiflòra, Campánula g.*	See page 352.		Early July to Oct.
"Blue" 52	PITCHER'S SAGE	*Sálvia azùrea var. grandiflòra *S. Pítcheri*	See page 413.		Late Aug., Sept.
"Pale blue" often bet. 43 & 44	GRASS-LEAVED SCABIOUS	*Scabiòsa graminifòlia	See page 262.		June to Oct.
"Bluish" near 39 paler	WOODLAND SCABIOUS	*Scabiòsa sylvática	See page 262.		Early June to late Sept.
"Blue" 52 dull	STOKES' ASTER	**Stokèsia cyànea	See page 414.		Aug. to early Oct.
Parti-colored	FRENCH OR CROZY CANNA	**Cánna vars.	See page 357.		July to late Sept.
Parti-colored	ITALIAN CANNA	**Cánna vars.	See page 358.		"
"White" & 34	PARKINSON'S CHECKERED MEADOW SAFFRON	*Cólchicum Párkinsoni	See page 414.		Aug., Sept.
Parti-colored	DAHLIA	**Dáhlia vars.	See page 359.		July to late Oct.
6 shading from 19 to 14	GREAT-FLOWERED GAILLARDIA	**Gaillárdia aristàta *G. grandiflòra*	See page 263.		June to Nov.
Parti-colored	SWORD LILY	**Gladìolus vars.	See page 365.		July to Oct.
Various	FRENCH OR CROZY CANNA	**Canna vars.	See page 357.		July to late Sept.
Various	ITALIAN CANNA	**Canna vars.	See page 358.		"
Various	HARDY CHRYSANTHEMUM	*Chrysànthemum vars.	**Single Hardy Chrysanthemums.** Pretty daisy-like flowers, effective but less known than the button form. Require protection when in bloom. Prop. by seed and cuttings. Any good garden soil. Hort. Some good	2-3 ft. *Sun*	Sept., Oct.

FRINGED GENTIAN. *Gentiana crinita.*

SINGLE AND DOUBLE HARDY CHRYSANTHEMUMS

Color	English Name	Botanical Name and *Synonyms*	Description	Height and *Situation*	Time of Bloom
			vars. are:—**Boston;* (effect color nos. 10 to 13 redder, centre 4), yellowish orange. *Constance;* (sulphur-white with centre color no. 4), yellowish white with dull yellow centre. **Miss Rhoe;* (color no. 6 lightly suffused with 13), rich yellow tinged with orange. **Sir Walter Raleigh;* (color nos. 10 to 14), yellow merging into terra cotta. **Northumberland;* (effect 19 deep, centre 4), vermilion with yellow centre.		
			Anemone-flowered Pompon Chrysanthemums. Pretty single flowers resembling Japanese Anemones, less hardy than the Button Pompons, but grown in open air successfully. Protect in winter. Prop. by seed and cuttings. Rich mellow soil. Hort. Some good vars. are:—*Antonicus,* bright yellow. *Clara Owen,* pale straw color. *Descartes,* crimson-red. **Emily Rowbottom;* (white with centre color no. 2), white with pale yellow centre. **Reine des Anemones;* (white tinged with color no. 36 pale, centre 4 dull), pinkish white with large yellow centre. *Rose Marguerite;* (color bet. nos. 40 & 41, centre dull 4), dull magenta with large yellow centre.		
			Aster or Large-flowering Pompon Chrysanthemums. Brilliantly beautiful large double flowers, larger and looser petaled than the old-fashioned button-shaped Pompons and not as hardy. Effective in rows or masses in border or edge of shrubbery. Prop. by seed and cuttings. Any good garden soil. Hort. See Plate, page 442. The following are some of the best kinds:		
			White vars.: *Ashbury,* sulphur-white. *Hester,* pearl-white. ***Prince of Wales,* pure white. **Queen of the Whites,* pure white. **Sœur Mélanie,* pure white, rather ragged petals. *St. Anselm,* pure white.		
			Yellow vars.: **Allen Town;* (color no. bet. 6 pale & 7), semi-double. *Bohemia;* (color no. 3), clear yellow. Petals rather large and semi-double. **Douckelaori;* (color no. 3), clear yellow. *Fred J.;* (color no. 6 very light), warm light yellow. **Globe d'Or;* (color no. 3 light), clear yellow. A standard var. *Sir Michael;* (color no. 3), lemon-yellow. *Sunshine;* (color no. 3), clear yellow. Petals somewhat ragged. *Zenobia;* (color no. 3), rather large and loose petals.		

Color	English Name	Botanical Name and *Synonyms*	Description	Height and *Situation*	Time of Bloom
			Orange, Red and Terra Cotta vars.: *Alice Carey,* bright orange. ***Cowenton;* (color no. 13 light streaked with 4), terra cotta and yellow. **Montclair;* (color no. 13 streaked with 4), terra cotta and yellow. *Mrs. Porter;* (color no. 2 pale suffused with color no. bet. 13 & 14), bronze. **Mrs. Vincent;* (color no. 35), maroon-red. **Patterson;* (color no. 3 suffused with 13 light), "old gold." *Sadie,* "bronze-orange." **Sunset;* (color no. 16 suffused with 13 redder), rich terra cotta. *The Czar;* (color no. 7 pale, much suffused with 13 & 14), "golden bronze." **Pink vars.:** *Arabella,* "crimson- salmon." *Blenheim,* "silver-pink." *Cerise Queen,* "cerise-pink." *Constance,* light pink. *Daybreak,* "daybreak-pink." *Duluth,* "salmon-pink." *Empress,* rose-pink. *Gloire de France,* "silver-pink." **Hijos;* (color no. 22 shading to cream-white), primrose-pink. Large. *Lady de Vaul,* "violet-pink." *Madeline,* pink. *Salem,* rose-pink. **Sheridan;* (color no. 22 suffused with 25 redder), deep rose-pink. *Crola;* (color no. bet. 32 & 39), rosy pink suffused with color no. 36), pink. **Viola;* (color no. bet. 32 & 39), rosy pink. **Lilac and Magenta vars.:** **Fred Peele;* (color no. 32), light lilac. *Hamlet;* (color nos. 27 & 33), crimson-magenta. **King Philip;* (color no. 33 dull and dark), dark crimson-magenta. *Little Pet;* (color no. 33), crimson-magenta. Very small button. **Button or Small-flowering Pompon Chrysanthemums.** Charming and effective double flowers, the last to succumb to winter, blooming after the frost and even in the snow. Decorative massed in the border or edge of shrubbery. Prop. by seed and cuttings. Any good garden soil. Hort. See Plate, page 442. The following are a few of the best vars.: **White vars.:** *Anna Mary,* creamy white. **James Boon,* pure white. *L'Ami Conderchet,* cream-white. Very small but numerous flowers. *Norwood,* pure white. *Paragon,* pure white. **Snowdrop,* pure white. Very small, **Yellow vars.:** *Fashion,* "maize-yellow." *Golden Fleece,* pure yellow. *Mignon;* (color no. 4), rich golden yellow. Very small button. **Savannah;* (color no. 3 intense), intense yel-		

Color	English Name	Botanical Name and *Synonyms*	Description	Height and *Situation*	Time of Bloom
			low, small button. *Tennyson,* pure yellow. **Orange vars.:** *Agalia,* light orange. **Golden Pheasant;* (color no. 13), deep orange-yellow. **Goldfinch;* (color nos. 6 & 13 light), "golden yellow shaded crimson." Effect orange. *Henrietta;* (color no. 13), bright yellowish terra cotta. **Red vars.:** **Black Douglas,* dark maroon. *Dundee,* "scarlet-maroon." *Erminie;* (effect color no. 13 redder), bright orange-scarlet. *Little Bob,* brownish crimson. **Northumberland;* (effect color no. 19 deep, centre 4), cardinal. Pretty single flower. *Ruby Queen;* (color no. 34), garnet-red with yellow centre. *Rufus;* (color no. 14 redder), pure terra cotta. **Pink vars.:** *Austin,* "pink." *Blushing Bride,* "light pink." **Dawn;* (color no. 36 shading to 33), pale shading to deeper shell-pink. *Dinizulu;* (color no. 38 to white), bluish pink. *Jeanette,* deep rose. **Rhoda;* sulphur-white.		
Various	**DAHLIA**	****Dáhlia vars.**	See page 359.		July to late Oct.
Various	**SWORD LILY**	****Gladìolus vars.**	See page 365.		July to Oct.
Various	**PERENNIAL PHLOX**	****Phlóx paniculàta vars.** *P. decussàta*	See page 370.		"
Various	**MOUNTAIN FLEECE**	***Polýgonum amplexicaùle** *P. multiflòrum, P. oxyphýllum, P. speciòsum*	Tufted plant with rose-red or creamy white flowers in spikes. Numerous heart-shaped leaves. Suitable for the border. Prop. by division, sometimes by seed. Any garden soil. Himalaya.	2-3 ft. *Sun or half shade*	Sept., early Oct.
Various	**PANSY, HEART'S-EASE**	***Viola trícolor**	See page 68.		Mid. Apr. to mid. Sept.

THE BEST HERBACEOUS PLANTS*

Botanical Name.	English Name.	Time of Bloom.	Color.	Height.	Situation.	Page.
Acanthus mollis & var. latifolius	BEAR'S BREECH & VAR.	July, Aug.	Various, purple	3-4 ft.	*Sun*	354, 337
Acanthus spinosissimus	VERY PRICKLY BEAR'S BREECH	"	Various	"	"	357
Acanthus spinosus	PRICKLY BEAR'S BREECH	"	"	"	"	357
Achillea Ptarmica var. "The Pearl"	DOUBLE SNEEZEWORT	June to Oct.	White	$1\frac{1}{2}$-$2\frac{1}{2}$ ft.	"	167
Aconitum autumnale	AUTUMN ACONITE	Mid. July to mid. Sept.	Blue	3-5 ft.	*Sun or shade*	345
Aconitum Napellus & var. album	TRUE MONKS-HOOD & VAR.	Mid. July to early Sept.	Blue, white	3-4 ft.	"	27, 274
Aconitum uncinatum	WILD MONKSHOOD	Mid. June to Sept.	Purple	3-5 ft.	*Sun or half shade*	233
Adonis vernalis	SPRING ADONIS	Mid. Apr. to June	Yellow	8-15 in.	"	23
Ajuga Genevensis	ERECT BUGLE	May	Blue	6-8 in.	*Sun or shade*	141
Ajuga reptans	BUGLE	Early May to mid. June	"	3-4 in.	"	141
Allium Moly	GOLDEN GARLIC	Mid. Apr. to June	Yellow	1 ft.	"	24
Althæa rosea	HOLLYHOCK	July, Aug.	Various	5-8 ft.	*Sun*	357
Alyssum saxatile	ROCK MADWORT	Mid. Apr. to late May	Yellow	1 ft.	"	24
Alyssum saxatile var. compactum	COMPACT ROCK MADWORT	Mid. Apr. to June	"	"	"	24
Amsonia Tabernæmontana	AMSONIA	Late May, early June	Blue	2-3 ft.	"	142
Anemone Japonica & vars.	JAPANESE WINDFLOWER & VARS.	Late Sept. to early Nov.	Rose, white, pink	2-5 ft.	*Sun or half shade*	431, 416
Anemone sylvestris	SNOWDROP WINDFLOWER	Late Apr. to mid. July	White	1-$1\frac{1}{2}$ ft.	"	11
Aquilegia cærulea	LONG-SPURRED COLUMBINE	Mid. May to July	Blue	"	*Sun*	145
Aquilegia chrysantha	GOLDEN-SPURRED COLUMBINE	Late May to late Aug.	Yellow	3-4 ft.	"	96
Aquilegia formosa var. hybrida	HYBRID CALIFOR-NIAN COLUMBINE	Mid. May to July	Red	1-$1\frac{1}{4}$ ft.	"	114
Aquilegia glandulosa	ALTAIAN COLUMBINE	May, June	Blue	1-$1\frac{1}{2}$ ft.	"	145

* All plants in this list are prefixed in the previous text by a double star, indicating their excellence.

Botanical Name.	English Name.	Time of Bloom.	Color.	Height.	Situation.	Page.
Aquilegia vulgaris var. nivea	WHITE COLUMBINE	Mid. May to July	White	2-3 ft.	*Sun*	72
Arabis albida	WHITE ROCK CRESS	Early Apr. to June	"	6-8 in.	"	11
Armeria maritima	CUSHION PINK	Mid. May to mid. June	Pink	3-6 in.	"	118
Armeria maritima var. Laucheana	LAUCHE'S THRIFT	Late Apr. to mid. June	"	"	"	39
Asclepias tuberosa	BUTTERFLY WEED	Early July to early Aug.	Orange	2-3 ft.	"	317
Aster alpinus & var. albus	BLUE ALPINE ASTER & VAR.	Late May to late June	Violet, white	3-10 in.	*Sun or half shade*	128, 75
Aster lævis	SMOOTH ASTER	Sept., early Oct.	Blue	4 ft.	"	438
Aster Novæ Angliæ	NEW ENGLAND ASTER	Mid. Aug. to late Sept.	Purple	3-7 ft.	*Sun*	404
Aster Shortii	SHORT'S ASTER	Late Sept. to late Oct.	Violet	2-4 ft.	*Sun or shade*	435
Aster Tartaricus	TARTARIAN ASTER	Late Sept., Oct.	Purple	7 ft. or less	*Sun or half shade*	436
Aster turbinellus	PRAIRIE ASTER	Mid. Sept. to Nov.	Mauve	2-4 ft.	*Sun*	436
Astilbe Chinensis	CHINESE GOAT'S BEARD	July, early Aug.	Pink	1½-2 ft.	*Half shade*	329
Astilbe decandra	FALSE GOAT'S BEARD	Early June to early July	White	3-6 ft.	"	168
Astilbe Japonica	JAPANESE FALSE GOAT'S BEARD	Mid. June to mid. July	"	1-3 ft.	"	168
Aubrietia deltoidea	PURPLE ROCK CRESS	Early Apr. to late May	Violet	2-10 in.	*Sun or half shade*	43
Baptisia australis	BLUE WILD INDIGO	Late May to mid. June	Blue	4-4½ ft.	*Sun*	277
Bocconia cordata	PLUME POPPY	Early July to early Aug.	White	3-8 ft.	"	277
Boltonia glastifolia	WOAD-LEAVED BOLTONIA	Aug., Sept.	"	4-5 ft.	"	376
Boltonia latisquama	BROAD-SCALED BOLTONIA	Late July to Oct.	Lavender	4 ft.	"	337
Bulbocodium vernum	SPRING MEADOW SAFFRON	Apr.	Purple	4-6 in.	"	44
Campanula Carpatica & var. alba	CARPATHIAN HAIRBELL & VAR.	Late June to late Aug.	Violet, white	9 18 in.	"	237, 171
Campanula Carpatica var. turbinata alba	WHITE TURBAN BELLFLOWER	July, Aug.	White	6-12 in.	"	278
Campanula glomerata	CLUSTERED BELLFLOWER	June, July	Purple	1-2 ft.	"	237
Campanula latifolia var. macrantha	LARGE-BLOSSOMED BELLFLOWER	Early June, July	"	3-5 ft.	*Sun or shade*	238

Botanical Name	English Name.	Time of Bloom.	Color.	Height.	Situation.	Page.
Campanula Medium	**CANTERBURY BELLS**	Late June, July	**Various**	1½-4 ft.	*Sun*	264
Campanula nobilis	**NOBLE BELLFLOWER**	Mid. June to Aug.	"	2 ft.	"	264
Campanula persicifolia	**PEACH-LEAVED BELLFLOWER**	Early June to mid. July	**Violet**	1½-3 ft.	"	238
Campanula persicifolia var. alba	**WHITE PEACH-LEAVED BELLFLOWER**	Early June to early July	**White**	1½-3 ft.	"	171
Campanula persicifolia vars. Backhousei & Moerheimi	**BACKHOUSE'S & MOERHEIM'S PEACH-LEAVED BELLFLOWERS**	June, July	"	1½-3 ft.	"	171
Campanula pyramidalis	**CHIMNEY CAMPANULA**	July, Aug.	**Blue**	4-6 ft.	"	346
Campanula rotundifolia	**ENGLISH HAIRBELL**	June to late Aug.	**Violet**	6-12 in.	"	238
Campanula Van Houttei	**VAN HOUTTE'S BELLFLOWER**	Early June to mid. July	"	2 ft.	"	238
Canna vars.	**FRENCH & ITALIAN CANNA**	July to late Sept.	**Various**	3-4½ ft.	"	357-358
Centaurea macrocephala	**CENTAURY**	Mid. July to Sept.	**Yellow**	2½-3 ft.	"	295
Centaurea montana	**MOUNTAIN BLUET**	June to Sept.	**Purple**	12-20 in.	"	239
Centaurea montana vars. alba & rosea	**WHITE & ROSY MOUNTAIN BLUETS**	Late May to early July	**White, rose**	9-20 in.	"	75, 121
Chionodoxa Luciliæ & var. gigantea	**GLORY OF THE SNOW**	Mid. Mar. to early May	**Blue**	3-8 in.	*Sun or half shade*	7
Chrysanthemum vars.	**HARDY CHRYSANTHEMUM**	Sept., Oct.	**Various**	2-3 ft.	*Sun*	440
Chrysanthemum coccineum	**RED CHRYSANTHEMUM**	June, July	"	1-2 ft.	"	264
Chrysanthemum maximum	**LARGE-FLOWERED WHITEWEED**	"	**White**	1 ft.	"	172
Chrysanthemum uliginosum	**GIANT DAISY**	Aug., Sept.	"	4-5 ft.	"	376
Clematis heracleæfolia var. stans	**ERECT HERACLEUM-LEAVED CLEMATIS**	Sept., early Oct.	**Blue**	"	"	439
Clematis recta	**WHITE HERBACEOUS VIRGIN'S BOWER**	Early June to mid. July	**White**	2-3 ft	"	173
Colchicum autumnale	**MEADOW SAFFRON**	Aug., Sept.	**Purple**	3-4 in.	*Sun or half shade*	407
Colchicum speciosum	**HANDSOME MEADOW SAFFRON**	Early Sept. to early Oct.	**Violet**	6-12 in.	"	436

BEST HERBACEOUS PLANTS

Botanical Name.	English Name.	Time of Bloom.	Color.	Height.	Situation.	Page.
Convallaria majalis	LILY-OF-THE-VALLEY	Mid. May to mid. June	White	8 in.	*Half shade or shade*	75
Coreopsis grandiflora	LARGE-FLOWERED TICKSEED	June to Sept.	Yellow	1-2 ft.	*Sun*	196
Coreopsis lanceolata	LANCE-LEAVED TICKSEED	June to Sept.	"	"	"	196
Crocus vars.	CROCUS	Mid. Mar. to late Apr.	Various	6-8 in.	*Sun or half shade*	8
Crocus biflorus	SCOTCH CROCUS	"	Parti-colored	"	"	8
Crocus Susianus	CLOTH OF GOLD CROCUS	"	Yellow	3 in.	"	4
Crocus vernus	SPRING CROCUS	"	Various	4-5 in.	"	8
Cypripedium pubescens	LARGE YELLOW LADY'S SLIPPER	May, June	Yellow	1-2 ft.	*Shade*	99
Cypripedium spectabile	SHOWY LADY'S SLIPPER	June	White	1-2½ ft.	*Half shade or shade*	174
Dahlia vars.	CACTUS, DECORATIVE, FANCY, QUILLED SHOW & SINGLE DAHLIAS	Late July to late Oct.	Various	3-6 ft.	*Sun*	359, 360, 363, 364
Daphne Cneorum	GARLAND FLOWER	Late Apr., May	Pink	6-12 in.	*Sun or half shade*	39
Delphinium elatum	BEE LARKSPUR	June to Sept.	Blue	2-6 ft.	"	253
Delphinium formosum	ORIENTAL LARKSPUR	June, July	"	2-3 ft.	"	253
Delphinium grandiflorum & var. album	GREAT-FLOWERED LARKSPUR & VAR.	July, Aug.	Blue, white	1-2 ft.	*Sun*	347, 282
Delphinium grandiflorum var. Chinense	CHINESE LARKSPUR	June, July	Various	1-2 ft.	"	267
Delphinium hybridum	HYBRID LARKSPUR	"	Blue	3-4 ft.	*Sun or half shade*	254
Dianthus barbatus	SWEET WILLIAM	"	Various	10-18 in.	*Sun*	267
Dianthus cinnabarinus	CINNAMON PINK	July, Aug	Pink	1 ft.	"	330
Dianthus cruentus	DARK RED PINK	June, July	Red	1-1½ ft.	"	213
Dianthus deltoides	MAIDEN PINK	May, June	Pink	6-9 in.	"	121
Dianthus latifolius	BROAD-LEAVED PINK	June to Sept.	"	6-12 in.	"	224
Dianthus "Miss Simkins"	PINK MISS SIMKINS	Late May to late June	White	4-6 in.	"	76
Dianthus plumarius	SCOTCH PINK	"	Pink	1 ft.	"	122
Dianthus plumarius vars.	GARDEN PINK	June	"	8-12 in.	"	227
Dianthus plumarius vars. alba plena & "White Witch"	DOUBLE WHITE GARDEN PINKS	"	White	8-12 in.	"	174

BEST HERBACEOUS PLANTS

Botanical Name.	English Name.	Time of Bloom.	Color.	Height.	Situation.	Page.
Dianthus Seguierii	**SEGUIER'S PINK**	Late June, July	**Rose**	1 ft.	*Sun*	227
Dicentra eximea	**WILD BLEEDING HEART**	Early June to Aug.	"	1-2 ft.	*Half shade*	227
Dicentra spectabilis	**BLEEDING HEART**	Late Apr. to mid. July	"	"	*Half shade best*	40
Dictamnus albus & var. rubra	**GAS PLANT**	June, July	**White, pink**	2-3 ft.	*Sun or half shade*	174, 227
Digitalis purpurea & var. alba	**COMMON FOX-GLOVE & VARS.**	June, early July	**Various, white**	2-3 ft.	"	267, 174
Echinacea purpurea	**PURPLE CONEFLOWER**	July, Aug.	**Pink**	2-3½ ft.	*Sun*	330
Echinops Ritro	**RITRO GLOBE THISTLE**	"	**Blue**	2-3 ft.	"	348
Eranthis hyemalis	**COMMON WINTER ACONITE**	Mar., Apr.	**Yellow**	3-8 in.	*Sun or half shade*	4
Erigeron speciosus	**SHOWY FLEABANE**	June, July	**Lilac**	1½-2 ft.	*Sun*	240
Eryngium amethystinum	**AMETHYST SEA HOLLY**	June to early Sept.	**Blue**	1-3 ft.	"	254
Eryngium planum	**FLAT-LEAVED ERYNGO**	July, Aug.	"	"	"	348
Erythronium albidum	**WHITE DOG-TOOTH VIOLET**	Late Apr., May	**White**	6 in.	*Half shade*	15
Erythronium Americanum	**COMMON ADDER'S TONGUE**	Late Apr. to late May	**Yellow**	"	"	27
Erythronium Dens-Canis	**COMMON DOG-TOOTH VIOLET** of Europe	Late Apr., May	**Lilac**	4-6 in.	"	44
Erythronium grandiflorum	**LARGE-FLOW-ERED DOGTOOTH VIOLET**	Late Apr. to mid. May	**Yellow**	6 in.	*Half shade or shade*	27
Fritillaria Imperialis	**CROWN IMPERIAL**	Mid. Apr. to mid. May	**Various**	2-3 ft.	*Sun or half shade*	56
Fritillaria Meleagris	**GUINEA-HEN FLOWER**	Late Apr. to late May	"	10-12 in.	*Sun or shade*	59
Fritillaria Meleagris var. alba	**WHITE GUINEA-HEN FLOWER**	Late Apr., May	**White**	1 ft.	"	15
Funkia lancifolia	**LANCE-LEAVED DAY LILY**	Late July to early Sept.	**Lilac**	1-2 ft.	*Half shade best*	340
Funkia subcordata var. grandiflora	**LARGE-FLOWERED SUBCORDATE PLANTAIN LILY**	Late Aug., Sept.	**White**	"	*Half shade*	380
Gaillardia aristata	**GREAT-FLOWERED GAILLARDIA**	June to Nov.	**Parti-colored**	1½-3 ft	*Sun*	263
Galanthus Elwesii	**GIANT SNOWDROP**	Mar., Apr.	**White**	6-12 in.	*Sun or half shade*	3
Galanthus nivalis	**COMMON SNOWDROP**	"	"	4-6 in.	"	3

BEST HERBACEOUS PLANTS

Botanical Name.	English Name.	Time of Bloom.	Color.	Height.	Situation.	Page.
Galanthus plicatus	PLAITED SNOWDROP	Mar. to early May	White.	4-8 in.	*Sun or half shade*	3
Galtonia candicans	CAPE HYACINTH	July, Aug.	"	3-5 ft.	"	284
Geranium sanguineum	BLOOD-RED CRANESBILL	Late May to mid. July	Crimson	1½-2 ft.	"	115
Gladiolus vars.	SWORD LILY	July to Oct.	Various	3-4½ ft.	*Sun*	365
Gypsophila paniculata	BABY'S BREATH	July, Aug.	White	2-3 ft.	"	285
Helenium autumnale	SNEEZEWEED	Aug. to late Sept.	Yellow	2-5 ft.	"	386
Helenium autumnale vars. grandiflorum, pumilum & superbum	SNEEZEWEED	Aug., Sept.	Yellow	1-6 ft.	"	386
Helenium Hoopesii	HOOPES'S SNEEZEWEED	Late May to late June	Yellow	1-3 ft.	"	102
Helenium nudiflorum var. grandicephalum striatum	STRIPED PURPLE-HEADED SNEEZEWEED	Aug.	Orange	3-4 ft.	"	393
Helianthus mollis	HAIRY SUNFLOWER	July, Aug.	Yellow	2-5 ft.	"	300
Helianthus rigidus var. "Miss Mellish"	STIFF SUNFLOWER MISS MELLISH	Late Aug., Sept.	"	6 ft.	"	387
Hemerocallis Dumortierii	DUMORTIER'S DAY LILY	June, July	Orange	1-2 ft.	*Sun or half shade*	210
Hemerocallis flava	LEMON LILY	June, early July	Yellow	3 ft.	*Half shade*	199
Hemerocallis Middendorfii	MIDDENDORF'S YELLOW DAY LILY	Late June, July	"	1-3 ft.	"	199
Hemerocallis Thunbergii	THUNBERG'S YELLOW DAY LILY	"	"	3-4 ft.	*Sun or half shade*	199
Heuchera sanguinea	CORAL BELLS	June to late Sept.	Red	1-1½ ft.	"	213
Hypericum Kalmianum	KALM'S ST. JOHN'S-WORT	Aug.	Yellow	2-4 ft.	*Shade*	388
Hypericum Moserianum	GOLD FLOWER	July, Aug.	"	2 ft.	*Sun or half shade*	303
Hyacinthus orientalis	DUTCH HYACINTH	Late Apr., May	Various	8-18 in.	*Sun*	59
Iberis sempervirens	EVERGREEN CANDYTUFT	May, early June	White	9-15 in.	*Sun or half shade*	78
Iberis Tenoreana	TENORE'S CANDYTUFT	"	"	9-12 in.	*Half shade*	78
Iris cristata	CRESTED DWARF IRIS	Late May to July	Lilac	4-9 in.	*Sun*	135
Iris Florentina	FLORENTINE FLAG	May, early June	White	1-2 ft.	*Half shade*	78
Iris Germanica vars.	GERMAN IRIS	Late May to July	Purple & lavender	1½-3 ft.	*Sun*	135, 156

BEST HERBACEOUS PLANTS

Botanical Name.	English Name	Time of Bloom.	Color.	Height.	Situation.	Page.
Iris lævigata	JAPANESE IRIS	June, July	Various	2-3 ft.	*Sun*	267
Iris neglecta	NEGLECTED IRIS	Late May to early June	Blue	1 ½-2 ft.	"	149
Iris pallida	GREAT PURPLE FLAG	Late May to July	Violet	2-4 ft.	*Sun or half shade*	135
Iris plicata	PLAITED FLAG	"	Lilac	"	"	135
Iris Sibirica & var. alba	SIBERIAN FLAG & VAR.	Late May to mid. June	Violet, white	2-3 ft.	*Sun*	136, 78
Iris Sibirica var. orientalis	EASTERN SIBERIAN IRIS	June, early July	Blue	1-2½ ft.	"	257
Iris xiphioides	ENGLISH IRIS	Late June, July	Purple	1-2 ft.	"	242
Iris Xiphium	SPANISH IRIS	Mid. June to July	Blue	"	"	258
Kniphofia Pfitzerii	EVERBLOOMING FLAME FLOWER	Early Aug. to early Oct.	Orange	3-4 ft.	*Sun or half shade*	393
Lathyrus vernus	SPRING BITTER VETCH	Mid. Apr. to late May	Violet	12-15 in.	*Sun*	47
Lathyrus vernus var. albus	WHITE SPRING BITTER VETCH	Late Apr., May	White	1-2 ft.	*Sun or half shade*	16
Lespedeza Sieboldi	VON SIEBOLD'S BUSH CLOVER	Early Sept. to mid. Oct	Purple	4-6 ft.	*Sun*	437
Liatris pycnostachya	DENSE-SPIKED BLAZING STAR	Aug. to mid. Sept.	"	3-5 ft.	*Half shade*	408
Liatris scariosa	SCARIOUS BLAZING STAR	Aug., Sept.	Lavender	1-5 ft.	*Sun*	408
Liatris spicata	DENSE BUTTON SNAKEROOT	Mid. July to early Sept.	Purple	2-5 ft.	*Sun or half shade*	341
Lilium auratum	GOLD-BANDED LILY	Mid. July to mid. Aug.	White	2-4 ft.	"	286
Lilium Browni	BROWN'S LILY	July, Aug.	Parti-colored	3-4 ft.	*Half shade*	354
Lilium Canadense	WILD YELLOW LILY	June, July	Various	1-4 ft.	*Sun or half shade*	268
Lilium candidum	MADONNA LILY	"	White	3-5 ft.	"	180
Lilium elegans	THUNBERGIAN LILY	"	Orange	1-2 ft.	"	214
Lilium elegans var. "Alice Wilson"	THUNBERGIAN LILY ALICE WILSON	July	Yellow	"	*Sun or shade*	304
Lilium elegans var. alutaceum	YELLOW THUNBERGIAN LILY	"	"	8-10 in.	*Sun or half shade*	304
Lilium elegans var. fulgens	SHINING THUNBERGIAN LILY	Mid. July to early Aug.	Apricot	1-3½ ft.	*Sun or shade*	318
Lilium elegans var. Wallacei	WALLACE'S THUNBERGIAN LILY	July	"	1-2½ ft.	*Sun or half shade*	318
Lilium maculatum	SPOTTED LILY	June, July	Orange	3-4 ft.	"	211
Lilium pardalinum	PANTHER LILY	July	Red	2-3 ft.	*Half shade*	323
Lilium Parryi	PARRY'S LILY	June, July	Yellow	2-6 ft.	*Shade*	200

BEST HERBACEOUS PLANTS

Botanical Name.	English Name.	Time of Bloom.	Color.	Height.	Situation.	Page.
Lilium speciosum & vars. Melpomene & rubrum	SPOTTED LILY & VARS.	Aug., Sept.	Pink	2-4 ft.	*Sun or half shade*	401
Lilium superbum	AMERICAN TURK'S CAP LILY	Early July to early Aug.	Orange	3-6 ft.	"	321
Lilium tenuifolium	SIBERIAN CORAL LILY	Late June, July	Scarlet	1-2 ft.	*Sun*	214
Lilium testaceum	NANKEEN LILY	Mid. June to mid. July	Buff	2-6 ft.	*Sun or half shade*	203
Lilium tigrinum	TIGER LILY	Mid. June to Sept.	Orange	2-5 ft.	"	321
Linum perenne	PERENNIAL FLAX	Mid. May to Aug.	Blue	1-1½ ft.	"	149
Lobelia cardinalis	CARDINAL FLOWER	Aug. to mid. Sept.	Crimson	2-4 ft.	*Sun or shade*	396
Lobelia syphilitica	GREAT LOBELIA	Mid. Aug. to late Sept.	Blue	1-3 ft.	*Sun*	413
Lobelia syphilitica var. alba	GREAT WHITE LOBELIA	Mid. Aug. to Oct.	White	2-3 ft.	"	381
Lupinus polyphyllus & var. albiflorus	TALL BLUE-FLOWERED PERENNIAL LUPINE & VAR.	June, July	Purple, white	2-5 ft.	"	242, 180
Lychnis Chalcedonica	MALTESE CROSS	Early June to mid. July	Scarlet	2-3 ft.	*Sun or shade*	214
Lychnis Chalcedonica vars. alba & alba plena	SINGLE & DOUBLE WHITE MALTESE CROSS	June to early Aug.	White	"	*Sun*	289
Lychnis Viscaria var. splendens	BRILLIANT GERMAN CATCHFLY	June	Red	6-20 in.	"	217
Lysimachia clethroides	JAPANESE LOOSESTRIFE	Mid. June to late July	White	2-3 ft.	"	180
Monarda didyma	OSWEGO TEA	Mid. June to early Sept.	Red	1½-2½ ft.	*Sun or shade*	217
Muscari botryoides & var. album	COMMON GRAPE HYACINTH	Apr., May	Blue, white	6-9 in.	*Sun or half shade*	52, 16
Muscari comosum var. monstrosum	FEATHERED HYACINTH	"	Blue	1 ft.	"	52
Myosotis palustris var. semperflorens	EVER-FLOWERING FORGET-ME-NOT	May to Sept.	Blue	8 in.	*Shade*	153
Narcissus incomparabilis & vars.	STAR DAFFODIL & VARS.	Mid. Apr. to mid. May	Yellow	12-15 in.	*Half shade best*	28
Narcissus poeticus	PHEASANT'S EYE	May	White	"	"	82
Narcissus Pseudo-Narcissus vars.	COMMON DAFFODIL	Late Apr., May	Yellow	12-18 in.	"	31
Œnothera fruticosa	SUNDROPS	June, July	"	1-3 ft.	*Half shade*	204
Œnothera glauca var. Fraseri	FRASER'S EVENING PRIMROSE	June to Sept.	"	2-3 ft.	*Sun*	204

Botanical Name.	English Name.	Time of Bloom.	Color.	Height.	Situation.	Page.
Œnothera Missouriensis	**MISSOURI PRIMROSE**	June to early Aug.	**Yellow**	10 in.	*Half shade*	204
Pæonia vars.	**HERBACEOUS PEONY**	June	**Various**	3-4 ft.	*Sun or half shade*	268
Pæonia albiflora & vars.	**WHITE-FLOWERED PEONY & VARS.**	Late May to mid. June	"	2-4 ft.	"	85, 159
Pæonia officinalis & vars.	**COMMON GARDEN PEONY**	Mid. May to mid. June	**Crimson**	2-3 ft.	*Half shade*	116
Papaver alpinum	**ALPINE POPPY**	Mid. May to early June	**White**	6 in.	*Sun*	86
Papaver nudicaule	**ICELAND POPPY**	Late Apr. to July, late Aug. to Oct.	**Yellow**	9-15 in.	"	35
Papaver nudicaule var. album	**WHITE ICELAND POPPY**	Late Apr. to mid. June, late Aug., Sept.	**White**	"	"	19
Papaver orientale & vars. "Blush Queen," "Silver Queen"	**ORIENTAL POPPY & VARS.**	Early June to early July	**Scarlet, pink, white**	2-3½ ft.	"	218, 230, 184
Pentstemon barbatus var. Torreyi	**TORREY'S BEARDED PENTSTEMON**	Early July to early Aug.	**Scarlet**	4-5 ft.	"	325
Pentstemon diffusus	**DIFFUSE PENTSTEMON**	Early June to early July	**Purple**	1-2 ft.	"	245
Pentstemon lævigatus var. Digitalis	**FOXGLOVE BEARD-TONGUE**	Early June to mid. July	**White**	3-4 ft.	"	187
Pentstemon secundiflorus	**ONE-SIDED PENTSTEMON**	June, July	**Lavender**	12-18 in.	"	245
Pentstemon spectabilis	**SHOWY PENTSTEMON**	Early June to mid. July	**Blue**	2-2½ ft.	"	261
Phlox amœna	**HAIRY PHLOX**	Late Apr., May	**Various**	4-6 in.	"	63
Phlox divaricata	**WILD SWEET WILLIAM**	May	**Lilac**	10-18 in.	"	136
Phlox paniculata vars.	**PERENNIAL PHLOX**	July to Oct.	**Various**	2-3½ ft.	"	370
Phlox subulata	**MOSS PINK**	Late Apr., May	"	4-6 in.	"	63
Phlox subulata var. "The Bride"	**MOSS PINK THE BRIDE**	Late Apr., to late May	**White**	"	"	19
Platycodon grandiflorum	**BALLOON FLOWER**	Early July to Oct.	**Blue**	1-3 ft.	*Sun or shade*	352
Platycodon grandiflorum var. album	**WHITE BALLOON FLOWER**	June to Oct.	**White**	"	*Sun or half shade*	187
Polemonium cæruleum	**AMERICAN JACOB'S LADDER**	Mid. May to July	**Purple**	"	*Half shade*	139

Botanical Name.	English Name.	Time of Bloom.	Color.	Height.	Situation.	Page.
Polemonium humile	DWARF JACOB'S LADDER	June, July	Blue	6 in.	*Half shade*	261
Primula Japonica	JAPANESE PRIMROSE	Late May to Aug.	Purple	1-2 ft.	"	139
Primula officinalis	ENGLISH COWSLIP	Late Apr. to late May	Yellow	6-12 in.	"	35
Primula Sieboldi	VON SIEBOLD'S PRIMROSE	Late Apr., May	Various	6-12 in.	*Sun or half shade*	63
Primula Stuartii	STUART'S PRIMROSE	Late May, June	Yellow	9-15 in.	*Half shade*	105
Rudbeckia laciniata var. flore-pleno	GOLDEN GLOW	Late July to late Sept.	"	2-10 ft.	*Sun*	309
Rudbeckia subtomentosa	SWEET CONEFLOWER	Mid. Aug. to Oct.	"	3-5 ft.	*Sun or shade*	391
Rudbeckia triloba	THIN-LEAVED CONEFLOWER	July, Aug.	"	2-5 ft.	"	309
Salvia pratensis	MEADOW SAGE	June, early July	Violet	2-3 ft.	*Sun*	262
Scabiosa Caucasica	PINCUSHION FLOWER	June, July	Blue	1½-2 ft.	"	262
Scilla Hispanica	SPANISH SQUILL	Late May, June	Various	12-18 in.	"	160
Scilla Hispanica vars. carnea & rosea	FLESH & ROSE COLORED SPANISH SQUILL	"	Pink	"	*Sun or shade*	126
Scilla hyacinthoides & vars.	HYACINTH SQUILL & VARS.	Aug.	Blue, white, flesh-color	1-1½ ft.	*Sun or half shade*	414, 383, 402
Scilla Sibirica & var. alba	SIBERIAN SQUILL	Mid. Mar. to early May	Blue, white	2-6 in.	"	7, 4
Sedum spectabile	SHOWY SEDUM	Late Aug. to mid. Sept.	Rose	1½-2 ft.	*Sun*	402
Silphium laciniatum	COMPASS PLANT	Mid. July to mid. Sept.	Yellow	6 ft.	"	313
Spiræa Aruncus	GOAT'S BEARD	June, early July	White	3-5 ft.	*Sun or shade*	190
Spiræa astilboides	ASTILBE-LIKE MEADOW SWEET	"	"	2 ft.	*Sun or half shade*	190
Spiræa palmata & var. elegans	PALMATE-LEAVED MEADOW SWEET & VAR.	Late June, July	Carmine, pink	1-4 ft.	*Half shade*	233
Statice latifolia	BROAD-LEAVED SEA LAVENDER	Late July to early Sept.	Purple	1½-2 ft.	*Sun*	344
Statice Limonium	COMMON SEA LAVENDER	July, Aug.	Blue	1½ ft.	"	353
Stokesia cyanea	STOKES'ASTER	Aug. to early Oct.	"	1-1½ ft.	"	414
Trillium grandiflorum	LARGE-FLOWERED WAKEROBIN	May, early June	White	9-12 in. or more	*Half shade*	94

Botanical Name.	English Name.	Time of Bloom.	Color.	Height.	Situation.	Page.
Tritonia crocosmæflora & vars.	CROCUS-FLOWERED BLAZING STAR	July to Oct.	Scarlet, yellow	2-4 ft.	*Sun*	326, 314
Tritonia Pottsii	POTTS' BLAZING STAR	"	Orange	2-3 ft.	"	321
Tritonia rosea	REDDISH BLAZING STAR	July, Aug.	Red	1 ft.	"	326
Trollius Asiaticus	ORANGE GLOBE FLOWER	Late Apr. to late May, early Aug. to Oct.	Yellow	1½-2 ft.	*Sun or half shade*	35
Trollius Europæus	MOUNTAIN GLOBE THISTLE	Early May to early June	"	6-15 in.	*Half shade best*	109
Tulipa vars.	SINGLE & DOUBLE EARLY BEDDING TULIPS	Late Apr. to late May	Various	10-14 in.	*Sun*	64
Tulipa vars.	SINGLE & DOUBLE LATE BEDDING TULIPS	Mid. to late May	"	12-18 in.	"	163
Tulipa carinata	KEELED TULIP	May	Scarlet	12-15 in.	*Sun*	117
Tulipa "Darwin"	DARWIN TULIP	Late May to early June	"	1½-2 ft.	*Sun or half shade*	163
Tulipa "Duc van Thol"	DUC VAN THOL TULIP	Late Apr., early May	Various	6 in.	*Sun or half shade*	68
Tulipa Gesneriana & vars.	COMMON GARDEN TULIP & VARS.	Mid. May to early June	Red	6-24 in.	*Sun*	118, 164
Tulipa Gesneriana var. Dracontia	PARROT OR DRAGON TULIP	Mid. May to June	Various	12-18 in.	*Half shade*	164
Tulipa Greigi	GREIG'S TULIP	Late Apr. to mid. May	Red	3-8 in.	*Sun*	39
Tulipa retroflexa	REFLEXED TULIP	Early to late May	Yellow	12-18 in.	"	110
Tulipa vitellina	VITELLINE TULIP	May	"	1-2 ft.	"	110
Verbascum Olympicum	OLYMPIAN MULLEIN	July	"	6-10 ft.	"	314
Veronica Chamædrys	ANGEL'S EYES	Late May, June	Blue	1-1½ ft.	"	154
Veronica gentianoides	GENTIAN-LEAVED SPEEDWELL	Late Apr. to late May	"	½-2 ft.	*Sun or half shade*	55
Veronica incana	HOARY SPEEDWELL	Mid. June to late July	"	1-2 ft.	*Sun*	262
Veronica longifolia var. subsessilis	SUBSESSILE LONG-LEAVED SPEEDWELL	Early Aug. to mid. Sept.	Purple	2-3 ft.	"	410
Veronica rupestris	ROCK SPEEDWELL	Mid. May to late June	"	4-5 in.	"	154
Veronica spicata	SPIKE-FLOWERED SPEEDWELL	"	Blue	2-2½ ft.	"	262

BEST HERBACEOUS PLANTS

Botanical Name.	English Name.	Time of Bloom.	Color.	Height.	Situation.	Page.
Veronica spicata var. alba	WHITE SPIKE-FLOWERED SPEEDWELL	Early June, July	White	2-2½ ft.	*Sun*	193
Veronica Teucrium	HUNGARIAN SPEEDWELL	Late May to early June	Blue	½-1 ft.	"	155
Viola cornuta	HORNED VIOLET	Late Apr. until frost	Violet	5-8 in.	*Sun or half shade*	48
Viola cornuta var. alba	WHITE HORNED VIOLET	"	White	"	"	23
Viola cornuta var. lutea major	YELLOW HORNED VIOLET	"	Yellow	"	"	35
Viola tricolor	PANSY	Mid. Apr. to mid. Sept.	Various	"	"	68
Yucca Filamentosa	ADAM'S NEEDLE	June, July	White	6 ft.	*Sun*	194

A FEW WATER PLANTS OR AQUATICS

Color	English Name	Botanical Name and *Synonyms*	Description	Height and *Situation*	Time of Bloom
"Yellow-ish green"	**STRIPED SWEET FLAG**	**Acorus gramíneus var. variegàtus**	The chief attraction of this plant is its striking foliage striped with white, in thick grassy clumps. Prop. by division. Japan.	8-12 in. *Sun or shade*	June, July
"White"	**CAPE POND WEED, WATER HAWTHORN**	**Aponogèton distàchyum**	Lovely fragrant flowers. Prop. by seed. Thrives in tubs with 2 ft. of water or in pools from 2 to 4 ft. deep. Cape of Good Hope.	*Sun or half shade*	June to Sept.
"Purple"	**WATER SHIELD**	**Brasenia peltàta**	Small flowers. Thrives in still water from 2 to 6 ft. deep.	*Sun*	May to Sept.
"Yellow" 5	**DOUBLE MARSH MARIGOLD**	**Cáltha palústris var. flòre-plèno** *C. p. var. monstròsa-plèna*	Double form of our native plant. Pretty in ponds and also useful in the bog-garden. Flowers $1\frac{1}{2}$ in. broad. Good for cutting. Hort. The single var. is also desirable. See Plate, page 459.	1-2 ft. *Sun or half shade*	Apr., May
Various	**JAPANESE IRIS**	****Ìris lævigàta** *I. Kæmpferi*	One of the most beautiful and effective plants. Forms vigorous clumps. Flowers large and flat, sometimes 10 in. across, ranging in color from white to deep blue and plum color, often mottled or deeply veined. Narrow erect leaves. Beautiful in shallow water or in the bog-garden. Prop. by seed and division. Any good soil. Water during flowering season. For names of some good varieties see page 267. E. Siberia; Japan.	2-3 ft. *Sun*	June, July
"Yellow" 5 & 2	**COMMON OR YELLOW WATER FLAG**	**Ìris Pseudácorus**	Forms luxuriant clumps having many stems which bear large, broad-petaled flowers veined with brown. Long stiff gray-green leaves. Beautiful for the margin of water. Prop. by division. Europe.	$1\frac{1}{2}$-3 ft. *Sun*	Late May to late June
"Bright purple" 55 lighter	**LARGER BLUE FLAG**	**Ìris versícolor**	Native Iris. Flowers marked with white, yellow and purple. Leaves slightly grayish. Good for bogs, ponds, also for dry positions. Prop. by division. Canada; Northern U. S. A.	1-3 ft. *Sun or half shade*	Late May, June
"Yellow" 4	**FRINGED BUCKBEAN**	****Limnánthemum nymphoìdes**	Rampant aquatic bearing a profusion of large flowers and mottled leaves. Spreads rapidly and is difficult to get rid of if grown in too much space. Prop. by division or freshly gathered seed sown in mud. Europe; Asia.	2-4 in. *Sun*	Late May to Aug.

MARSH MARIGOLD. *Caltha palustris.*

FALSE LOTUS. *Nelumbium speciosum.*

AQUATICS

Color	English Name	Botanical Name and *Synonyms*	Description	Height and *Situation*	Time of Bloom
"Yellow" 2	**AMERICAN LOTUS OR NELUMBO, WATER CHINKAPIN**	***Nelúmbium lùteum** *Nelúmbo lùtea*	Beautiful and striking plant with very large showy blossoms and round bluish green leaves on long stalks. Effective when massed in ponds and slow streams. When transplanting in spring never disturb the plants until young growth is evident. Prop. in spring when growth begins by division or by seed. Rich soil. N. Amer.	2-6 ft. *Sun*	July, Aug.
"Pink" bet. 22 & 26	**INDIAN OR FALSE LOTUS**	***Nelúmbium speciòsum** *Nelúmbo Índica, Nelúmbo nucífera*	Fragrant showy flowers just rising above the grayish brown long-stalked leaves. Plant 4-5 in. in masses in artificial basins or ponds and slow streams. Do not transplant in spring until young growth is evident. Prop. in spring when growth begins by seed or division. Rich soil. Asia; Australia. See Plate, page 460. Var. *kermesina,* pale pink blossoms. Var. *rosea,* rose-pink blossoms. Var. *rosea plena,* double rose var. Hort. All excellent in small numbers for artificial basins and massed in ponds and slow streams.	6-8 ft. *Sun*	July to early Sept.
"White"	**MAGNOLIA LOTUS**	***Nelúmbium speciòsum var. álbum** *Nelúmbo nucífera var. álba*	A good variety of this effective plant. Var. *alba striata,* flowers striped. Plant groups of 3-4 in artificial basins or mass in ponds and slow streams. Prop. in spring when growth begins by seed or division. Rich soil. Hort.	6-8 ft. *Sun*	July to early Sept.
"Yellow"	**COMMON SPATTER-DOCK, LARGE YELLOW POND LILY**	****Nùphar ádvena** *Nymphǽa ádvena*	Distinguished more for its large erect or floating leaves than for the cup-shaped flowers, occasionally purplish, which have a disagreeable odor. Plant in the margins of slow streams or muddy ponds. Prop. by division. Rich muddy soil. N. Amer.	2-3 in. *Sun or half shade*	Late May to Sept.
"Yellow"	**SMALL YELLOW POND LILY**	****Nùphar Kalmiànum** *Nymphǽa Kalmiàna*	Peculiar aquatic with small cup-shaped flowers and many round leaves, some beneath the water, others floating. Excellent near the margins of slow streams or in muddy ponds. Prop. by division. N. Amer.	2-3 in. *Sun or half shade*	"
"Yellow"	**EUROPEAN YELLOW LILY**	****Nùphar lùteum**	Small round slightly fragrant flowers with fleshy petals. Large arrow-shaped leaves floating or erect. Excellent near the margins of slow streams or in muddy ponds. Prop. by division. Europe.	2-3 in. *Sun or half shade*	"
"White"	**WHITE WATER LILY**	****Nymphǽa álba**	One of the hardiest aquatics. Charming cup-shaped flowers which close about four o'clock. Large round floating leaves, sometimes reddish. Excellent in small ponds. Plant in spring or summer beneath about 2 ft. of water, in well-enriched loam. Europe; Asia.	4-8 in. *Sun*	"

Color	English Name	Botanical Name and *Synonyms*	Description	Height and *Situation*	Time of Bloom
"White"	**PUREST WHITE WATER OR POND LILY**	***Nymphæa álba var. candidíssima** *N. candidissima*	Superior to the type, bearing large pure white flowers on very thick stalks. Excellent for small ponds. Plant in spring or summer beneath about 2 ft. of water in well-enriched loam.	2-4 in. *Sun*	Late May to Sept.
"Red" 26	**ANDREW'S WATER OR POND LILY**	***Nymphæa Andreàna**	Dull flowers. Spotted leaves. Good for large artificial basins. Prop. by division. Good rich soil. Hort.	2-3 in. *Sun*	June to Sept.
"Reddish pink"	**ARETHUSA WATER OR POND LILY**	***Nymphæa Arethusa**	An improvement on N. Laydekeri var. fulgens. Stronger grower and with larger flowers. Good for spacious artificial basins. Plant beneath 2 ft. of water in rich soil.	2-3 in. *Sun*	"
"Orange" 20 more orange	**AURORA WATER OR POND LILY**	***Nymphæa Aurora**	Yellowish blossoms gradually changing to red. Pretty in ponds. Prop. by division. Good rich soil. Hort.	2-4 in. *Sun*	"
"Purple carmine"	**ELLIS' WATER OR POND LILY**	***Nymphæa Ellisiàna**	Very showy and pleasing. Bright flowers. Prop. by division. Good rich soil. Hort.	2-4 in. *Sun*	"
"Yellow"	**YELLOW WATER LILY**	***Nymphæa flàva**	Resembles N. Mexicana but less vigorous. Large pale flowers open from eleven until four o'clock. Dark floating foliage. Good in small ponds. Protect slightly in winter. Plant in spring or summer beneath 2 or more feet of water in well-enriched loam. S. Eastern U. S. A.	4-5 in. *Sun*	June. July
"Yellow"	**TAWNY WATER OR POND LILY**	****Nymphæa fùlva**	Large floating flowers and spotted leaves. Effective in ponds. Prop. by division. Plant beneath 2 ft. of water in rich soil. Hort.	1-3 in. *Sun*	June to Sept.
"Brilliant white"	**GLADSTONE'S WATER OR POND LILY**	***Nymphæa Gladstoniàna**	Vigorous large flowers and attractive bold foliage. Excellent for spacious ponds. Prop. by division. Plant in spring or summer beneath 2 ft. of water in rich loam. Hort.	2-4 in. *Sun*	"
Various	**LAYDEKER'S WATER OR POND LILY**	***Nymphæa Laydekeri vars.**	An interesting group. Adapted for cultivation in tubs and fountain basins. Prop. by division. Good rich soil. Var. *fulgens;* (color no. near 28), charming brilliant reddish pink flowers. Var. *liliacea;* fragrant lilac flowers and spotted leaves. Var. *purpurata;* (color no. 33 darker), crimson flowers. Var. *rosea;* (color no. 33 lighter), rather small pink flowers.	2-3 in. *Sun*	"

AQUATICS

Color	English Name	Botanical Name and *Synonyms*	Description	Height and *Situation*	Time of Bloom
"White"	MARLIAC'S WATER OR POND LILY	*Nymphæa Marliàcea var. álbida	Vigorous plant, perhaps the best white Water Lily. Flowers large and brilliant. tinted with pink. Good for large artificial basins. Prop. by division. Plant in spring or summer beneath 2 ft. of water in well-enriched soil. Hort.	2-4 in. *Sun*	June to Sept.
"Pink" white to pale 36	MARLIAC'S FLESH-COLORED WATER OR POND LILY	*Nymphæa Marliàcea var. càrnea	Vigorous plant. Pale pink flowers with a slight fragrance. Good in large artificial basins. Prop. by division. Plant in spring or summer beneath 2 ft. of water in rich loam. Hort.	2-4 in. *Sun*	"
"Yellow" 2	CANARY WATER LILY	*Nymphæa Marliàcea var. chromatélla *N. tuberòsa var. flavéscens*	Excellent free-blooming variety. The dark green spotted leaves often rise high above the water. Good for large artificial basins. Prop. by division. Plant in spring or summer beneath 2 ft. of water in well-enriched loam. Hort.	2-4 in. *Sun*	"
"Reddish purple" white to 36	MARLIAC'S FLAMING WATER OR POND LILY	*Nymphæa Marliàcea var. flámmea	Effective plant. Striking flowers in abundance. Good for large artificial basins. Prop. by division. Plant beneath 2 ft. of water in rich soil. Hort.	2-4 in. *Sun*	"
"Reddish purple" 27 light	MARLIAC'S FIERY WATER OR POND LILY	*Nymphæa Marliàcea var. ígnea	Strong grower. Very similar to var. flammea. Flowers striking. Good in large artificial basins. Plant beneath 2 ft. of water. Prop. by division. Rich soil. Hort.	2-4 in. *Sun*	"
"Rose" white to 36	MARLIAC'S PINK WATER OR POND LILY	*Nymphæa Marliàcea var. ròsea	Robust plant. An improved form of N. Marliacea var. carnea. Dark green foliage, reddish when young. Good for large artificial basins. Prop. by division. Rich soil. Hort.	2-4 in. *Sun*	"
"White"	SWEET-SCENTED WATER OR POND LILY, WATER NYMPH	*Nymphæa odoràta *Castàlia odoràta*	One of the best Water Lilies. Fragrant flowers, open only in the morning. Round leathery leaves. Effective when single or in isolated groups in large ponds. Prop. by division. Plant in spring or summer in enriched loam beneath 2 or more ft. of water. Eastern U. S. A.	2-4 in. *Sun*	"
Various	SWEET-SCENTED WATER OR POND LILY	*Nymphæa odoràta vars. *Castàlia odoràta vars.*	Var. *Caroliniana;* large fragrant narrow-petaled blossoms of a delicate pink shade. Hort. Var. *exquisita;* large deep rose flowers. Hort. Var. *Luciana;* (color no. 26), rose-colored flowers. Hort. Var. *sulphurea;* (color no. 2), yellow flowers a few in. above the water, open only in the morning. Small floating leaves. Var. *sulphurea grandiflora;* large flowers. All these vars. are effective singly or in isolated groups in large ponds. Plant in spring or summer in enriched loam beneath 2 or more ft. of water. Prop. by division.	2-4 in. *Sun*	"

Color	English Name	Botanical Name and *Synonyms*	Description	Height and *Situation*	Time of Bloom
"Pink" 29	**CAPE COD WATER OR POND LILY**	***Nymphæa odoràta var. ròsea** *N. odoràta var. rùbra, Castàlia odoràta var. ròsea*	Rather small flowers which gradually fade to white. Round leaves dark red when small. Effective single or massed in large ponds. Plant in spring or summer in rich loam, beneath 2 or more feet of water. S. Eastern Mass. to N. J.	2-4 in. *Sun*	June to Sept.
"Yellow" 26	**ROBINSON'S WATER OR POND LILY**	**Nymphæa Robinsòni** *N. Robinsoniàna*	Large flowers tinted with red which float on the surface of the water. Spotted leaves. Good for large artificial basins. Prop. by division. Good rich soil. Hort.	1-3 in. *Sun*	"
"Rose & carmine" 24 to 2	**SEIGNORETI'S WATER OR POND LILY**	***Nymphæa Seignoréti**	Pleasing Water Lily. Delicate flowers tinted with red. Spotted leaves. Pretty in ponds. Prop. by division. Good rich soil. Hort.	2-6 in. *Sun*	"
"White"	**SMALL WHITE WATER LILY**	***Nymphæa tetrágona** *N. pýgmæa, Castàlia tetrágona*	The smallest Water Lily, with charming flowers about 2 in. across, which open in the afternoon. Dark green horse-shoe-shaped leaves. Pretty in tubs and small basins. Prop. by seed. Asia; N. Amer.	1-2 in. *Sun*	"
"Yellow" 2	**HELVOLA WATER LILY**	***Nymphæa tetrágona var. Helvola**	Flowers, small, produced in great abundance. Leaves small and dark green above. Well adapted to tubs and small fountain basins. Prop. by seed. Plant 2 ft. under water.	1-2 in *Sun*	"
"White"	**TUBEROUS WHITE WATER LILY**	****Nymphæa tuberòsa** *N. renifórmis, Castàlia tuberòsa*	Luxuriant in growth. Large flowers about 6 or 9 in. wide, open only in the morning. Large roundish leaves, somewhat veiny. Excellent for large ponds. Prop. by division. Plant in spring or summer in loam under shallow water. U. S. A.	4-6 in. *Sun*	July, Aug.
"White"	**RICHARDSON'S TUBEROUS WHITE WATER LILY**	***Nymphæa tuberòsa var. Richardsonii**	Especially vigorous plant. Very double flowers rising well above the water. Large clean bright green leaves. Excellent for large ponds. Prop. by division. Very rich soil. Hort.	4-6 in. *Sun*	"
"Pink" 29	**PINK TUBEROUS WATER LILY**	****Nymphæa tuberòsa var. ròsea**	Plant of rank growth. Pink flowers, rising above the water. Excellent for large ponds. Prop. by division. Requires rich soil. Hort.	4-6 in. *Sun*	"
"Yellow"	**GOLDEN CLUB**	****Oróntium aquáticum**	Vigorous plant. Tiny flowers in narrow club-like spikes. Dark green oblong leaves on long stalks. Difficult to eradicate when once established. Plant in bogs or in fairly swift streams 1 ft. or more deep. Eastern U. S. A.	½-2 ft. *Sun*	May

AQUATICS

Color	English Name	Botanical Name and *Synonyms*	Description	Height and *Situation*	Time of Bloom
"Green"	ARROW ARUM	**Peltándra Virgínica *P. undulàta*	Sub-aquatic foliage plant. Attractive on account of its thick dark green arrow-shaped leaves. Excellent for bogs and shallow water. N. Amer.	2-3 ft. *Sun*	June
"Purplish blue" 46	PICKEREL WEED	**Pontedèria cordàta	Graceful sub-aquatic with tall clumps of dense flower spikes and beautiful arrow-shaped leaves. Wild in shallow water everywhere. Excellent in bogs or in shallow water near the margins of ponds. Prop. by division. N. Amer.	1½-4 ft. *Sun*	July, Aug.
"White"	OLD WORLD ARROW-HEAD	**Sagittària sagittæfòlia var. flòre-plèno *Sagittària Japónica*	Delicate double buttercup-like flowers. Large broad arrow-shaped leaves. Excellent in shallow ponds. Prop. by seed, oftener by division. Plant beneath water 1 ft. deep. Hort.	2-3 ft. *Sun*	"
"Light brown"	GREAT BULRUSH	**Scírpus lacústris var. zebrìna *S. Tabernæmontàna var. zebrìna*	Sub-aquatic. Peculiar flowers in spikes on large round stalks. Leaves striped with green and white. Good for margin of ponds. Moist soil. Hort.	3-6 ft. *Sun*	"
"Brown"	BROAD-LEAVED CAT-TAIL, COMMON REED MACE	**Týpha latifòlia	Graceful plant for marshes or shallow water. Peculiar brownish flowers in dense spikes and very narrow ribbon-like leaves, long and glossy. Excellent in bogs and shallow ponds. Amer.; Europe; Asia.	4-8 ft. *Sun*	June, July
"Green white"	INDIAN OR WILD RICE, WATER OATS	*Zizània aquática	A beautiful kind of reed-like grass. Useful for margins of streams on account of the graceful habit of its leaves. Prop. by seed, sown annually. N Amer.	7-9 ft. *Sun*	Aug.

SOME BOG-GARDEN OR MARSH PLANTS

Color	English Name	Botanical Name and *Synonyms*	Description	Height and *Situation*	Time of Bloom
"Yellowish green"	**SWEET FLAG**	**Ácorus Cálamus**	Marsh plant. Flowers insignificant. Foliage erect and rush-like. Rootstock odorous and edible. Prop. by division. Europe. Var. *variegatus*. Foliage striped with green and gold. More frequently cultivated than the type.	2 ft. *Sun or shade*	June, early July
"Reddish"	**GIANT REED**	**Arúndo Dònax**	A plant, decorative because of its striking foliage. Reddish flowers insignificant. Prop. by seed and division. Any soil. S. Europe. Var. *variegata*, (var. *versicolor*). Foliage variegated.	10 ft. *Sun*	
"Yellow" 5	**DOUBLE MARSH MARIGOLD**	**Cáltha palústris var. flòre-plèno** *C. p. var. monstròsa-plèna*	Double form of our native plant. Flowers 1½ in. broad. Good for cutting. Hort. The single var. is also good. See Plate, page 459.	1 ft. *Sun or half shade*	Apr., May
	COMMON HORSE-TAIL OR SCOURING RUSH	**Equisètum hyemàle**	Unique primeval looking plant with slender hollow reed-like stems and no apparent leaves or flowers. Prop. by spores. U. S. A.	1-2 ft. *Sun or shade*	
"Purple pink" bet. 45 & 25	**JOE-PYE OR TRUMPET WEED**	**Eupatòrium purpùreum**	Plant of coarse growth. Flowers in immense flat clusters, 18 in. across. Foliage in whorls. Naturalize on banks of streams or in wet meadows. Prop. by cuttings. Any soil. N. Amer.	5-7 ft. *Sun or half shade*	Late July to early Sept.
"Lilac" 43 deep	**LARGE PURPLE-FRINGED ORCHIS**	**Habenària fimbriàta** *H. grandiflòra*	An Orchid. Grows in wet meadows. Fragrant flowers prettily fringed, rarely white, rise in spikes above inconspicuous foliage. Bog-garden. Leaf-mold and sand kept moist by mulch of leaves. N. Amer.	1-1½ ft. *Half shade*	June, July
Various often bet. 39 & 41	**SMALLER PURPLE-FRINGED ORCHIS**	**Habenària psycòdes** *Órchis psycòdes*	Closely related to H. fimbriata, but with smaller blossoms. Very fragrant fringed flowers, rose, lilac or crimson, rarely white, in tall spikes 4-10 in. long. Showy in bog-garden. Leaf-mold and sand kept moist by mulch of leaves. N. Amer.	2-3 ft. *Half shade*	July, Aug.
Various	**JAPANESE IRIS**	****Ìris lævigàta** *I. Kæmpferi*	One of the most beautiful and effective plants. Forms vigorous clumps. Flowers large and flat, sometimes 10 in. across, ranging in color from white to deep blue and	2-3 ft. *Sun*	June, July

PITCHER PLANT. *Sarracenia purpurea.*

CARDINAL FLOWER. *Lobelia cardinalis.*

BOG PLANTS

Color	English Name	Botanical Name and *Synonyms*	Description	Height and *Situation*	Time of Bloom
			plum color, sometimes mottled or deeply veined. Narrow erect leaves. Beautiful in masses beside water. Prop. by seed and division. Any good soil. Water during flowering season. For names of good vars. see page 267.		
"Yellow" 5 & 2	**COMMON OR YELLOW WATER FLAG**	**Ìris Pseudácorus**	Forms luxuriant clumps having many stems which bear large broad-petaled flowers veined with brown. Long stiff gray-green leaves. Beautiful for the margin of water. Prop. by division. Europe.	1½-3 ft. *Sun*	Late May to late June
"Bright purple" 55 lighter	**LARGER BLUE FLAG**	**Ìris versícolor**	Native Iris. Flowers marked with white, yellow and purple. Leaves slightly grayish. Good for margin of ponds and also for dry positions. Prop. by division. Canada; Northern U. S. A.	1-3 ft. *Sun or half shade*	Late May, June
Various 1 spotted brown or 11 shading to 14	**CANADA OR WILD YELLOW LILY**	**Lílium Canadénse**	Well-known native species found in moist meadows and bogs. Spotted flowers, varying from yellow to red, droop in a circle, surmounting the graceful stems around which the leaves grow in whorls. Easily naturalized. Bulbous. Prop. by offsets or scales. Light well-drained soil. Avoid direct contact with manure. Eastern N. Amer.	1-4 ft. *Sun or half shade*	June, July
"Reddish orange" bet. 19 & 20	**AMERICAN TURK'S CAP LILY**	****Lílium supérbum**	Native in meadows and marshes. Delicate drooping flowers having pointed reflexed petals, spotted within, in a pyramidal panicle of about twenty. Bulbous. Prop. by offsets, scales, or very slowly by seed. Any well-drained soil. Avoid direct contact with manure. Eastern N. Amer.	3-6 ft. *Sun or half shade*	Early July to early Aug.
"Crimson" 20	**CARDINAL FLOWER, INDIAN PINK**	****Lobèlia cardinàlis**	Brilliant flowers in spikes on erect unbranching stems. Leaves narrow mostly on lower part of stalk. Naturalize near water or plant in shaded border. Prop. by seed or division. Rich soil, preferably moist. Wet places, Eastern N. Amer. See Plate, page 468.	2-4 ft. *Sun or shade*	Mid. July to Sept.
"Blue" 46	**GREAT LOBELIA, BLUE CARDINAL FLOWER**	****Lobèlia syphilítica**	Tubular flowers in long leafy spikes on slightly hairy stalks. Foliage large, smooth or hairy. Good in bog-garden or border. Moist soil. Wet places, Eastern U. S. A. See Plate, page 468.	1-3 ft. *Sun*	Mid. Aug. to late Sept.
"White"	**GREAT WHITE LOBELIA**	****Lobèlia syphilítica var. álba**	Handsome variety. Flowers in long spikes. Leaves almost stemless on the flower stalks. Damp grounds or bogs. Prop. by seed or cuttings. Hort.	2-3 ft. *Sun*	Mid. Aug. to Sept.

BOG PLANTS

Color	English Name	Botanical Name and *Synonyms*	Description	Height and *Situation*	Time of Bloom
"Purple" 40	**SPIKED OR PURPLE LOOSESTRIFE**	***Lýthrum salicària**	Erect bushy plant. Delicate starry flowers with wavy petals in wandlike spikes on long graceful leafy stems. Pretty for water-side. Prop. by division. Any moist soil. Temperate Zone.	2-3 ft. *Sun*	Late July to late Aug.
"Rose"	**PINK SPIKED LOOSESTRIFE**	****Lýthrum salicària var. ròseum supérbum** *L. ròseum supérbum*	Tall spikes of flowers on long graceful stems covered with willowy foliage. Beautiful on banks of streams and ponds. Prop. by division or cuttings. Moist soil preferable. Hort.	4-6 ft. *Sun or half shade*	"
"Violet" 47	**SQUARE-STEMMED MONKEY FLOWER**	**Mímulus ríngens**	Snapdragon-like flowers with open throats. Bog-garden or margin of water. Prop. by seed. Any soil, abundantly watered. N. Amer.	1-2 ft. *Sun*	July, Aug.
"Brownish violet"	**EULALIA**	**Miscánthus Sinénsis** *Eulàlia Japónica*	Striking kind of grass with plume-like spikes of flowers. Prop. by seed and division. Any soil. Japan. Vars. *gracillimus,* (*Eulalia gracillima univittata, E. Japonica gracillima*). Leaves narrower. *Variegatus,* foliage variegated. *Zebrinus,* leaves with cross bands of gold. Prop. the last two vars. by division.	4-9 ft. *Sun*	Oct.
"Blue" 57	**EVER-FLOWERING FORGET-ME-NOT**	****Myosòtis palústris var. sempérflorens**	Dwarf plant of spreading habit, called *semperflorens* from its long season of bloom. Flowers in loose clusters. Good for damp shady spots of rock-garden. Prop. by seed and cuttings. Moist soil. Hort.	8 in. *Shade*	May to Sept.
"Pale green"	**ROYAL OR FLOWERING FERN**	***Osmúnda regàlis**	Distinctive in appearance. Smooth pale green foliage and conspicuous fruit in flower-like clusters. Thrives in rich moist soil, even with water standing 2 or 3 in. deep. Prop. by spores. Eastern Amer.	2-4 ft. *Sun or shade*	"
"Purplish blue" 46	**PICKEREL WEED**	***Pontedèria cordàta**	Among the most beautiful of aquatics. Many small flowers in spikes. Glossy heart-shaped leaves. Pretty in bog-garden or best in water 6-12 in. deep. Prop. by division at any time. N. Amer.	1½-4 ft. *Sun or half shade*	July to Sept.
"Rosy purple"	**MEADOW BEAUTY, DEER GRASS**	***Rhéxia Virgínica**	Numerous showy flowers about 1½ in. wide, several on the stem, conspicuous golden anthers. For bogs or moist spots in wild garden where it will form a bed of bloom. Prop. by seed or division. Peaty soil. Eastern U. S. A.	9-12 in. *Sun*	July to Oct.
"White"	**COMMON ARROW-HEAD**	**Sagittària latifòlia** *S. sagittæfòlia var. variábilis, S. variábilis*	Flowers similar to Buttercups in whorls of 3 around leafless sta.ks. Arrow-shaped leaves. For bogs, edges of ponds and shallow water. Prop. by seed and division. N. Amer.	4-6 ft. *Sun*	July to early Sept.

BOG PLANTS

Color	English Name	Botanical Name and *Synonyms*	Description	Height and *Situation*	Time of Bloom
"Reddish brown"	GREAT TRUMPET LEAF, PITCHER PLANT OP SIDE-SADDLE FLOWER	**Sarracènia Drummóndii**	Curious wild bog plant. Odd-shaped unobtrusive flowers. Horn-like leaves 2 ft. long, marked with pale yellow and purple. Bog-garden. Protect in winter. Prop. by division. Moist soil. S. Eastern U. S. A.	2-3 ft. *Sun*	June
"Deep purple"	PITCHER PLANT, SIDE-SADDLE FLOWER	**Sarracènia purpùrea**	Curious wild bog plant. A solitary flower, sometimes brownish, of unusual shape, less conspicuous than the purple-veined pitcher-like leaves. Bog garden or water-side. Prop. by division. Moist soil. Atlantic States. See Plate, page 467.	8-12 in. *Sun*	"
"Greenish"	PENNSYLVANIA OR SWAMP SAXIFRAGE	**Saxífraga Pennsylvánica**	A profusion of flowers in elongated panicles. Large leaves in clumps at the base of the plant. Prop. by division, offshoots and stolons. Good for bogs. Swamps, Eastern U. S. A.	3 ft. *Half shade*	May, June
"Orange" 6	JAPANESE GROUNDSEL	**Senècio Japónicus** *Erythrochæte palmatifida, Liguläria Japónica*	Effective flowers on tall stems, sometimes branching. Leaves very broad, 1 ft. across, deeply lobed. Excellent for foliage effect in border. Prop. by seed, division or cuttings. Moist soil. Japan.	4-5 ft. *Sun*	Late July to mid. Aug.
"White"	GOAT'S BEARD	**Spiræa Arúncus** *Arúncus sylvêster*	Erect branching herb growing wild in rich woods. Abundant flowers in plumy panicles. Handsome compound foliage. Invaluable for rough places and for grouping with foliage plants. Prop. by division. Any soil. N. Europe; Asia; N. Amer.	3-5 ft. *Sun or shade*	June, early July
"Carmine" 26 deep	PALMATE-LEAVED MEADOW SWEET	**Spiræa palmàta** *Filipêndula purpùrea, Ulmària purpùrea*	Vigorous plant. Broad clusters of brilliant flowers borne on erect stems. Tufted root leaves, palmately divided. Prop. by seed and division. A fairly rich moist soil. Japan.	2-4 ft. *Half shade*	June, July
"Cream white"	POISONOUS ZYGADENUS	**Zygadènus venenòsus**	Slender plant. Small flowers in short racemes on erect stems. Leaves mostly about the root. Bulb poisonous. Plant in wild or bog-garden. Slight winter protection necessary. Prop. by seed, more frequently by offsets. Moist soil preferable. Western U. S. A.	1½-2 ft. *Sun*	Mid. May to mid. June

SOME PLANTS CONSPICUOUS FOR THEIR FOLIAGE

Color	English Name	Botanical Name and *Synonyms*	Description	Height and *Situation*	Time of Bloom
"Cream white"	**VARIEGATED GOUTWEED OR ASHWEED**	**Ægopòdium podogrària var. variegàtum**	A common foliage plant. Spreads quickly by creeping rootstocks, and makes attractive mats of white margined foliage. Hard to get rid of when established. Prop. by division. Europe.	12-15 in. *Sun*	Late May to late June
"Purple"	**VARIEGATED HAIR GRASS**	**Aìra cærùlea var. variegàta** *Molìnia cærùlea var. variegàta*	Rather stiff grass plant of ornamental foliage. Flowers in spike-like panicles. Leaves striped green and gold. Excellent for carpeting. Central Europe; Asia.	1-2½ ft. *Sun*	
"White"	**INDIAN WILD SARSAPARILLA**	**Aràlia Cachemírica** *A. Cashmeriàna, A. macrophýlla*	A strong-growing plant. Umbels of flowers forming long loose panicles. Small round black or purplish berries. Leaves compound. Good for shady places under trees and for subtropical effects. Prop. by cuttings. Mts. of India.	5-8 ft. *Half shade*	
"White"	**CORDATE WILD SARSAPARILLA**	**Aràlia cordàta** *A. èdulis*	Umbels of flowers forming long loose panicles. Small round black or purplish berries. Leaves divided into somewhat heart-shaped leaflets. Good for subtropical effects. Prop. by cuttings. Japan.	4-6 ft. *Half shade*	
"Greenish white"	**SPIKENARD**	**Aràlia racemòsa**	Flower umbels in loose panicles. Bears small round black berries. Foliage compound. Good under trees. Prop. by cuttings. Northern U. S. A.	3-4 ft. *Sun*	Late July to late Aug.
"Green"	**VARIEGATED OAT-GRASS**	**Arrhenathèrum bulbòsum var. variegàtum**	A pretty dwarf tufted grass. Leaves striped green and white. Good for the edge of border. Prop. by division. Ordinary garden soil. Hort.	6-8 in. *Sun*	
"White"	**SOUTHERNWOOD, OLD MAN**	**Artemísia Abrótanum**	A plant of shrubby growth. Drooping flowers on long slender panicles. Grown for its dark green fragrant foliage. Prop. by division or cuttings. Thrives even in poor soil. Europe.	3-5 ft *Sun*	
"Whitish"	**LEWIS' SOUTHERNWOOD**	**Artemísia Ludoviciàna**	Very hoary plant with panicles of small tubular blossoms. Silvery colored foliage is more effective than flowers. Margin of border. Prop. by division. Thrives in poor soil. Plains of Western U. S. A.	2 ft. *Sun*	

FOLIAGE PLANTS

Color	English Name	Botanical Name and *Synonyms*	Description	Height and *Situation*	Time of Bloom
"Whitish yellow"	**ROMAN WORM-WOOD**	**Artemísia Póntica**	An effective bushy foliage plant. Panicles of small globe-shaped drooping flowers. Leaves finely divided and whitish beneath. Prop. by division. Thrives in poor soil. Europe.	1 ft. *Sun*	
"Yellow"	**OLD WOMAN**	**Artemísia Stelleriàna**	A very hoary shrubby plant. Small globe-shaped flowers in compact racemes. Attractive for its silvery effect. Prop. by division. Thrives in poor soil. Coast of Mass.; Asia.	2 ft. *Sun*	
"Green"	**JAPANESE BAMBOO**	**Arundinària Japónica** *Bambùsa Metáke*	Grass-like plant. Foliage large, dark and glossy, whitish underneath. Grows well in cities. Needs plenty of room, sheltered position and also some winter protection. Prop. preferably by division in spring. Rich loamy soil. Japan.	6-10 ft. *Half shade*	
"Green"	**SIMON'S BAMBOO**	**Arundinària Simòni** *A. Naribìra, Bambùsa Simòni, B. viridi-strìata*	Imposing plant which takes some years to establish. Leaves 8-12 in. long and very narrow. Needs sheltered position, some winter protection and pruning. Prop. preferably by spring division. Rich loamy soil. India; China.	10-20 ft. *Sun*	
"Reddish"	**GIANT REED**	**Arúndo Dònax**	Grown for foliage. Erect stalks with plumy racemes of small flowers, and graceful large pointed leaves. Effective in clumps. Prop. by seed and division. Any soil. S. Europe. Var. *variegata,* (*A. D. var. versicolor*), variegated foliage. Hort.	10 ft. *Sun*	
"Bronze"	**LARGE-LEAVED PLUME POPPY**	***Boccònia microcarpa**	Plumy clusters of bronze-tinted flowers at the summit of stems clad with large handsome leaves like those of B. cordata. Naturalize on edge of lawns or shrubbery, etc. Prop. by division, generally by suckers. Rich soil is essential. China.	9 ft. *Sun*	July
"Red" bet. 19 & 20	**NIGRICANS CANNA**	**Cánna nigricans**	One of the most desirable of the tall Cannas on account of its foliage, which is very dark. Plant in formal garden or shrubbery border. After frost, dig up the roots and store in dry cellar until spring. Prop. by division of rootstock. Light soil, rich, deep and moist.	4-6 ft. *Sun*	
"Green"	**VARIE-GATED ORCHARD GRASS OR COCK'S-FOOT**	**Dáctylis glomeràta var. variegàta**	Dwarf grass in inconspicuous flower clusters. Foliage marked with silver. Much used for edgings. Prop. by division. Any good soil. Hort.	1½-2 ft. *Sun*	Aug.

FOLIAGE PLANTS

Color	English Name	Botanical Name and *Synonyms*	Description	Height and *Situation*	Time of Bloom
"Green"	SEA LYME GRASS	Élymus arenàrius	Vigorous strong-rooting ornamental grass. Flowers valueless. Plant near shrubbery or on sandy banks. Temperate Zone.	2-5 ft. *Sun*	
"Greenish"	WOOL GRASS, PLUME GRASS, RAVENNA GRASS	Eriánthus Ravénnæ *Sáccharum Ravênnæ*	Ornamental grass. Very long leaves tinged with violet and with white line down the centre. Handsome in clumps like the Pampas Grass. Prop. by division. Any soil. S. Europe.	4-7 ft. *Sun*	Aug.
	BLUE FESCUE GRASS	Festùca glaùca *F. ovìna var. glaùca*	Ornamental grass. Inconspicuous flowers in panicles grown for the dense tufts of very narrow bluish leaves. Used for edgings or for contrast with darker foliage. Prop. by division. Europe.	18-20 in. *Sun or half shade*	June, July
"White"	GIANT PARSLEY OR GIANT COW PARSNIP	Heraclèum villòsum *H. gigánteum*	Large bold plant for subtropical effects. Small flowers in dense clusters 1 ft. or more across. Leaves very large. Easy to naturalize in rough places. Prop. by seed and division. Rich moist loam. Europe.	8-10 ft. *Sun*	July, early Aug.
"Brownish violet"	EULALIA	Miscánthus Sinénsis *Eulàlia Japónica*	An excellent grass plant with feathery panicles of flowers which have long silky hairs. Good for ornamental beds or in the shrubbery border. Prop. by seed and division. Any soil. Japan. Vars. *gracillimus*, (*Eulalia gracillima univittata, Eulalia Japonica gracillima*), narrower leaves. *Variegatus*, variegated foliage. *Zebrinus*, has cross bands of yellow on the leaves. The last two Japanese vars. are prop. better by division.	4-9 ft. *Sun*	Oct.
"White"	RIBBON GRASS, GARDENER'S GARTERS	Phálaris arundinàcea var. variegàta *P. arundinàcea var. pícta*	A grass with flowers in spikes. Grown for its ribbon-like foliage, which is striped with white. Good for edging beds. Northern N. Amer.	2-4 ft. *Sun*	Aug.
"Brownish"	GOLDEN BAMBOO	Phyllóstachys aùrea *Bambùsa aùrea*	Graceful grass plant with small light-colored leaves and stems of yellowish green. Effective if planted against setting of dark foliage. Prop. usually by spring division. Rich loamy soil. Slight protection necessary. Japan.	10-15 ft. *Half shade*	
"Brownish"	BLACK BAMBOO	Phyllóstachys nìgra *Bambùsa nìgra*	One of the best-known species. Branches turn black after a year. Delicate, paper-like foliage of medium size. Prop. usually by spring division. Rich loamy soil; sheltered position, slight protection. Orient.	10-20 ft. *Half shade*	

FOLIAGE PLANTS

Color	English Name	Botanical Name and *Synonyms*	Description	Height and *Situation*	Time of Bloom
"Green"	**OVAL-LEAVED BAMBOO**	**Phyllóstachys ruscifòlia** *P. Kumâsaca, Bambùsa ruscifòlia, B. viminàlis*	Very angular habit of growth; small dark oval leaves. Prop. usually by spring division. Rich loamy soil; sheltered position; slight winter protection. Japan.	1½-2 ft. *Half shade*	
"Green"	**GLAUCOUS BAMBOO**	**Phyllóstachys viridi-glaucéscens** *Bambusa viridi-glaucêscens*	Graceful and easily cultivated Bamboo, covering large space. Running rootstock. Small yellowish branches, medium sized leaves. Prop. usually by spring division. Rich loamy soil; sheltered position; slight winter protection. China.	10-18 ft. *Half shade*	
"Greenish white"	**MEDICINAL RHUBARB**	**Rhèum officinàle**	Effective foliage plant with very large leaves, sometimes 3 ft. broad. Flowers numerous. Plant near shrubberies, etc. Prop. by division. Rich deep soil. Thibet; W. China.	5-6 ft. *Sun*	
"Greenish white"	**PALMATE RHUBARB**	**Rhèum palmàtum** *R. sanguíneum*	Foliage plant of tropical effect. Flowers in elongated clusters. Large roundish heart-shaped leaves, deeply lobed. Plant near shrubberies, etc., isolated or in masses. Prop. by seed and division. Deep rich soil. China. Var. *Tanghuticum*, (*Rheum Tanghuticum*), is more vigorous than the type. It increases rapidly. Leaves longer, and less deeply divided. Hort.	6-8 ft. *Sun*	
"White"	**VARIEGATED JAPANESE ROHDEA**	**Ròhdea Japónica var. variegàta**	Foliage plant. Bell-shaped flowers in compact spikes. Beautiful variegated leaves, 10-12 in. long, in upright rosettes, rising above the flowers. Deep rich soil. Japan.	9-12 in. *Half shade*	June
"Yellow"	**LAVENDER COTTON**	**Santolìna chamæ-cyparíssus** *S. incàna*	Shrubby downy plant of compact habit with inconspicuous greenish yellow flowers. Valuable for its silvery foliage, evergreen and fragrant. Desirable in groups or in the border. Prop. by cuttings. Slight protection. Europe; Asia.	1½-2 ft. *Sun*	July, Aug.
"Brown"	**VARIEGATED BULRUSH OR SEDGE**	**Scírpus Holoschœnus var. variegàtus**	Ornamental, rush-like foliage plant. Minute brown flowers in clustered spikes. Leaves striped with yellowish white. Effective in bog-garden or beside water. Prop. by seed, division or suckers. Moist or dry soil. Hort.	1-1½ ft. *Sun*	June, July
"Green"	**FEATHER GRASS**	**Stìpa pennàta**	Looks like ordinary grass except when in bloom. Feathery plumes of flowers used in 'everlasting' bouquets. Border. Prop. by seed and division. Deep sandy loam. Steppes of Europe and Siberia.	20 in *Sun*	July, Aug.

FOLIAGE PLANTS

Color	English Name	Botanical Name and *Synonyms*	Description	Height and *Situation*	Time of Bloom
Various	**VARIEGATED COMMON COMFREY**	**Sýmphytum officinàle var. variegàtum** *S. officinàle var. luteo-marginàtum*	Broad leaves deeply margined with yellow or cream color. When cultivated for ornamental foliage the flower stems should be cut off. Europe.	3-4 ft. *Sun or half shade*	June, July
"Green"	**GAMA OR SESAME GRASS**	**Trípsacum dactyloìdes** *T. Dáctylis, T. Violàceum*	Ornamental grass of tufted habit. Terminal spikes of flowers. Plant in wild garden or among other ornamental grasses. Prop. by seed, preferably by rootstock cuttings. Moist situation. Southern and Central U. S. A.	4-7 ft. *Sun*	"
"Purple"	**SHRUB-YELLOW ROOT**	**Xanthorrhìza apiifòlia** *Zanthorhìza apiifòlia*	Shrubby yellow stemmed plant. Flowers in slender drooping racemes under the pinnate leaves. Grown for its foliage, which becomes golden yellow in autumn. Prop. by seed and division. Any garden soil, preferably moist. Eastern U. S. A.	1-2 ft. *Shade*	May, early June

PANICLED CLEMATIS. *Clematis paniculata.*

CHINESE WISTARIA. *Wistaria Chinensis.*

A FEW SELECTED VINES AND CLIMBERS

Color	English Name	Botanical Name and *Synonyms*	Description	Height and *Situation*	Time of Bloom
"White"	**ASIATIC ACTINIDIA**	**Actínidea polýgama**	Strong rapid grower which makes long shoots each season. A deciduous twining shrub with inconspicuous fragrant flowers in clusters. Handsome glossy foliage. Especially good for screens, trellises or arbors. Prop. by cuttings and layers. Rich soil. E. Asia.	20 ft. or more *Sun or shade*	June
"Purplish" 42	**AKEBIA**	**Akèbia quinàta**	A graceful twining shrub with pretty clusters of small fragrant flowers and charming evergreen foliage. Forms a thick screen. Good for trellises, pergolas, etc. Prop. by seed, root-division, cuttings or layers. Well-drained soil. Japan; China.	12 ft. or more *Sun*	May, early June
"Greenish"	**VIRGINIA CREEPER**	**Ampelópsis quinquefòlia** *A. hederàcea, Vìtis quinquefòlia*	Rapidly climbing shrub of free and luxuriant habit bearing inconspicuous clusters of flowers followed by handsome dark blue berries. Beautiful divided foliage, turning brilliantly in fall. Invaluable for covering walls, dead trees, buildings, etc. Prop. by seed, generally by hardwood cuttings. Any soil. N. Amer.	12-20 ft. *Sun or half shade*	July
"Greenish"	**JAPANESE OR BOSTON IVY**	**Ampelópsis tricuspidàta** *A. Róylei, A. Veìtchi, Vìtis incónstans*	One of the best wall creepers, resisting dust, etc. Climbing shrub clinging closely and having dense growth of glossy foliage which turns bronze or scarlet in fall, and bunches of berries. Much used in cities. It will grow well in a northern exposure. Prop. by seed, greenwood cuttings and layers. Any soil. China; Japan.	30-40 ft. *Sun or shade*	Late June, early July
"Purplish brown" 21	**DUTCHMAN'S PIPE**	**Aristolòchia macrophýlla** *A. Sìpho*	Striking climbing shrub with odd inconspicuous flowers and large round dark leaves, 10 in. across. Useful for screens, porches, etc. Prop. by cuttings. Any good loamy soil. U. S. A.	12-20 ft. *Sun or half shade*	Late May, early June
"Greenish white"	**ORIENTAL BITTER SWEET**	**Celástrus orbiculàtus** *C. articulàtus*	Shrubby climber with clusters of small flowers, succeeded by clusters of bright orange-yellow berries with conspicuous crimson seeds which are hidden until the leaves fall. Prop. by fall-sown seed, cuttings of the root or layers. Any soil. China; Japan.	12-15 ft. *Sun or shade*	June

Color	English Name	Botanical Name and *Synonyms*	Description	Height and *Situation*	Time of Bloom
"Greenish white"	**FALSE BITTER SWEET**	**Celástrus scándens**	Shrubby climber with terminal clusters of small flowers, followed by bright yellow berries with conspicuous crimson seeds which last all winter. Prop. by fall-sown seed, cuttings of the root or layers. Any soil. N. Amer.	20 ft. *Sun or shade*	June
"Lilac"	**BLUISH CLEMATIS**	**Clématis cærùlea** *C. azùrea. C. pàtens*	Large spreading blossoms of a beautiful shade when grown in a northern exposure. Rather slow-growing. Requires rich deep soil and plenty of rotten manure. Needs plenty of water during dry weather. Prop by cuttings or graftings. Rich loamy soil, well-drained and enriched. Japan.	8-10 ft. *Half shade*	June, July
"Purple" 48 intense	**JACKMAN'S CLEMATIS**	**Clématis Jáckmani**	A beautiful species bearing a profusion of large, spreading, deep purple flowers. Support is necessary and winter mulching advisable. Prop. by cuttings or graftings. Rich loamy soil, well-drained and enriched. Needs plenty of water. Hort. Var. *Gypsy Queen* has deep violet flowers. Var. *Star of India* has purple flowers striped with red. Var. *magnifica, (Clematis magnifica)*, has purple flowers crimson tinted and striped with red. Hort.	5-6 ft. *Half shade*	Late June to early Sept.
"White"	**WHITE JACKMAN'S CLEMATIS**	**Clématis Jáckmani var. álba**	This variety is similar to the type, differing only in having white blossoms. Support is necessary and winter mulching advisable. Prop. by cuttings or graftings. Rich loamy soil, well-drained and enriched. Needs plenty of water in summer. Hort.	5-6 ft. *Half shade*	"
"Lavender"	**GREAT-FLOWERED VIRGIN'S BOWER**	**Clématis lanuginòsa**	This species is remarkable for the size of its blossoms, which are borne in succession throughout the summer. Support is necessary and winter mulching advisable. Prop. by cuttings or grafts. Rich loamy soil, well-drained and enriched. There are many vars. China. Var. *candida, (C. candida)*, differs from the type, having larger flowers and larger leaves. Var. *excelsior, (C. excelsior)*, has double pale purple flowers with red stripes. Hort.	5-6 ft. *Half shade*	June to Sept.
"Grayish white"	**GREAT-FLOWERED VIRGIN'S BOWER OTTO FROEBEL**	**Clématis lanuginòsa var. "Otto Froebel"**	This variety is similar in habit to the type. The flowers are large and bluish tinted. Support is necessary and winter mulching advisable. Prop. by cuttings or grafts. Rich loamy soil, well-drained and enriched. Hort.	5-6 ft. *Half shade*	"

VINES AND CLIMBERS

Color	English Name	Botanical Name and *Synonyms*	Description	Height and *Situation*	Time of Bloom
"White"	**MOUNTAIN CLEMATIS**	**Clématis montàna**	A beautiful species. Strong-growing climber with large showy flowers. Requires protection in winter. Prop. by cuttings or graftings. Rich soil. Mediterranean Region.	**15-20 ft.** *Sun*	June
"White"	**PANICLED CLEMATIS**	**Clématis paniculàta**	A vigorous climber which covers a large space the first season. Remarkably plentiful in blossoms, and delightfully fragrant. Small starry flowers. Invaluable for covering porches, arbors, etc. Prop. by seed, cuttings or graftings. Rich light loamy soil. Japan. See Plate, page 477.	**20-25 ft.** *Sun*	**Sept.**
"Bluish purple"	**PURPLE VIRGIN'S BOWER**	**Clématis verticillàris** *Atrâgene Americàna*	A species with drooping flowers 2-3 in. across. Prop. by seed, cuttings or grafts. Rich loamy soil, well-drained and enriched Eastern N. Amer.	**8-10 ft.** *Sun*	May, June
"Red"	**RED LEATHER FLOWER**	**Clématis Vióna var. coccínea** *C. coccínea*	More desirable than the type. Scarlet or rosy red pitcher-shaped flowers. Grayish foliage. Winter mulching desirable. Prop. by seed and cuttings. Rich soil, light and loamy. Texas.	**8-10 ft.** *Sun*	Early June, July, late Sept.
"White"	**TRAVELLER'S JOY**	**Clématis Vitálba**	One of the most vigorous species of Clematis. A profusion of small fragrant flowers in panicles. Grows quickly and covers arbors rapidly. Support is necessary. Prop. by cuttings or graftings. Rich loamy soil with good drainage. Europe; Africa.	**20-30 ft.** *Sun*	July to Sept.
"White"	**JAPANESE SPINDLE TREE**	**Euónymus radìcans** *E. Japónicus var. radìcans*	A bushy climbing evergreen much valued for its dark glossy foliage. Forms a dense covering over walls, rocks, fences, etc. Prop. by cuttings of half-ripe wood. A warm exposure and ordinary soil. There are vars. with variegated foliage. Japan.	**10-12 ft.** *Sun or shade*	
"Greenish"	**ENGLISH IVY**	**Hédera Hèlix**	Climbing or trailing evergreen subshrub with inconspicuous flowers and beautiful large dark green leaves. There are many vars., all of which do especially well on the north side of buildings. Protect in winter. Prop. by half-ripe cuttings. Preferably rich damp soil. Europe; Africa; Asia.	**30-40 ft.** *Shade*	June, July
"Greenish yellow"	**COMMON HOP**	**Hùmulus Lùpulus**	Vigorous free-growing twining perennial which bears greenish yellow catkins. Leaves rough and hairy. Useful for its rapid growth in covering trellises, old fences, etc. Prop. by seed or division in spring. Any soil, preferably rich loam. Europe; N. Amer.; Asia.	**25-30 ft.** *Sun*	Mid. July to early Aug.

Color	English Name	Botanical Name and *Synonyms*	Description	Height and *Situation*	Time of Bloom
"White"	**MAN-OF-THE-EARTH, WILD POTATO VINE, PERENNIAL MOON-FLOWER**	**Ipomœa panduràta**	Hardy perennial vine. Funnel-shaped purple-throated flowers in clusters. Dense foliage. Good for covering fences or stumps in wild garden, etc. Prop. by seed, division or cuttings. Any soil. Canada; Eastern U. S. A.	2-12 ft. *Sun*	July, Aug.
"Purplish rose" 31 darker	**TWO-FLOWERED EVER-LASTING PEA**	**Láthyrus grandiflòrus**	Perennial climber of the Pea order, not so vigorous as L. latifolius. Flowers, size of Sweet Pea and largest of species, grow in pairs. Habit free and neat. Good covering for rocks, banks, stumps, etc. Prop. by seed and division. Any garden soil. S. Europe.	4-6 ft. *Sun or shade*	June, July
"Rose" 40 lighter & 37	**EVER-LASTING OR PERENNIAL PEA**	**Láthyrus latifòlius**	Vigorous perennial climber. Numerous large loose clusters of pea-shaped flowers continuous in bloom even after a frost. Grayish foliage. Excellent for covering rocks, stumps, banks, etc. Good for cutting. Very easily cultivated. Prop. by seed, division or cuttings. Any soil. Europe. Var. *splendens;* (color no. 31 or 40) is the best form of L. latifolius. Very showy and free-growing. Brilliant purplish pinkish flowers, occasionally red. The luxuriant foliage makes a good background.	4-8 ft. *Sun or shade*	Mid. July to early Sept.
"White"	**WHITE EVER-LASTING OR PERENNIAL PEA**	**Láthyrus latifòlius var. álbus**	Vigorous perennial climber. Numerous large pea-shaped flowers in loose clusters. Gray-green compound foliage. Excellent for covering stumps, banks, etc. Prop. by seed and division. Any soil. Hort.	4-8 ft. *Sun or shade*	Mid. July to Sept.
"Old rose" 25	**SHOWY WILD PEA**	**Láthyrus venòsus**	Wild on shady banks. Strong-stemmed perennial vine. Blossoms in many-flowered clusters. Good for cutting. Useful in covering rough places. Prop. by seed and division. Any good garden soil. N. Amer.	2-3 ft. *Sun or shade*	Mid. June to July
"Yellowish white"	**ITALIAN HONEY-SUCKLE, FRAGRANT WOODBINE**	**Lonícera Caprifòlium** *C. hortênse, C. perfoliàtum*	Climbing plant with fragrant purplish tinted flowers. Useful for trellises and walls. Prop. by fall-sown seed or ripe wood cuttings. Any good soil. Europe; Asia; naturalized in N. Amer.	12-15 ft. *Sun*	May, June
"White"	**JAPANESE OR CHINESE HONEY-SUCKLE**	**Lonícera Japónica**	Climbing vine with dainty fragrant flowers and semi-evergreen leaves. Charming when grown over walls or shrubs. Prop. by fall-sown seed or ripe wood cuttings. Any good soil. China; Japan; naturalized in N. Amer. Var. *aureo-reticulata,* (*L. reticulata aurea, L. brachypoda reticulata*), has smaller yellow-veined leaves. Hort.	15 ft. *Sun*	June, July

Color	English Name	Botanical Name and *Synonyms*	Description	Height and *Situation*	Time of Bloom
"White"	HALL'S JAPANESE HONEY-SUCKLE	**Lonícera Japónica var. Halliàna** *L. flexuòsa Halliàna, Caprifòlium Hallìanum*	Climbing vine differing from the type mainly in its time of bloom. Semi-evergreen foliage. Prop. by fall-sown seed or ripe wood cuttings. Any good soil. China; Japan.	15 ft. *Sun*	Sept.
"Scarlet" 18	TRUMPET HONEY-SUCKLE	**Lonícera sempérvirens** *Caprifòlium sempérvirens*	Beautiful native climber with terminal clusters of lovely bright flowers, sometimes yellow. Handsome foliage, evergreen in the Southern States. Prop. by fall-sown seed or ripe wood cuttings. Any good soil. U. S. A.	8-10 ft. *Sun*	May to Sept.
"Yellow"	SULLI-VANT'S HONEY-SUCKLE	**Lonícera Sullivántii** *L. flàva*	Climbing plant with close spikes of purplish tinted flowers. Thick grayish leaves. Attractive in the fall on account of its bright red berries. Prop. by fall-sown seed or ripe wood cuttings. Any good soil. N. Amer.	4-5 ft. *Sun*	May, June
"Purple"	CHINESE BOX THORN	**Lýcium Chinénse**	Branching shrub which if trained to a support will grow 15 ft. high, the long slender branches being laden with ornamental scarlet fruit. Good rich soil. China.	12-15 ft. *Sun*	June to Sept.
"White"	MATRI-MONY VINE, BOX THORN	**Lýcium vulgàre** *L. flàccidum, L. halimi-fòlium*	Branching shrub with a profusion of bright red berries following funnel-like flowers which vary to purple. Grayish foliage. Useful for covering walls, etc. Prop. by seed, cuttings of hard wood, layers or suckers. Plant in any position not too damp. Europe.	12-15 ft. *Sun*	Late May to late Sept.
"White"	BALTIMORE BELLE ROSE	**Ròsa "Baltimore Belle"**	A variety of the Prairie Rose. It has clusters of rather small double flowers and is not so hardy as the type. A useful climber for trellis or porch. Prop. generally by cuttings. Any rich soil. Hort.	6 ft. *Sun*	June, July
"Rosy carmine" bet. 27 & 34	CARMINE PILLAR ROSE	**Ròsa "Carmine Pillar"**	A lovely climbing Rose having delicate single flowers of a peculiarly fine color. Foliage good. An excellent pillar rose and good for walls, arbors, etc. It is best to cover in winter. Prop. by cuttings or graftings. Rich soil. Hort.	10-12 ft. *Sun*	June
"Crimson"	CRIMSON RAMBLER ROSE	**Ròsa "Crimson Rambler"**	A climber bearing a profusion of rather small dazzling double flowers. A vigorous plant which is a great favorite and makes quick growth, covering a large space in the first season. Prop. generally by cuttings. Any good soil. Hort.	15-20 ft. *Sun*	June, July

Color	English Name	Botanical Name and *Synonyms*	Description	Height and *Situation*	Time of Bloom
"Crimson"	**THE DAWSON ROSE**	**Ròsa "Dawson"**	A cross between Rosa multiflora and General Jacqueminot. A rampant grower, covered with a myriad of small single roses in clusters. Pretty clean foliage. Very attractive for arbors, trellises, etc. Prop. by cuttings and graftings. Good rich soil. Hort.	10-15 ft. *Sun*	June
"Pink" 30	**DÉBUTANTE ROSE**	**Ròsa "Débutante"**	A "perpetual flowering" climbing Rose bearing a profusion of flowers in clusters. Excellent for pillars. Prop. generally by cuttings. Rich heavy soil, either loamy or clayey.	10-15 ft. *Sun*	June
"White"	**BRAMBLE ROSE**	**Ròsa multiflòra** *R. intermèdia, R. polyántha, R. Wichùræ*	A rampant grower covered with clusters of small single fragrant flowers. Free vigorous habit. Useful for pillar work and pretty in the shrubbery, where it forms a large bush. Prop. generally by cuttings. Requires good rich soil. China; Japan.	6-10 ft. *Sun*	"
"Deep rose" 30	**PRAIRIE ROSE**	**Ròsa setígera**	Climber with clusters of flowers 2 in. across, which gradually fade to white. Very effective, the flowers being exceptionally showy. Prop. generally by cuttings. Any rich soil. Eastern N. Amer.	6 ft. *Sun*	June, July
"Bright pink" near 30	**SWEETHEART ROSE**	**Ròsa "Sweetheart"**	A perpetual flowering climbing Rose, bearing a profusion of double flowers in clusters. Good for covering pillars and porches. Very effective and pleasing. Prop. generally by cuttings. Heavy soil, either loamy or clayey. Hort.	10-12 ft. *Sun*	"
"Pink" 20 to 30 & darker	**THE FARQUHAR ROSE**	**Ròsa "The Farquhar"**	One of the most rapid growers. Flowers in large clusters. The splendid shiny foliage lasts in good condition throughout the season. Prop. by cuttings. Hort.	15-20 ft. *Sun*	"
"White"	**WICHURIANA ROSE**	**Ròsa Wichuriàna**	A strong-growing Rose of prostrate habit with single flowers 1½-2 in. across and brilliantly glossy leaves. Forms long shoots in a season and is pretty on rocks or sunny banks, etc. Prop. by division and cuttings. Any ordinary soil. There are many pretty vars. in different colors. Japan.	12-20 ft. *Sun*	July, Aug.
"White"	**CLIMBING HYDRANGEA**	**Schizophrágma hydrangeoìdes**	A very effective climbing shrub which resembles the Hydrangea. Small flowers in large clusters and pretty bright green deciduous foliage which forms a pleasing contrast to the young red shoots. Useful for covering walls, etc. Prop. by seed, cuttings of green wood or layers. Rich soil fairly moist. Japan.	30 ft. *Sun or half shade*	July

VINES AND CLIMBERS

Color	English Name	Botanical Name and *Synonyms*	Description	Height and *Situation*	Time of Bloom
"Scarlet" bet. 16 & 17	**CHINESE TRUMPET CREEPER**	**Técoma grandiflòra** *T. Chinênsis, Bignònia Chinênsis, Cámpsis adrepens*	Shrubby climber, not as high growing as T. radicans but with larger flowers. Terminal clusters of gorgeous blossoms about 2 in. broad. Prop. by seed, cuttings or layers. Rich and moist soil is best. China; Japan.	4-6 ft. *Sun*	Late July, Aug.
"Scarlet orange" 19 shading to 10	**TRUMPET VINE, CREEPER OR HONEYSUCKLE**	**Técoma radìcans** *Bignònia radìcans, Cámpsis radìcans*	Shrubby climber with terminal clusters of showy tubular flowers, yellow inside, and handsome dark green compound leaves. Very effective when in flower. Good for covering walls, fences, old trees, etc. Prop. by seed, cuttings or layers. Rich moist soil is best. Eastern U. S. A.	8-12 ft. *Sun*	July, Aug.
"Purple" 44	**CHINESE WISTARIA**	**Wistària Chinénsis** *W. consequàna, W. polystàchya, W. Sinênsis*	A deservedly favorite climber with large dense pendent clusters of pea-shaped flowers and beautiful light green foliage. This wonderful vine is of a sturdy growth and very florescent. It needs support. Prop. most easily by layers. Rich soil is preferable. China. See Plate, page 478.	20-30 ft. *Sun*	May
"White"	**WHITE CHINESE WISTARIA**	**Wistària Chinénsis var. albiflòra**	A lovely variety though not so hardy as the type. The clusters are longer and more graceful. Plant in a warm sheltered spot. Prop. by layers. Rich soil.	20-30 ft. *Sun*	"
"Purple" 44	**LOOSE-CLUSTERED WISTARIA**	**Wistària multijùga** *W. Chinênsis var. multijùga, W. grandiflòra*	The drooping clusters of this species are much larger and looser than those of the Chinese Wistaria. Pale green foliage. Prop. most easily by layers. Rich soil is preferable. Japan.	20-30 ft. *Sun*	"

A FEW OF THE BEST FERNS

English Name	Botanical Name and *Synonyms*	Description	Height and *Situation*
MAIDENHAIR FERN	***Adiántum pedàtum**	Graceful clusters of delicate fronds on dark wiry stems. Excellent for cutting. Set plants a foot apart in masses. Easily cultivated in rich soil with plenty of water and good drainage. Prop. by spores. U. S. A.	8-14 in. *Shade*
CHRISTMAS OR SHIELD FERN	***Aspídium acrostichoìdes** *Dryópteris acrostichoìdes*	A tough evergreen with deep green polished fronds, coarsely divided. Excellent for cutting. Plant several 1 ft. apart in good well-drained garden soil. Easily transplanted. Prop. by spores. Eastern Amer.	1 ft. *Shade*
MALE SHIELD FERN	***Aspídium Fìlix-mas** *Dryópteris Fìlix-mas*	Beautiful and uncommon evergreen. Thrives under trees. Plant in masses a foot apart. Rich soil mixed with sand and leaf-mold. Prop. by spores. N. Europe; Canada; Col.	1-1½ ft. *Shade*
GOLDIE'S WOOD FERN	**Aspídium Goldiànum** *Dryópteris Goldieàna*	Our tallest wood fern with deep green fronds occasionally more than a foot wide. Plant these ferns at intervals of 18 in. in moist soil. Prop. by spores. Eastern Amer.	2-4 ft. *Shade*
MARGINAL SHIELD OR EVERGREEN WOOD FERN	***Aspídium marginàle** *Dryópteris marginàlis*	Thick fronds, 3-5 in. wide, remain deep green all the year round. Thrives anywhere but prefers rich soil. Plant at intervals of 1 foot. Prop. by spores. N. Amer.	1-2 ft. *Shade*
SPINULOSE WOOD FERN	***Aspídium spinulòsum** *Dryópteris spinulòsa*	An evergreen with minutely divided fronds about 4 in. wide. These ferns thrive when planted 18 in. apart in rich soil with plenty of muck or leaf-mold. Prop. by spores. Northern N. Amer.; Europe.	15 in. *Shade*
NARROW-LEAVED SPLEENWORT	**Asplènium angustifòlium**	Pale delicate foliage easily beaten down by wind or rain. Plant in background about 18 in. apart. Rich moist soil preferable. Prop. by spores. N. Amer.	1-4 ft. *Shade*
EBONY SPLEENWORT	***Asplènium ebèneum** *A. platyneùron*	An evergreen species with delicate leaves, 1 to 2 in. wide on dark erect stems. Good for rock-garden. Plant 18 in. apart in dry soil mixed with some leaf-mold and a little lime. Prop. by spores. N. Amer.	6-15 in. *Sun or half shade*
LADY FERN	***Asplènium Fìlix-fœmina**	A striking species with finely toothed fronds. Set out plants in rich moist soil nearly 2 ft. apart. Prop. by spores. Europe; N. Amer.	2-3 ft. *Sun or shade*
MAIDENHAIR OR DWARF SPLEENWORT	***Asplènium Trichómanes**	Dainty evergreen with delicate fronds about 1½ in. wide. Excellent for rock-garden. Thrives in any cleft of rock, if given a bare foot-hold. Plant 8 in. apart. Leaf-mold beneficial. Prop. by spores. Northern Hemisphere.	3-6 in. *Shade*

English Name	Botanical Name and *Synonyms*	Description	Height and *Situation*
WALKING LEAF OR FERN	***Camptosòrus rhyzophýllus**	Excellent for carpeting the rock-garden. Small prostrate fern with tapering evergreen leaves. The fronds root at their tips, if they touch the ground, thus forming by "steps" new plants. Set out plants 6 in. apart in dry soil with a sprinkling of leaf-mold. Prop. by rooting of leaves and spores. Eastern N. Amer.	4-12 in. *Half shade*
BULBLET BLADDER FERN	**Cystópteris bulbífera**	Thrives on lime-stone cliffs near falling water. Slender tapering foliage, pale in color. Set out plants 8-12 in. apart, in moist soil, in front of stronger ferns. Prop. by spores. Eastern N. Amer.	1-2 ft. *Shade*
HAY-SCENTED OR GOSSAMER FERN, HAIRY DICKSONIA	***Dicksònia punctilóba,** *D. pilosiúscula, Dennstædtia punctilóbula*	Excellent for massing in open ground. Finely cut foliage fragrant when dry or crushed. Plant at intervals of 8 in. in dry or moist soil, well drained. Prop. by spores. Eastern N. Amer.	2-3 ft. *Sun or half shade*
CLIMBING, CREEPING OR HARTFORD FERN	***Lygòdium palmàtum**	The only climbing fern in America. Delicate heart-shaped fronds of a pale green on slender trailing stems. Needs some support. Plant in moist soil with leaf-mold and a mulch of leaves for a few years. Prop. by spores. Florida; Tenn.	1-3 ft. *Shade*
SENSITIVE FERN	**Onoclèa sensíbilis**	Thrives anywhere. Withers rapidly after being picked. Coarse foliage peculiar in shape and delicate in color. Set out plants a foot apart. Prop. by spores. Eastern N. Amer.	1-2 ft. *Sun or shade*
OSTRICH FERN	***Onoclèa Struthiópteris** *Struthiópteris Germánica*	One of the tallest and most stately ferns. Graceful feathery foliage, dark green and usually erect. Effective as background for smaller ferns, as its growth is of tropical luxuriance. A gross feeder, requiring much manure. Prop. by spores. Northeastern N. Amer.	2-5 ft. *Sun or shade*
CINNAMON FERN	***Osmúnda cinnamòmea**	A striking plant. Fronds 8 in. wide turn to cinnamon-brown. Thrives in rich moist soil with plenty of leaf-mold. Prop. by spores. Eastern Amer.	2-5 ft. *Sun or shade*
CLAYTON'S OR INTERRUPTED FERN	**Osmúnda Claytoniàna**	Easily cultivated. Resembles the Cinnamon Fern with graceful foliage curving outward like a feather duster. Plant at intervals of about 2 ft. in dry stony soil. Prop. by spores. Eastern Amer.	2-4 ft. *Shade*
ROYAL OR FLOWERING FERN	***Osmúnda regàlis**	Distinctive in appearance. Smooth pale green foliage and conspicuous fruit in flower-like clusters. Thrives in rich moist soil, even with water standing 2 or 3 in. deep. Prop. by spores. Eastern Amer.	2-4 ft. *Sun or shade*
BROAD OR HEXAGON BEECH FERN	***Phegópteris hexagonóptera**	Triangular leaves broader than their length, and much divided. Establish plants about 8 in. apart in any garden soil, preferably rather dry. Prop. by spores. Eastern N. Amer.	7-12 in. *Shade*

FERNS

English Name	Botanical Name and *Synonyms*	Description	Height and *Situation*
COMMON BRAKE OR EAGLE FERN	***Ptèris aquilìna**	Strikingly tall and vigorous plant. Leathery foliage, dull green in a sunny exposure, but a fresh color when growing in the shade. Any soil. Prop. by spores. N. Amer.	1-5 ft. *Sun or shade*
COMMON POLYPODY, ROCK OR SNAKE FERN	***Polypòdium vulgàre**	Evergreen. One of the best for the rock-garden, and useful for cutting. Finely cut fronds. Grows on decayed tree-trunks and on flat shelves in rocky ledges. Plant at intervals of 4-6 in. in any soil, not very moist. Prop. by spores. Europe; U. S. A.	4-12 in. *Half shade*
RUSTY WOODSIA	**Woódsia Ilvénsis**	Suitable for rock-garden. Grows in compact tufts on exposed rocky ledges. Plant at intervals of about 8 in. in moist well-drained soil. Prop. by spores. Amer.	2-6 in. *Sun or half shade*
NARROW-LEAVED OR NET-VEINED CHAIN FERN	**Woodwárdia angustifòlia**	Minutely toothed fronds 3-4 in. wide grown in compact tufts. The plant thrives in wet bogs, in moist or dry garden soil, well-drained. Prop. by spores. Eastern U. S. A.	1 ft. *Shade*

SUPPLEMENTARY LISTS

SOME PLANTS WHICH WILL BLOOM IN SHADE

Ajuga alpina
Ajuga reptans
Asarum Europæum
Convallaria majalis
Epimedium alpinum
Funkia Fortunei
Funkia lancifolia
Funkia lancifolia var. alba marginata
Funkia lancifolia var. undulata
Funkia ovata
Funkia subcordata
Funkia subcordata var. grandiflora
Funkia Sieboldiana
Hepatica acutiloba
Hepatica triloba
Orobus vernus
Pachysandra terminalis
Pachysandra terminalis var. variegata
Vancouveria hexandra
Vinca minor

SOME PLANTS WHICH WILL GROW IN POOR SOIL

Achillea serrata
Ajuga alpina
Alyssum saxatile
Aquilegia Canadensis
Arabis albida
Cerastium tomentosum
Dianthus deltoides, plumarius, and many others
Eryngium campestre
Euphorbia Myrsinites
Geranium sanguineum, and maculatum
Helianthemum vulgare
Iberis sempervivens
Lamium album
Linaria vulgaris
Opuntia Missouriensis, and Rafinesquii
Orobus lathroides
Phlox subulata
Potentilla tridentata
Saxifraga Pennsylvanica
Sedum acre and stoloniferum
Sempervivum tectorum, and arvernense
Verbascum Thapsus
Veronica rupestris
Viola cucullata

A FEW PLANTS HAVING ESPECIALLY LONG BLOOMING SEASONS

Campanula Carpatica
Delphiniums, if cut back
Heuchera sanguinea
Lathyrus latifolius
Lychnis coronaria
Scabiosa graminifolia
Scabiosa sylvatica
Verbascums, if cut back
Viola cornuta
Viola tricolor

SOME PERENNIALS FLOWERING THE FIRST SEASON FROM SEED SOWN IN SPRING

Bupthalmum salicifolium
Campanula Carpatica
Campanula Carpatica var. alba
Campanula Carpatica var. rotundifolia
Centaurea montana
Chrysanthemum maximum
Delphinium exaltatum
Delphinium formosum
Delphinium grandiflorum
Dianthus plumarius, and many other species
Erigeron Coulteri
Erigeron glabellus
Gaillardia aristata var. grandiflora
Geum atrosanguineum
Inula ensifolia
Linaria Dalmatica
Papaver nudicaule
Pentstemon campanulatus

SUPPLEMENTARY LISTS

FIFTY OF THE BEST ANNUAL PLANTS

1 Ageratum Mexicanum
2 Alonsoa incisifolia
3 Amarantus cordatus
4 Antirrhinum majus, many varieties
5 Arctotis grandis
6 Bartonia aurea
7 Brachycome iberidifolia
8 Browallia elata var. grandiflora
9 Calendula officinalis, many varieties
10 Calistephus hortensis (China Aster)
11 Celosia cristata, many varieties
12 Celosia plumosa
13 Centaurea Americana
14 Centaurea cyanus
15 Chrysanthemum carinatum, many varieties
16 Clarkia elegans
17 Cleome pungens
18 Collinsia bicolor
19 Convolvulus tricolor
20 Coreopsis Drumondii
21 Cosmos bipinnatus, and varieties
22 Delphinium consolida, and varieties
23 Dianthus Chinensis
24 Eschscholtzia Californica
25 Gaillardia picta var. Lorenziana
26 Gilia tricolor
27 Helianthus cucumerifolius var. Stella
28 Helipterum roseum
29 Helichrysum bracteatum, and varieties
30 Iberis umbellata, and varieties
31 Ipomœa purpurea
32 Lathyrus odoratus, many varieties
33 Lupinus hirsutus, and other species
34 Matthiola incana, and var. annua
35 Mirabilis jalapa
36 Papaver Rhœas, and somniferum in variety
37 Petunia hybrida
38 Phacelia viscida and Whitlavia
39 Phlox Drumondii
40 Reseda odorata, many varieties
41 Salpiglossis sinuata, and varieties
42 Salvia splendens
43 Scabiosa atropurpurea, and varieties
44 Schizanthus pinnatus, and varieties
45 Tagetes patula, and erecta
46 Tropæolum majus, and minus
47 Thunbergia alata
48 Torenia Fournieri
49 Verbena hybrida
50 Zinnia elegans

TWENTY-FIVE OF THE BEST HARDY HERBACEOUS PLANTS

1 Adonis vernalis
2 Anemone Japonica, and its varieties
3 Aquilegia cærulea, chrysantha, and chrysantha hybrids
4 Aster alpinus
5 Campanula Carpatica, persicifolia, and macrantha
6 Clematis recta
7 Delphinium elatum, formosum, and hybridum
8 Dicentra spectabilis
9 Iris Germanica, and its varieties
10 Iris lævigata, and its varieties
11 Lilium Hansoni, speciosum, tenuifolium, and many other kinds
12 Lychnis Chalcedonica
13 Œnothera fruticosa var. Youngii
14 Pæonia albiflora, officinalis, and many fine hybrids
15 Papaver orientale, and its varieties
16 Phlox paniculata, and its varieties
17 Phlox subulata
18 Platycodon grandiflorum
19 Pyrethrum (Chrysanthemum) roseum
20 Pyrethrum (Chrysanthemum) uliginosum
21 Rudbeckia speciosa
22 Sedum spectabile
23 Statice latifolia
24 Trollius Europæus
25 Veronica rupestris, and latifolia var. subsessilis

A FEW OF THE BEST ROCK-GARDEN PLANTS

Adonis vernalis
Alyssum saxatile
Aquilegia cærulea, Canadensis, vulgaris, and many others
Arabis albida
Armeria maritima
Aster alpinus
Aubretia deltoidea
Campanula Carpatica, turbinata, and rotundifolia
Dianthus deltoides, and many others
Epimedium alpinum
Geranium sanguineum
Gypsophila repens
Iberis sempervirens
Lathyrus vernus
Papaver alpinum, and nudicaule
Phlox subulata
Primula denticulata
Helianthemum vulgare
Heuchera sanguinea
Lychnis viscaria var. flore-pleno
Saxifraga cordifolia, crassifolia, and ligulata
Sedum spectabile and Sieboldii
Sempervivum arachnoideum, and sobiliferum
Tunica saxifraga
Viola cornuta

SUPPLEMENTARY LISTS

SOME OF THE BEST HARDY PLANTS FOR EDGINGS

Achillea tomentosa
Ajuga reptans
Arabis albida
Armeria maritima
Aubrietia deltoidea
Campanula Carpatica
Cerastium tomentosum
Dianthus deltoides
Phlox subulata
Phlox procumbens
Sedum stoloniferum
Stellaria Holostea
Veronica incana
Veronica rupestris
Viola cornuta

SOME HARDY PLANTS FOR BOLD OR SUBTROPICAL EFFECTS

Acanthus latifolius
Bocconia cordata
Cimicifuga racemosa
Dipsacus Fullonum
Echinops Ritro and exaltatus
Elymus arenarius
Eulalia Japonica
Helianthus orgyalis
Heracleum giganteum
Onopordon Acanthium
Polygonum cuspidatum
Polygonum sachalinense
Silphium perfoliatum
Spiræa Aruncus
Telekia cordifolia
Verbascum Olympicum
Yucca tomentosa

A FEW OF THE BEST PLANTS FOR WILD GARDENS

Anemone Canadensis
Aquilegia vulgaris, and Canadensis
Aster corymbosus, Novæ Angliæ, and many others
Campanula rapunculoides
Convallaria majalis
Coreopsis lanceolata
Coronilla varia
Crocus, in variety
Epilobium angustifolium
Erythronium Americanum
Galanthus nivalis
Helianthus mollis, and rigidus
Hemerocallis fulva
Hesperis matronalis
Malva Alcea
Lamium album
Narcissus pœticus, princeps, and many others
Scilla Sibirica and Hispanica
Tulipa sylvestris
Rudbeckia speciosa
Verbascum Thapsus
Viola cuculata, pedata, and pubescens
Lilium superbum, and Canadense
Solidago Canadensis, serotina, and many others
Geranium maculatum

INDEX OF BOTANICAL AND ENGLISH NAMES

INDEX

[Figures in bold-face type refer to pages bearing descriptive text, *i.e.*, the important references; entries set in italic type are synonyms; varieties entered with no page numbers following are subordinate to the entries under which they are indented; "illus." indicates pages on which illustrations of the respective entries occur.]

INDEX

INDEX

INDEX

INDEX

INDEX

INDEX

INDEX

INDEX

INDEX

INDEX

INDEX

INDEX

INDEX

INDEX

INDEX

INDEX

INDEX

INDEX

INDEX

INDEX

INDEX